D1428857

PETRI NET THEORY AND THE MODELING OF SYSTEMS

JAMES L. PETERSON
The University of Texas at Austin

PETRI NET THEORY AND THE MODELING OF SYSTEMS

PRENTICE-HALL, INC., Englewood Cliffs, N.J. 07632

Library of Congress Cataloging in Publication Data

PETERSON, JAMES LYLE.

Petri net theory and the modeling of systems.

Bibliography: p. 241

Includes index.

1. System design. 2. System analysis. 3. Petri nets. I. Title.

QA76.9.S88P47 003 80-24559

ISBN 0-13-661983-5

Editorial/production supervision by *Karen J. Clemments*
Interior design by *Karen J. Clemments*
Paste up by *Diane Koromhas*
Manufacturing buyers: *Joyce Levatino and Gordon Osbourne*

Printed in the United States of America

10 9 8 7 6 5 4 3 2 1

PRENTICE-HALL INTERNATIONAL, INC., *London*
PRENTICE-HALL OF AUSTRALIA PTY. LIMITED, *Sydney*
PRENTICE-HALL OF CANADA, LTD., *Toronto*
PRENTICE-HALL OF INDIA PRIVATE LIMITED, *New Delhi*
PRENTICE-HALL OF JAPAN, INC., *Tokyo*
PRENTICE-HALL OF SOUTHEAST ASIA PTE. LTD., *Singapore*
WHITEHALL BOOKS LIMITED, *Wellington, New Zealand*

Contents

Preface

Petri net theory has developed considerably from its beginnings with Dr. Petri's 1962 Ph.D. dissertation. However, much of the work on Petri nets is hard to obtain, being available only as reports and dissertations scattered among many sources. Despite the difficulty in learning about Petri nets, however, their use is constantly increasing. It is becoming expected that every computer scientist know some basic Petri net theory.

This book brings together the major parts of Petri net theory, presenting them in a coherent and consistent manner. The presentation and organization is suitable both for individual study by the practicing professional and for organized graduate study in computer science. Petri net theory can be applied to a vast number of areas (as shown in Chapter 3); a knowledge of the fundamentals of Petri net theory is becoming mandatory for the computer science, system analysis, and engineering professions.

For the student or professional who desires immediately applicable information on Petri nets, Chapters 1 through 4 and Chapter 7 are invaluable. These chapters are suitable for self-study, and provide a sufficient foundation in Petri net theory to allow immediate use in a wide range of areas.

This book can also be used as a text for a graduate seminar in Petri nets, for while the definitions and applications of the first four chapters can be easily learned, the remaining chapters take the student to the edge of current research. Each chapter includes exercises to provide practice with the concepts and reinforce the basics of the theory. In addition, the "Topics for Further Study" point the way for new research and study. Many of these topics could easily develop into theses and dissertations at both the Master's and Ph.D. level.

The basic concepts of Petri net theory can be understood with a minimal background. However, Petri nets, even more than most research topics, touch on many different aspects of computer science and mathematics. Full appreciation and understanding of current Petri net theory requires a good background in the study of formal languages and automata, operating systems, computer architecture, and linear algebra. An individual with an undergraduate degree in computer science or a year of graduate work in computer science should have the background necessary for research in Petri nets.

Obviously, more research has been done on Petri nets than can be presented here. We encourage further reading. The bibliography has been extensively researched in an effort to make it as complete as possible.

Specifically, we note that Dr. Petri has continued his research. What we refer to here as Petri net theory is, in his terminology, known as *Special Net Theory*. This is only a part of his *General Net Theory* [Petri 1973; Petri 1975; Petri 1976; Petri 1979a].

Acknowledgements

The creation of this volume benefited from the assistance of a number of people. Tilak Agerwala, Michel Hack, Tai-Yuan Hou, C. Matthias Laucht, Dino Mandrioli, Jerre Noe, Gary Nutt, and William Riddle helped with the technical content. J. C. Browne, K. Mani Chandy, Jim Daniel, Nancy Eatman, and R. T. Yeh, along with the Department of Computer Sciences and the Department of Mathematics of the University of Texas at Austin, and the Laboratory for Computer Science of the Massachusetts Institute of Technology provided the logistic support which allowed me the time and facilities to put together the manuscript.

Throughout the writing, editing, and revising process, my wife Jeanne has been a source of love and support.

The use of computer-based editing and typesetting procedures created new and unique problems in the production of this volume. I am grateful for the support, patience, and resolve of Prentice-Hall in this respect, and especially for the wisdom and professionalism of my editor, Karen Clemments.

J.L.P
Austin, Texas

PETRI NET THEORY
AND
THE MODELING
OF SYSTEMS

1

Introduction

Petri nets are a tool for the study of systems. Petri net theory allows a system to be modeled by a Petri net, a mathematical representation of the system. Analysis of the Petri net can then, hopefully, reveal important information about the structure and dynamic behavior of the modeled system. This information can then be used to evaluate the modeled system and suggest improvements or changes. Thus, the development of a theory of Petri nets is based on the application of Petri nets in the modeling and design of systems.

1.1 Modeling

The application of Petri nets is through *modeling*. In many fields of study, a phenomenon is not studied directly but indirectly through a *model* of the phenomenon. A model is a representation, often in mathematical terms, of what are felt to be the important features of the object or system under study. By the manipulation of the representation, it is hoped that new knowledge about the modeled phenomenon can be obtained without the danger, cost, or inconvenience of manipulating the real phenomenon itself. Examples of the use of modeling include astronomy (where models of the birth, death, and interaction of stars allow studying theories which would take long times and massive amounts of matter and energy), nuclear physics (where the radioactive atomic and subatomic particles under study exist for very short periods of time), sociology (where the direct manipulation of groups of people for study might cause ethical problems), biology (where models of biological systems require less space, time, and food to develop), and so on.

Most modeling uses mathematics. The important features of many physical phenomena can be described numerically and the relations between these features described by equations or inequalities. Particularly in the natural sciences and engineering, properties such as mass, position, momentum, acceleration, and forces are describable by mathematical equations. To successfully utilize the modeling approach, however, requires a knowledge of both the modeled phenomena and the properties of the modeling technique. Thus, mathematics has developed as a science in part because of its usefulness in modeling the phenomena of other sciences. For example, the differential calculus was developed in direct response to the need for a means of modeling continuously changing properties, such as position, velocity, and acceleration in physics.

The development of high-speed computers has greatly increased the use and usefulness of modeling. By representing a system as a mathematical model, converting that model into instructions for a computer, and running the computer, it is possible to model larger and more complex systems than ever before. This has resulted in considerable study of computer modeling techniques and of computers themselves. Computers are involved in modeling in two ways: as a computational tool for modeling and as a subject of modeling.

1.2 Features of Systems

Computer systems are very complex, often large, systems of many interacting components. Each component can be quite complex, as can its interactions with other components in the system. This is also true of many other systems. Economic systems, legal systems, traffic control systems, and chemical systems all involve many individual components interacting with other components, possibly in complex ways.

Thus, despite the diversity of systems which we want to model, several common points stand out. These should then be features of a useful model of these systems. One fundamental idea is that systems are composed of separate, interacting *components.* Each component may itself be a system, but its behavior can be described independently of other components of the system, except for well-defined interactions with other components. Each component has its own *state* of being. The state of a component is an abstraction of the relevant information necessary to describe its (future) actions. Often the state of a component depends on the past history of the component. Thus the state of a component may change over time. The concept of "state" is very important to modeling a component. For example, in a queueing system model of a bank, there may be several tellers and several customers. The tellers may be either idle (waiting for a customer to need service) or busy (serving a customer). Similarly, the customers may be idle (waiting for a teller to be free to serve them) or busy (being served by a teller). In a model of a hospital, the state of a patient might be critical, serious, fair, good, or excellent.

The components of a system exhibit *concurrency* or *parallelism.* Activities of one component of a system may occur simultaneously with other activities of other components. In a computer system, for example, peripheral devices, such as card readers, line printers, tape drives, and so on, may all operate concurrently under the control of the computer. In an economic system, manufacturers may be producing some products while retailers are selling other products, and consumers are using still other products, all at the same time.

The concurrent nature of activity in a system creates some difficult modeling problems. Since the components of the systems interact, it is necessary for *synchronization* to occur. The transfer of information or materials from one component to another requires that the activities of the involved components be synchronized while the interaction is occurring. This may result in one component waiting for another component. The timing of actions of different components may be very complex and the resulting interactions between components difficult to describe.

1.3 The Early Development of Petri Nets

Petri nets are designed specifically to model these types of systems: systems with interacting concurrent components. Petri nets have been developed from the early work of Carl Adam Petri [1962a]. In his doctoral dissertation, "Kommunikation mit Automaten," [Communication with automata], Petri formulated the basis for a theory of communication between asynchronous components of a computer system. He was particularly concerned with the description of the causal relationships between events. His dissertation was mainly a theoretic development of the basic concepts from which Petri nets have developed.

The work of Petri came to the attention of A. W. Holt and others of the Information System Theory Project of Applied Data Research, Inc. (ADR). Much of the early theory, notation, and representation of Petri nets developed from the work on the Information System Theory Project and was published in the final report of that project [Holt, et al.

1968] and in a separate report entitled "Events and Conditions" [Holt and Commoner 1970]. This work showed how Petri nets could be applied to the modeling and analysis of systems of concurrent components.

Petri's work also came to the attention of Project MAC at the Massachusetts Institute of Technology (M.I.T.). The Computation Structures Group, under the direction of Professor Jack B. Dennis, has been the source of considerable research and publication on Petri nets, publishing several Ph.D. dissertations and numerous reports and memos (see the bibliography). Two important conferences on Petri nets have been held by the Computation Structures Group: the Project MAC Conference on Concurrent Systems and Parallel Computation in 1970 at Woods Hole [Dennis 1970b] and the Conference on Petri Nets and Related Methods in 1975 at M.I.T.. Both of these conferences have helped to disseminate results and approaches in Petri net theory.

The use and study of Petri nets has spread widely in the last few years. A workshop on Petri nets was held in Paris in 1977 and an advanced course on General Net Theory in Hamburg in 1979. A special interest group on Petri nets has been formed in Germany. Research in and application of Petri nets is becoming widespread.

1.4 Applying Petri Net Theory

The practical application of Petri nets to the design and analysis of systems can be accomplished in several ways. One approach considers Petri nets as an auxiliary analysis tool. For this approach, conventional design techniques are used to specify a system. This system is then modeled as a Petri net and this Petri net model is analyzed. Any problems encountered in the analysis point to flaws in the design. The design must be modified to correct the flaws. This modified design can then be modeled and analyzed again. This cycle is repeated until the analysis reveals no unacceptable problems. This approach is diagrammed in Figure 1.1. Note that this approach can also be used to analyze an existing, currently operational system.

The conventional approach described above for using Petri nets in the design of a system requires constant conversion between the designed system and the Petri net model. An alternate approach has been suggested. In this more radical approach, the entire design and specification process is carried out in terms of Petri nets. Analysis techniques are applied only as necessary to create a Petri net design which is error-free. Then the problem is to transform the Petri net representation into an actual working system.

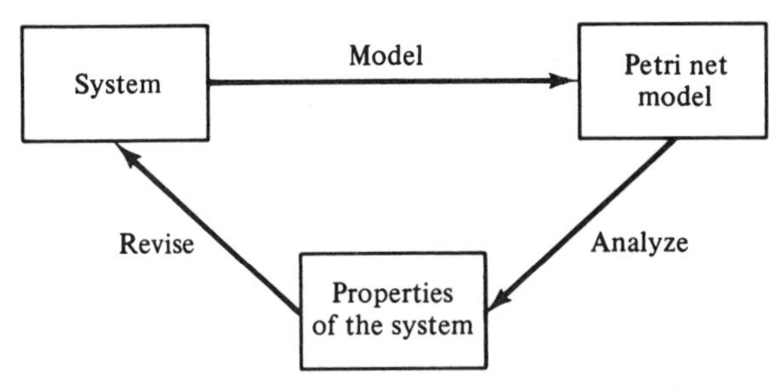

Figure 1.1 The use of Petri nets for the modeling and analysis of systems. The system is first modeled as a Petri net, and then this model is analyzed. The understanding of the system which results from the analysis will lead to a hopefully better system. Research is aimed at developing automatic techniques for the modeling and analysis of systems with Petri nets.

These two approaches to using Petri nets in the design process provide different types of problems for the Petri net researcher. In the first case, modeling techniques must be developed to transform systems into a Petri net representation; in the second case, implementation techniques must be developed to transform Petri net representations into systems. In both cases, we need analysis techniques to determine the properties of our Petri net model. Thus, our primary concern in the development of a theory of Petri nets is to study the properties of the Petri nets themselves.

1.5 Applied and Pure Petri Net Theory

The study of Petri nets has developed in two directions: *Applied Petri net theory* is concerned mainly with the application of Petri nets to the modeling of systems, the analysis of these systems, and the resulting insights into the modeled system. Successful work in this area requires a good knowledge of the application area and of Petri nets and Petri net techniques.

Pure Petri net theory is the study of Petri nets to develop the basic tools, techniques and concepts needed for the application of Petri nets. Although the motivation for Petri net research is based on applications, there is a need for a firm foundation of Petri net theory to be able to apply Petri nets. Much of the work on Petri nets at this point has concentrated on the fundamental theory of Petri nets, developing the tools and approaches which may someday be useful in the application of Petri nets to specific real-world problems. In this book, we present some of both areas of Petri net theory (pure and applied) but concentrate mainly on the basic theory. The applications which are given are intended mainly to demonstrate the versatility and power of Petri nets and to motivate the development of the analysis techniques.

We do not attempt to cover the entire range of Petri net topics in depth but rather hope to provide a firm foundation in terms, concepts, approaches, results, and history of Petri nets to allow a computer scientist or graduate student to be able to use and understand the growing body of Petri net literature and to be able to apply this theory to an even wider range of applications. We begin with some formal definitions and examples of Petri nets in Chapter 2 and then proceed with demonstrating their power and usefulness in the remainder of the book. The annotated bibliography provides references to most work on Petri nets.

1.6 Further Reading

The birth of Petri nets was Petri's dissertation [Petri 1962a], but most work in the United States is also based on the final report of the Information System Theory Project [Holt, et al 1968] which translated Petri's dissertation into English as well as extending the work considerably. The "Events and Conditions" paper by Holt and Commoner [1970] is also an important part of the early works. Petri presented a short paper to the 1962 IFIP Congress which was printed in the proceedings [Petri 1962b]. This paper is based on the ideas in his dissertation.

The approach presented in this book derives largely from the work of the Computation Structures Group at M.I.T. and has developed from the work of Dennis [1970a], Patil [1970a], and others, culminating in the work of Hack [1975c]. Keller has also been influential with his report on vector replacement systems [Keller 1972] and his view of modeling [Keller 1975a].

1.7 Topics for Further Study

1. Trace the origin and flow of the important ideas in Petri net theory. The starting point is most obviously Petri, but how did the ideas flow to the United States and to whom? How did they flow from there? Use published reports, papers, dissertations, and memos to determine precedence by date and citations in their references. You will probably want to interview some of the key people: Petri, Holt, Dennis, Patil, and so on.

2

Basic Definitions

In this chapter, we give formal definitions for the basic Petri net concepts. These basic concepts are used throughout our study of Petri nets and so are fundamental to a correct understanding of Petri nets.

Our formalisms are based on *bag* theory, an extension of set theory. If you are not familiar with bag theory, we suggest you read the appendix for the relevant concepts.

The definitions given here are similar in style to definitions in automata theory [Hopcroft and Ullman 1969]. In fact, they define a new class of machines, the Petri net automaton. As we shall see later (Chapters 5, 6, 7, and 8), this point of view can lead to some interesting results in formal language theory and automata theory.

2.1 Petri Net Structure

A Petri net is composed of four parts: a set of *places* P, a set of *transitions* T, an *input* function I, and an *output* function O. The input and output functions relate transitions and places. The input function I is a mapping from a transition t_j to a collection of places $I(t_j)$, known as the *input places* of the transition. The output function O maps a transition t_j to a collection of places $O(t_j)$ known as the *output places* of the transition.

The structure of a Petri net is defined by its places, transitions, input function, and output function.

DEFINITION 2.1 A *Petri net structure*, C, is a four-tuple, $C = (P,T,I,O)$.

$P = \{p_1, p_2, \ldots, p_n\}$ is a finite set of *places*, $n \geqslant 0$. $T = \{t_1, t_2, \ldots, t_m\}$ is a finite set of *transitions*, $m \geqslant 0$. The set of places and the set of transitions are disjoint, $P \cap T = \emptyset$. $I: T \rightarrow P^\infty$ is the *input* function, a mapping from transitions to bags of places. $O: T \rightarrow P^\infty$ is the *output* function, a mapping from transitions to bags of places.

The cardinality of the set P is n, and the cardinality of the set T is m. We denote an arbitrary element of P by p_i, $i = 1, \ldots, n$, and an arbitrary element of T by t_j, $j = 1, \ldots, m$.

Examples of Petri net structures are given in Figures 2.1, 2.2, and 2.3.

A place p_i is an *input place* of a transition t_j if $p_i \in I(t_j)$; p_i is an *output place* if $p_i \in O(t_j)$. The inputs and outputs of a transition are *bags* of places. A *bag* is a generalization of sets which allows multiple occurrences of an element in a bag. The appendix contains a description of bag theory. The use of bags, rather than sets, for the inputs and outputs of a transition allows a place to be a *multiple input* or a *multiple output* of a transition. The *multiplicity* of an input place p_i for a transition t_j is the number of occurrences of the place in the input bag of the transition, $\#(p_i, I(t_j))$. Similarly, the multiplicity of an output place p_i for a transition t_j is the number of occurrences of the place in the output bag of the transition, $\#(p_i, O(t_j))$. If the input and output functions are sets (rather than bags), then the multiplicity of each place is either zero or one.

The input and output functions can be usefully extended to map places into bags of transitions in addition to mapping transitions into bags of places. We define a transition t_j to be an input of a place p_i if p_i is an output of t_j. A transition t_j is an output of place p_i if p_i is an input of t_j.

DEFINITION 2.2 We extend the input function I and output function O as follows:

$$I: P \rightarrow T^\infty$$

$$O: P \rightarrow T^\infty$$

such that

$$\#(t_j, I(p_i)) = \#(p_i, O(t_j))$$

$$\#(t_j, O(p_i)) = \#(p_i, I(t_j))$$

$$C = (P, T, I, O)$$
$$P = \{p_1, p_2, p_3, p_4, p_5\}$$
$$T = \{t_1, t_2, t_3, t_4\}$$

$I(t_1) = \{p_1\}$ $\quad$ $O(t_1) = \{p_2, p_3, p_5\}$

$I(t_2) = \{p_2, p_3, p_5\}$ $\quad$ $O(t_2) = \{p_5\}$

$I(t_3) = \{p_3\}$ $\quad$ $O(t_3) = \{p_4\}$

$I(t_4) = \{p_4\}$ $\quad$ $O(t_4) = \{p_2, p_3\}$

Figure 2.1 A Petri net structure represented as a 4-tuple. The tuple consists of a set of places (P), a set of transitions (T), an input function ($I: T \rightarrow P^\infty$), and an output function ($O: T \rightarrow P^\infty$).

$$C = (P, T, I, O)$$
$$P = \{p_1, p_2, p_3, p_4, p_5, p_6\}$$
$$T = \{t_1, t_2, t_3, t_4, t_5\}$$

$I(t_1) = \{p_1\}$ $\quad$ $O(t_1) = \{p_2, p_3\}$

$I(t_2) = \{p_3\}$ $\quad$ $O(t_2) = \{p_3, p_5, p_5\}$

$I(t_3) = \{p_2, p_3\}$ $\quad$ $O(t_3) = \{p_2, p_4\}$

$I(t_4) = \{p_4, p_5, p_5, p_5\}$ $\quad$ $O(t_4) = \{p_4\}$

$I(t_5) = \{p_2\}$ $\quad$ $O(t_5) = \{p_6\}$

Figure 2.2 A Petri net structure.

$$C = (P, T, I, O)$$
$$P = \{p_1, p_2, p_3, p_4, p_5, p_6, p_7, p_8, p_9\}$$
$$T = \{t_1, t_2, t_3, t_4, t_5, t_6\}$$

$I(t_1) = \{p_1\}$ $\quad$ $O(t_1) = \{p_2, p_3\}$

$I(t_2) = \{p_8\}$ $\quad$ $O(t_2) = \{p_1, p_7\}$

$I(t_3) = \{p_2, p_5\}$ $\quad$ $O(t_3) = \{p_6\}$

$I(t_4) = \{p_3\}$ $\quad$ $O(t_4) = \{p_4\}$

$I(t_5) = \{p_6, p_7\}$ $\quad$ $O(t_5) = \{p_9\}$

$I(t_6) = \{p_4, p_9\}$ $\quad$ $O(t_6) = \{p_5, p_8\}$

Figure 2.3 A Petri net structure.

For the Petri net of Figure 2.1, the extended input and output functions are:

$$
\begin{array}{ll}
I(p_1) = \{\ \} & O(p_1) = \{t_1\} \\
I(p_2) = \{t_1, t_4\} & O(p_2) = \{t_2\} \\
I(p_3) = \{t_1, t_4\} & O(p_3) = \{t_2, t_3\} \\
I(p_4) = \{t_3\} & O(p_4) = \{t_4\} \\
I(p_5) = \{t_1, t_2\} & O(p_5) = \{t_2\}
\end{array}
$$

Exercises

1. Give the extended input and output functions for Figures 2.2 and 2.3.
2. Show that both an input and output function are not needed but that a Petri net can be defined from a set of places, a set of transitions, and an extended input (or output) function. To do this, show how an extended output function can be defined from an extended input function and vice versa.

2.2 Petri Net Graphs

Most theoretical work on Petri nets is based on the formal definition of Petri net structures given above. However, a *graphical* representation of a Petri net structure is much more useful for illustrating the concepts of Petri net theory. A Petri net graph is a representation of a Petri net structure as a bipartite directed multigraph.

A Petri net structure consists of places and transitions. Corresponding to these, a Petri net graph has two types of nodes. A *circle* O represents a place; a *bar* | represents a transition. Since the circles represent places, we call the circles places. Similarly, we call the bars transitions.

Directed arcs (arrows) connect the places and the transitions, with some arcs directed from the places to the transitions and other arcs directed from transitions to places. An arc directed from a place p_i to a transition t_j defines the place to be an input of the transition. Multiple inputs to a transition are indicated by multiple arcs from the input places to the transition. An output place is indicated by an arc from the transition to the place. Again, multiple outputs are represented by multiple arcs.

A Petri net is a *multigraph*, since it allows multiple arcs from one node of the graph to another. In addition, since the arcs are directed, it is a *directed* multigraph. Since the nodes of the graph can be partitioned into two sets (places and transitions), such that each arc is directed from an element of one set (place or transition) to an element of the other set (transition or place), it is a *bipartite* directed multigraph. We refer to it simply as a Petri net graph.

DEFINITION 2.3 A *Petri net graph* G is a bipartite directed multigraph, $G = (V,A)$, where $V = \{v_1,v_2, \ldots ,v_s\}$ is a set of vertices and $A = \{a_1,a_2, \ldots ,a_r\}$ is a bag of directed arcs, $a_i = (v_j,v_k)$, with $v_j,v_k \in V$. The set V can be partitioned into two disjoint sets P and T such that $V = P \bigcup T$, $P \bigcap T = \emptyset$, and for each directed arc, $a_i \in A$, if $a_i = (v_j,v_k)$, then either $v_j \in P$ and $v_k \in T$ or $v_j \in T$ and $v_k \in P$.

Figures 2.4, 2.5, and 2.6 are Petri net graphs equivalent to the Petri net structures of Figures 2.1, 2.2, and 2.3.

To demonstrate the equivalence of these two representations of a Petri net, the Petri net structure and the Petri net graph, we show how to transform one into the other. Assume we are given a Petri net structure $C = (P,T,I,O)$ with $P = \{p_1,p_2, \ldots ,p_n\}$ and $T = \{t_1,t_2, \ldots ,t_m\}$. Then we can define a Petri net graph as follows,

DEFINITION 2.4 Define $V = P \bigcup T$. Define A as a bag of directed arcs such that for all $p_i \in P$ and $t_j \in T$

$$\#((p_i,t_j),A) = \#(p_i,I(t_j))$$

$$\#((t_j,p_i),A) = \#(p_i,O(t_j))$$

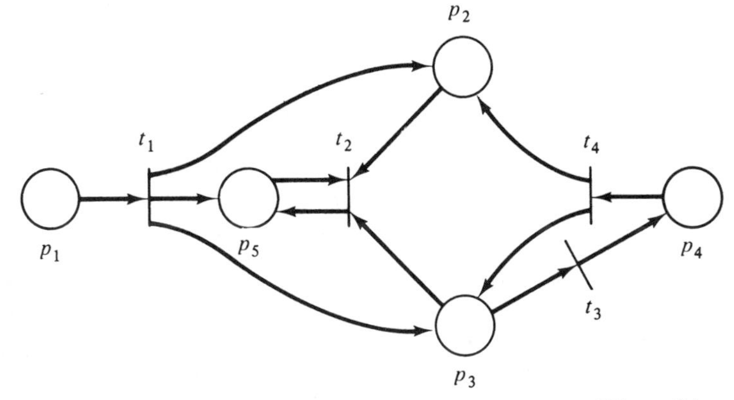

Figure 2.4 A Petri net graph equivalent to the structure of Figure 2.1.

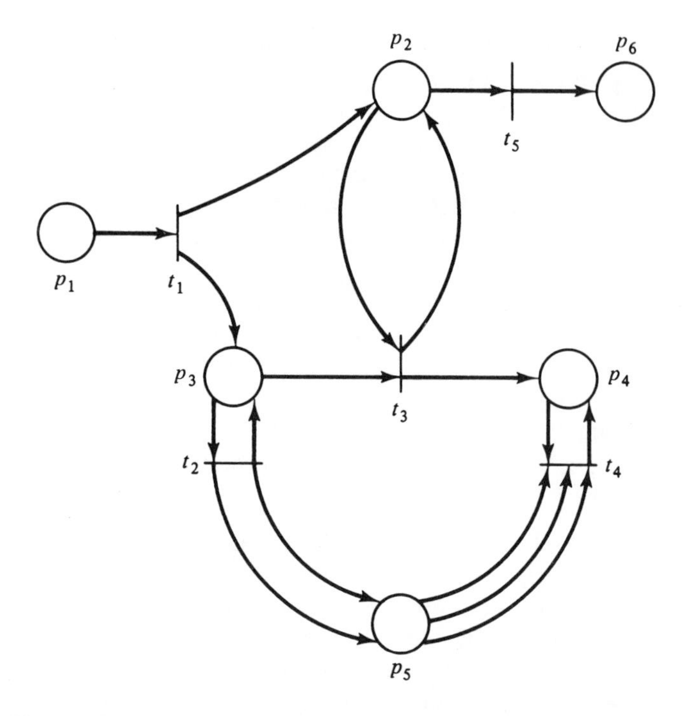

Figure 2.5 A Petri net graph equivalent to the structure of Figure 2.2.

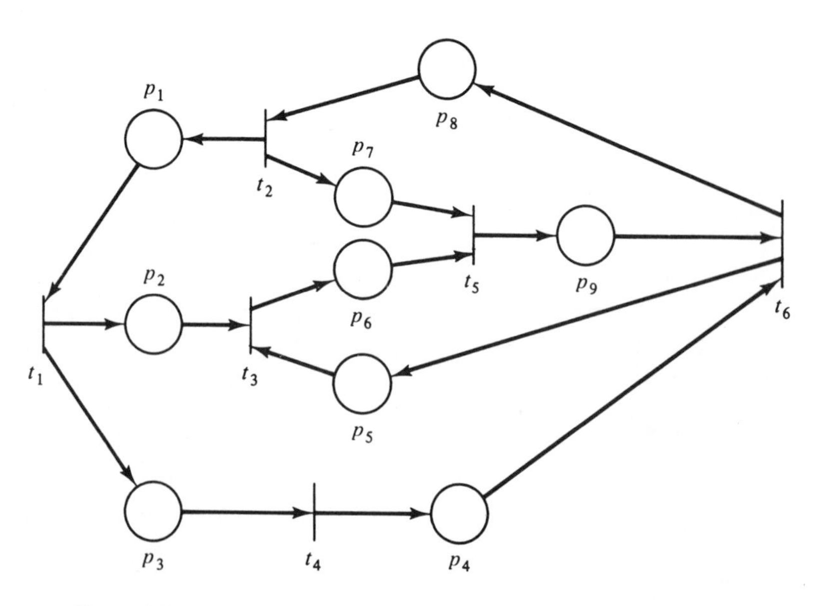

Figure 2.6 A Petri net graph equivalent to the structure of Figure 2.3.

$G = (V,A)$ is a Petri net graph which is equivalent to the Petri net structure $C = (P,T,I,O)$.

Conversion in the opposite direction (from a Petri net graph to a Petri net structure) is similar, and we leave the detailed description to the reader. One interesting problem arises in the translation from a Petri net graph to a Petri net structure: If the set of vertices is partitioned into the two sets S and R, which set should be the places and which set should be the transitions? Both possible selections allow a Petri net to be defined, although the two resulting structures have interchanged places and transitions.

The *dual* of a Petri net $C = (P,T,I,O)$ is the Petri net $\bar{C} = (T,P,I,O)$ which results from interchanging places and transitions. The graph structure is maintained, simply interchanging the circles and bars of the graph to indicate the change in places and transitions (but see Exercise 6). Figure 2.7 is the dual of the Petri net of Figure 2.4. Duality is a commonly used aspect of graph theory and would appear to be an interesting Petri net concept. However, no use has been made of the concept of the dual of a Petri net in Petri net research. This results, most likely, from the difficulty of defining the dual of a marked Petri net. Marked Petri nets are discussed next.

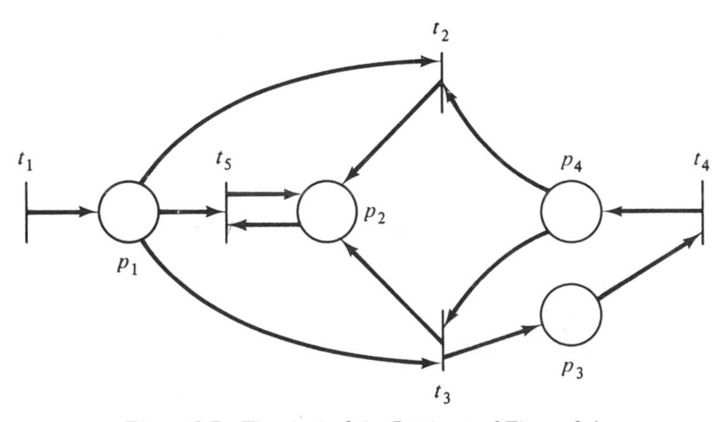

Figure 2.7 The dual of the Petri net of Figure 2.4.

Exercises

1. Give the dual of the Petri net graphs of Figures 2.5 and 2.6.
2. Give the Petri net graph of the following Petri net structure:

$$P = \{p_1, p_2, p_3, p_4\}$$
$$T = \{t_1, t_2, t_3, t_4, t_5\}$$
$$I(t_1) = \{\ \} \qquad O(t_1) = \{p_1\}$$
$$I(t_2) = \{p_1\} \qquad O(t_2) = \{p_2\}$$
$$I(t_3) = \{p_2, p_4\} \qquad O(t_3) = \{p_1, p_3\}$$
$$I(t_4) = \{\ \} \qquad O(t_4) = \{p_3\}$$
$$I(t_5) = \{p_3\} \qquad O(t_5) = \{p_4\}$$

3. Give the Petri net graph of the following Petri net structure:

$$P = \{p_1, p_2\}$$
$$T = \{t_1, t_2, t_3\}$$
$$I(t_1) = \{p_1\} \qquad O(t_1) = \{p_1, p_2\}$$
$$I(t_2) = \{p_1\} \qquad O(t_2) = \{p_2\}$$
$$I(t_3) = \{p_2\} \qquad O(t_3) = \{\}$$

4. Show that the dual of the dual of a Petri net structure C is the same structure C.
5. Define the class of Petri nets which are equal to their own duals. Can you give a simple characterization of this class of Petri nets?
6. If the dual of a Petri net structure, $C = (P,T,I,O)$ is defined as $\overline{C} = (T,P,I,O)$, the input and output functions must first be extended to map both P and T. Why? If $C = (P,T,I,O)$ with nonextended input and output functions, give the definition of $\overline{C} = (T,P,I',O')$ with nonextended input and output functions.
7. Give the Petri net structure corresponding to the Petri net graph of Figure 2.8. Give the Petri net structure for Figure 2.9.
8. Petri net graphs are multigraphs because a place may be a multiple input or output of a transition. This results in a graph with several arcs between the place and the transition. While this is satisfactory for small multiplicity (up to maybe three), it would be inconvenient for very large multiplicity. Thus, an alternative representation for structures with a large multiplicity is to use a *bundle* of arcs. A bundle is a special arc which is drawn very thick and labeled with its multiplicity. Figure 2.10 illustrates a transition with input multiplicity

7 and output multiplicity 11. Draw the Petri net graph for the following structure:

$$P = \{p_1, p_2, p_3, p_4\}$$

$$T = \{t_1, t_2, t_3, t_4\}$$

$$I(t_1) = \{\ \} \qquad O(t_1) = \{p_1, p_1, p_1, p_1, p_2\}$$

$$I(t_2) = \{p_2\} \qquad O(t_2) = \{p_1, p_1, p_1, p_1, p_1, p_1, p_3\}$$

$$I(t_3) = \{p_1, p_1, p_1, p_1, p_1, p_1\} \qquad O(t_3) = \{p_2, p_2, p_2, p_2, p_4, p_4\}$$

$$I(t_4) = \{p_3, p_4, p_4, p_2\} \qquad O(t_4) = \{\ \}$$

9. The *inverse* Petri net $-C$ for a Petri net $C = (P,T,I,O)$ is defined by interchanging the input and output functions, $-C = (P,T,O,I)$. What is the effect on the Petri net graph? How does this differ from the dual of a Petri net? Does it matter if the input and output functions have been extended? Draw the inverse Petri net of Figure 2.7.

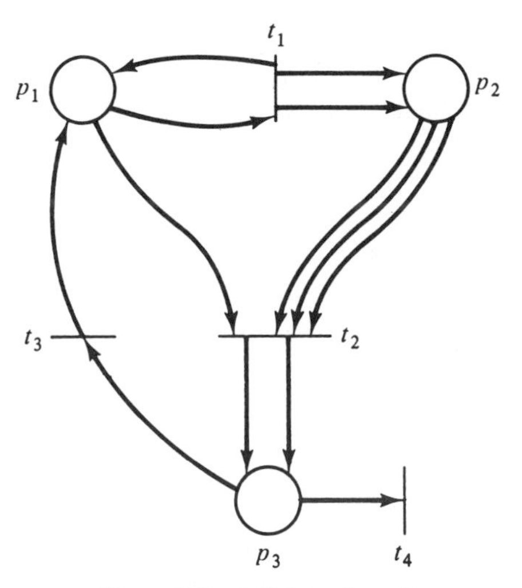

Figure 2.8 A Petri net graph.

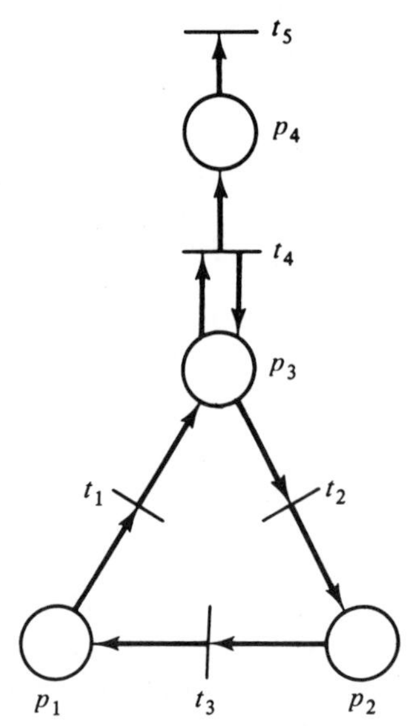

Figure 2.9 A Petri net graph.

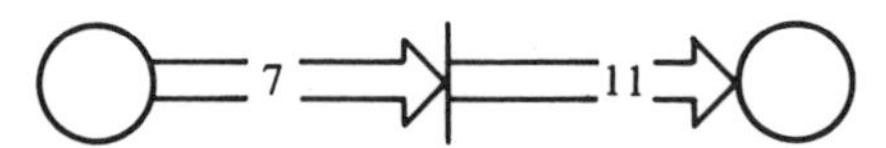

Figure 2.10 Bundles of arcs. For graphs with large multiplicity, a bundle is used, tagged with the multiplicity, rather than drawing multiple arcs.

2.3 Petri Net Markings

A *marking* μ is an assignment of *tokens* to the places of a Petri net. A token is a primitive concept for Petri nets (like places and transitions). Tokens are assigned to, and can be thought to reside in, the places of a Petri net. The number and position of tokens may change during the *execution* of a Petri net. The tokens are used to define the execution of a Petri net.

DEFINITION 2.5 A *marking* μ of a Petri net $C = (P,T,I,O)$ is a function from the set of places P to the nonnegative integers N. $\mu : P \rightarrow N$.

The marking μ can also be defined as an n-vector, $\mu = (\mu_1, \mu_2, \ldots, \mu_n)$, where $n = |P|$ and each $\mu_i \in N$, $i = 1, \ldots, n$. The vector μ gives for each place p_i in a Petri net the number of tokens in that place. The number of tokens in place p_i is μ_i, $i = 1, \ldots, n$. The definitions of a marking as a function and as a vector are obviously related by $\mu(p_i) = \mu_i$. The functional notation is somewhat more general and so is more commonly used.

A *marked Petri net* $M = (C, \mu)$ is a Petri net structure $C = (P, T, I, O)$ and a marking μ. This is also sometimes written as $M = (P, T, I, O, \mu)$.

On a Petri net graph, tokens are represented by small dots • in the circles which represent the places of a Petri net. Figures 2.11 and 2.12 show examples of a graph representation of a marked Petri net.

Since the number of tokens which may be assigned to a place of a Petri net is unbounded, there are an infinity of markings for a Petri net. The set of all markings for a Petri net with n places is simply the set of all n-vectors, N^n. This set, although infinite, is of course denumerable.

Exercises

1. For the marked Petri net of Figure 2.12, give the marking, both as a function and as a vector.
2. For the Petri net structure of Figure 2.2, draw the Petri net graph and indicate on the graph a marking $\mu = (1, 0, 1, 1, 0, 0)$.
3. Although the number of tokens in a Petri net which is drawn seldom exceeds 5 or 6, it is possible that a marking might have 10 or 20 or hundreds of tokens assigned to a place. In this case, it would be very inconvenient to actually draw hundreds of tokens in the circles

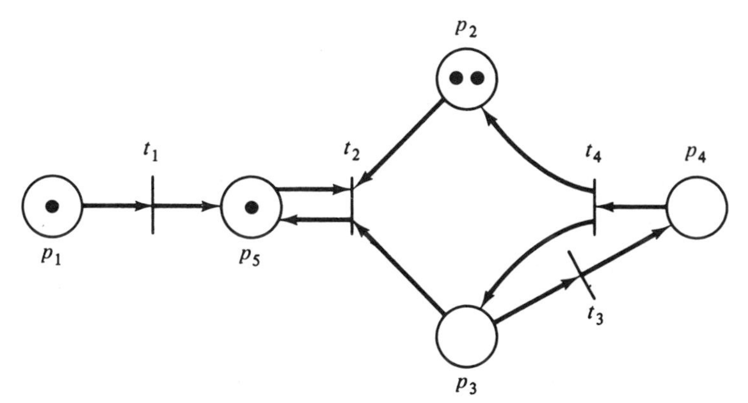

Figure 2.11 A marked Petri net. The Petri net structure is the same as Figures 2.1 and 2.4. The marking is $(1, 2, 0, 0, 1)$.

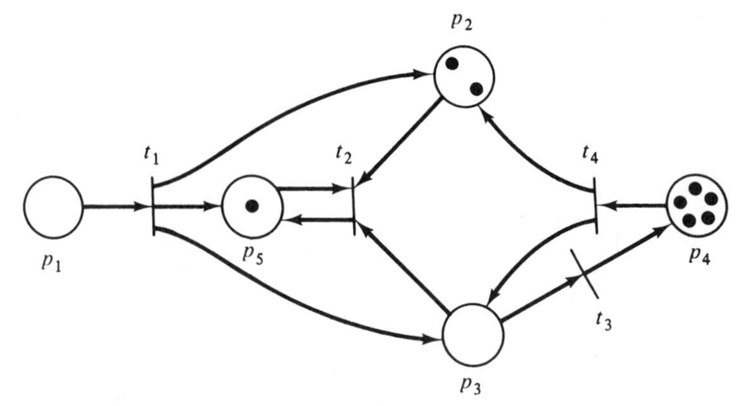

Figure 2.12 A marked Petri net. The structure is the same as the structure of Figure 2.11, but the marking is different.

representing the places of a Petri net graph. The convention in this case is to write the number of tokens in the place as in Figure 2.13. Using this convention, draw the Petri net of Figure 2.11 with a marking of $\mu = (137, 22, 2, 0, 14)$.

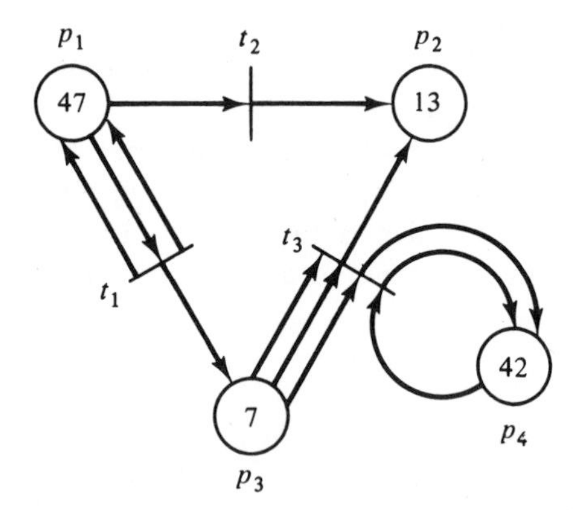

Figure 2.13 A Petri net graph with a very large marking, $(47, 13, 7, 42)$.

2.4 Execution Rules for Petri Nets

The *execution* of a Petri net is controlled by the number and distribution of tokens in the Petri net. Tokens reside in the places and control the execution of the transitions of the net. A Petri net executes by *firing* transitions. A transition *fires* by removing tokens from its input places and creating new tokens which are distributed to its output places.

A transition may fire if it is *enabled.* A transition is enabled if each of its input places has at least as many tokens in it as arcs from the place to the transition. Multiple tokens are needed for multiple input arcs. The tokens in the input places which enable a transition are its *enabling tokens.* For example, if the inputs to transition t_4 are places p_1 and p_2, then t_4 is enabled if p_1 has at least one token and p_2 has at least one token. For a transition t_7 with input bag $\{p_6, p_6, p_6\}$, place p_6 must have at least three tokens to enable t_7.

DEFINITION 2.6 A transition $t_j \in T$ in a marked Petri net $C = (P, T, I, O)$ with marking μ is *enabled* if for all $p_i \in P$,

$$\mu(p_i) \geqslant \#(p_i, I(t_j))$$

A transition *fires* by removing all of its enabling tokens from its input places and then depositing into each of its output places one token for each arc from the transition to the place. Multiple tokens are produced for multiple output arcs. A transition t_3 with $I(t_3) = \{p_2\}$ and $O(t_3) = \{p_7, p_{13}\}$ is enabled whenever there is at least one token in place p_2. Transition t_3 fires by removing one token from place p_2 and depositing one token in place p_7 and one token in place p_{13} (its outputs). Extra tokens in place p_2 are not affected by firing t_3 (although they may enable additional firings of t_3). A transition t_2 with $I(t_2) = \{p_{21}, p_{23}\}$ and $O(t_2) = \{p_{23}, p_{25}, p_{25}\}$ fires by removing one token from p_{21} *and* one token from p_{23} and then deposits one token in p_{23} and two tokens in p_{25} (since p_{25} has a multiplicity of two).

Firing a transition will in general change the marking μ of the Petri net to a new marking, μ'. Notice that since only enabled transitions may fire, the number of tokens in each place always remains non-negative when a transition is fired. Firing a transition can never try to remove a token which is not there. If there are not enough tokens in any input place of a transition, then the transition is not enabled and cannot fire.

DEFINITION 2.7 A transition t_j in a marked Petri net with marking μ may *fire* whenever it is enabled. Firing an enabled transition t_j results in a new marking μ' defined by

$$\mu'(p_i) = \mu(p_i) - \#(p_i, I(t_j)) + \#(p_i, O(t_j))$$

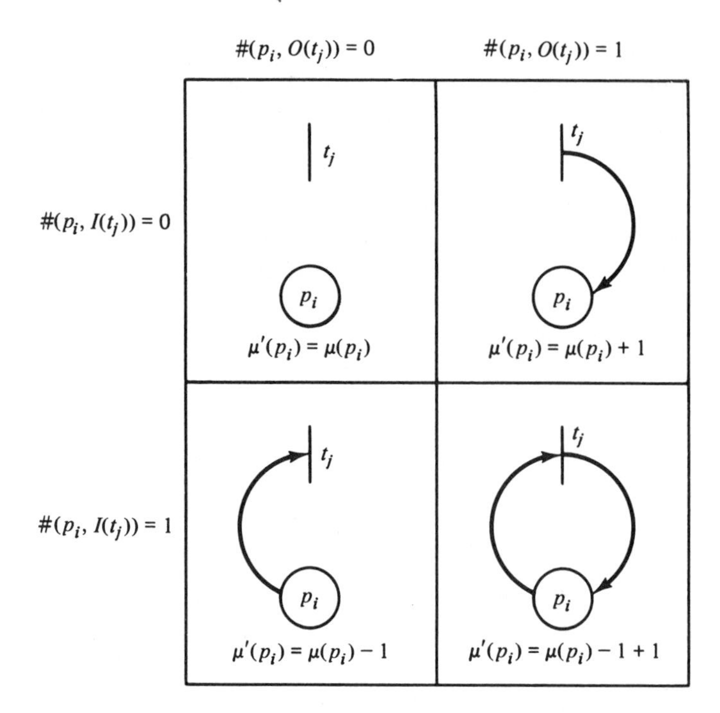

Figure 2.14 Illustrating how the marking of a place changes when a transition t_j fires. Each place may or may not be an input or output of the transition. This figure illustrates only the cases for a multiplicity of zero or one.

As an example, consider the marked Petri net of Figure 2.15. With this marking, three transitions are enabled, transitions t_1, t_3, and t_4. Transition t_2 is not enabled because there is no token in either place p_2 or p_3, which are both inputs of transition t_2. Since transitions t_1, t_3, and t_4 are enabled, any of them may fire. If transition t_4 fires, it removes a token from each input and deposits a token in each output. This removes the token in p_5, deposits one token in p_3, and increases the number of tokens in p_4 from two to three. Thus, the new marking which results from firing transition t_4 is shown in Figure 2.16.

In the marked Petri net of Figure 2.16, only transitions t_1 and t_3 are enabled. Firing transition t_1 will remove the token in p_1 and deposit tokens in p_2, p_3, and p_4 (two tokens in p_4 since it is a multiple output of transition t_1). This produces the marking of Figure 2.17. In this marked Petri net, transitions t_2 and t_3 are enabled. Firing transition t_3 produces the marking of Figure 2.18 where two tokens have been removed from p_4 and one has been added to p_5.

Transition firings can continue as long as there exists at least one enabled transition. When there are no enabled transitions, the execution *halts*.

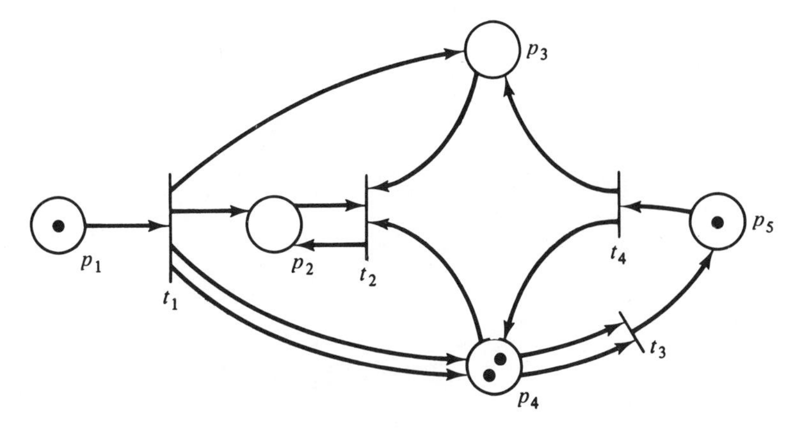

Figure 2.15 A marked Petri net to illustrate the firing rules. Transitions t_1, t_3, and t_4 are enabled.

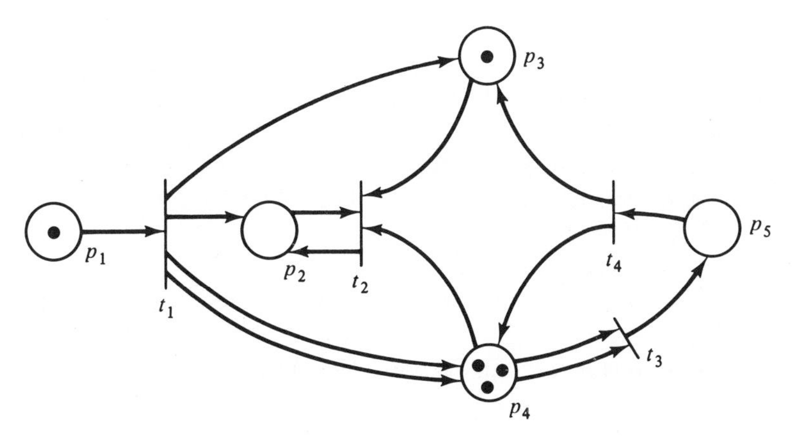

Figure 2.16 The marking resulting from firing transition t_4 in Figure 2.15.

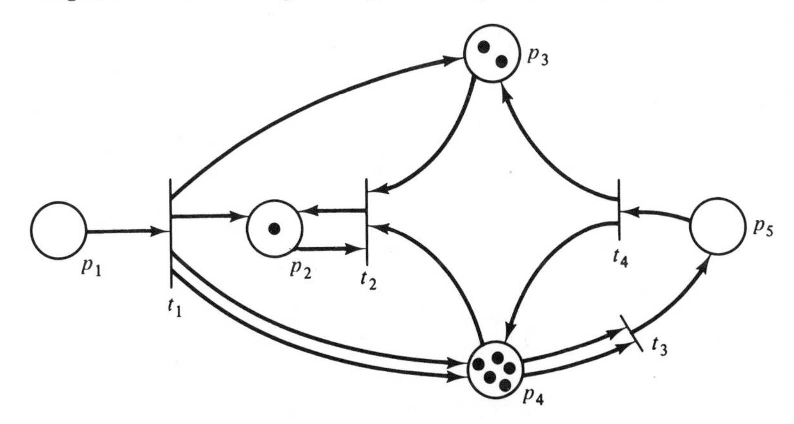

Figure 2.17 The marking resulting from firing transition t_1 in Figure 2.16.

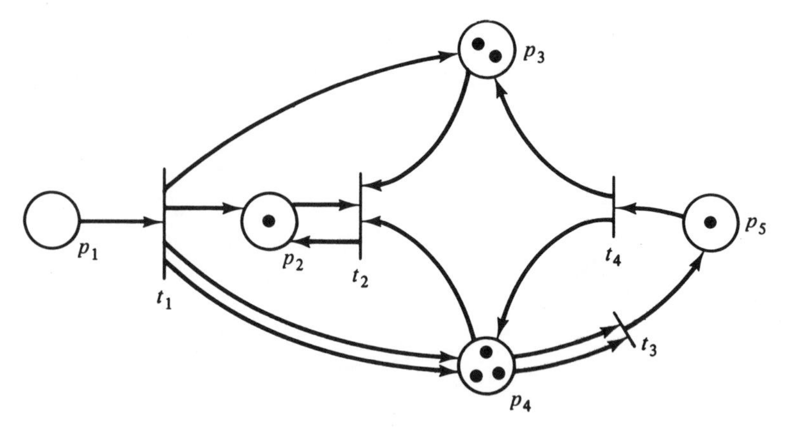

Figure 2.18 The marking resulting from firing transition t_3 in Figure 2.17.

Exercises

1. What transitions are enabled in the marked Petri net of Figure 2.11? Of Figure 2.12?
2. What marking results from firing transition t_1 in Figure 2.11? From firing transition t_4 in Figure 2.12? From firing first transition t_2 and then t_4 in Figure 2.12?
3. What transitions are enabled in Figure 2.13? For each of these transitions, what marking results from firing that transition in Figure 2.13?
4. Can transitions be fired in Figure 2.19? Which?

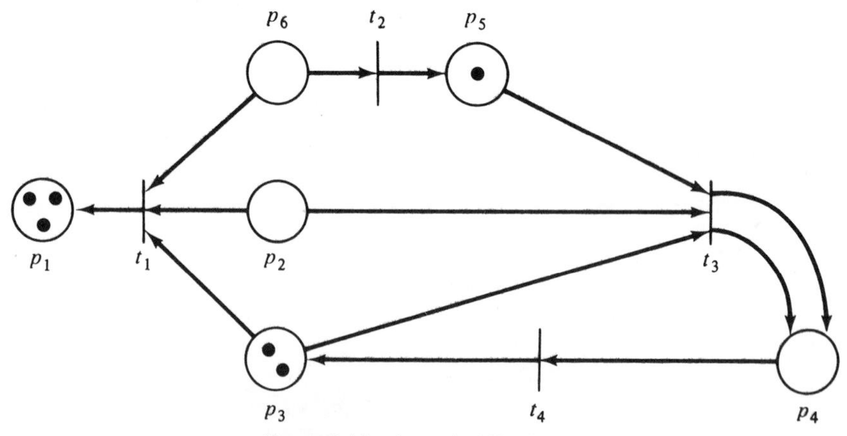

Figure 2.19 A marked Petri net.

2.5 Petri Net State Spaces

The *state* of a Petri net is defined by its marking. The firing of a transition represents a change in the state of the Petri net by a change in the marking of the net. The *state space* of a Petri net with n places is the set of all markings, that is, N^n. The change in state caused by firing a transition is defined by a change function δ called the *next-state function.* When applied to a marking (state) μ and a transition t_j this function yields the new marking (state) which results from firing transition t_j in marking μ. Since t_j can fire only if it is enabled, $\delta(\mu,t_j)$ is *undefined* if t_j is not enabled in marking μ. If t_j is enabled, then $\delta(\mu,t_j) = \mu'$, where μ' is the marking which results from removing tokens from the inputs of t_j and adding tokens to the outputs of t_j.

DEFINITION 2.8 The *next-state function* $\delta: N^n \times T \longrightarrow N^n$ for a Petri net $C = (P,T,I,O)$ with marking μ and transition $t_j \in T$ is defined if and only if

$$\mu(p_i) \geqslant \#(p_i,I(t_j))$$

for all $p_i \in P$. If $\delta(\mu,t_j)$ is defined, then $\delta(\mu,t_j) = \mu'$, where

$$\mu'(p_i) = \mu(p_i) - \#(p_i,I(t_j)) + \#(p_i,O(t_j))$$

for all $p_i \in P$.

Given a Petri net $C = (P,T,I,O)$ and an initial marking μ^0, we can execute the Petri net by successive transition firings. Firing an enabled transition t_j in the initial marking produces a new marking $\mu^1 = \delta(\mu^0,t_j)$. In this new marking, we can fire any new enabled transition, say t_k, resulting in a new marking $\mu^2 = \delta(\mu^1,t_k)$. This can continue as long as there is at least one enabled transition in each marking. If we reach a marking in which no transition is enabled, then no transition can fire, the next-state function is undefined for all transitions, and the execution must halt.

Two sequences result from the execution of a Petri net: the *sequence of markings* $(\mu^0,\mu^1,\mu^2,\ldots)$ and the *sequence of transitions* which were fired $(t_{j_0},t_{j_1},t_{j_2},\ldots)$. These two sequences are related by the relationship $\delta(\mu^k,t_{j_k}) = \mu^{k+1}$ for $k = 0, 1, 2, \ldots$. Given a transition sequence and μ^0, we can easily derive the marking sequence for the execution of the Petri net, and, except for a few degenerate cases, given the marking sequence, we can derive the transition sequence. Both of these sequences thus provide a record of the execution of the Petri net.

In a marking μ, a set of transitions will be enabled and may fire. The result of firing a transition in a marking μ is a new marking μ'. We say that μ' is *immediately reachable* from μ; that is, we can immediately get to state μ' from state μ.

> DEFINITION 2.9 For a Petri net $C = (P,T,I,O)$ with marking μ, a marking μ' is *immediately reachable* from μ if there exists a transition $t_j \in T$ such that $\delta(\mu,t_j) = \mu'$.

We can extend this concept to define the set of reachable markings for a given marked Petri net. If μ' is immediately reachable from μ and μ'' is immediately reachable from μ', then we say that μ'' is *reachable* from μ. We define the *reachability set* $R(C,\mu)$ of a Petri net C with marking μ to be all markings which are reachable from μ. A marking μ' is in $R(C,\mu)$ if there is any sequence of transition firings which will change marking μ into marking μ'. The "reachability" relationship is the reflexive transitive closure of the "immediately reachable" relationship.

> DEFINITION 2.10 The *reachability set* $R(C,\mu)$ for a Petri net $C = (P,T,I,O)$ with marking μ is the smallest set of markings defined by,
>
> 1. $\mu \in R(C,\mu)$.
> 2. If $\mu' \in R(C,\mu)$ and $\mu'' = \delta(\mu',t_j)$ for some $t_j \in T$, then $\mu'' \in R(C,\mu)$.

For the Petri net of Figure 2.20, and the marking $\mu = (1,0,0)$, two markings are immediately reachable: (0, 1, 0) and (1, 0, 1). From (0, 1, 0), no markings are reachable since no transition is enabled. However from (1, 0, 1), we can reach (0, 1, 1) and (1, 0, 2). Using techniques which are developed in Chapter 4, we can show that the reachability set $R(C,\mu)$ is $\{(1,0,n), (0,1,n) | n \geq 0\}$.

It is convenient to extend the next-state function to map a marking and a *sequence* of transitions into a new marking. For a sequence of

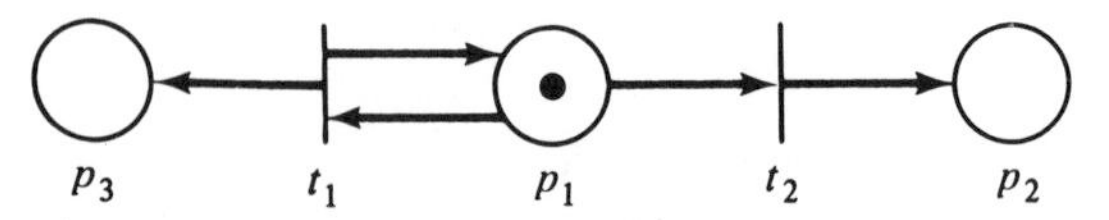

Figure 2.20 A marked Petri net.

transitions $t_{j_1} t_{j_2} \cdots t_{j_k}$ and marking μ, the marking $\mu' = \delta(\mu, t_{j_1} t_{j_2} \cdots t_{j_k})$ is the result of firing first t_{j_1}, then t_{j_2}, and so on until t_{j_k} is fired. (This is possible, of course, only if each transition is enabled when it is to be fired.)

DEFINITION 2.11 The *extended next-state function* is defined for a marking μ and a sequence of transitions $\sigma \in T^*$ by

$$\delta(\mu, t_j \sigma) = \delta(\delta(\mu, t_j), \sigma)$$
$$\delta(\mu, \lambda) = \mu$$

In general, we use this extended next-state function.

Exercises

1. For the marked Petri net of Figure 2.21 and the transition sequence $t_1 t_2 t_3 t_4 t_5$, give the corresponding marking sequence. For the marking sequence, (1, 0, 0), (0, 0, 1), (0, 0, 0), what is the corresponding transition sequence?

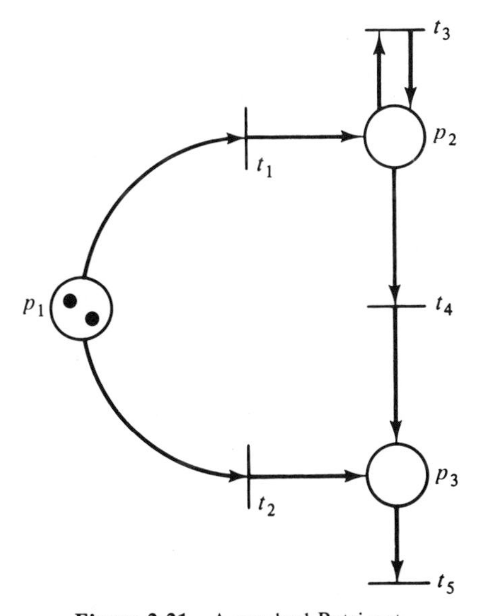

Figure 2.21 A marked Petri net.

2. It was said that there are a few degenerative cases in which a marking sequence does not define a unique transition firing sequence. Characterize the class of Petri nets for which this may be the case.
3. Show that $\bigcup_{\mu \in N^n} R(C,\mu) = N^n$.
4. Prove that if $\mu' \in R(C,\mu)$, then $R(C,\mu') \subseteq R(C,\mu)$.
5. Prove that $\mu' \in R(C,\mu)$ if and only if $R(C,\mu') \subseteq R(C,\mu)$.
6. Is $R(C,\mu) = \bigcup_{t_j \in T} \delta(\mu,t_j)$?
7. Is $R(C,\mu) = \bigcup_{t_j \in T} R(C,\delta(\mu,t_j))$?
8. Some of the literature on Petri nets refers to the reachability set of a marked Petri net as its *marking class*. More specifically, the *forward* marking class of a Petri net is what we have defined as the reachability set. The *backward* marking class is the reachability set of the inverse Petri net. (See Exercise 9 of Section 2.2.) The marking class of a marked Petri net is then the union of the forward and backward marking classes. Give formal definitions of the forward marking class, the backward marking class, and the marking class of a Petri net C with marking μ. Then show that the marking classes of a Petri net partition the set of all markings into disjoint equivalence classes. This requires showing that the relationship of having equal marking classes is reflexive, symmetric, and transitive.
9. It has been observed that Petri nets with their tokens and firing rules are very similar to board games, such as checkers, backgammon, nim, go, and so on. One can imagine a game for one to four players consisting of a board with a Petri net and a collection of plastic tokens. Tokens are distributed on the places of the Petri net, and players take turns selecting enabled transitions and firing them. The game would be sold with a set of 20 different Petri nets to provide variety. Starting with these ideas, develop a complete set of rules including
   (a) How is the initial distribution of tokens determined? (Each player starts with one token on the "home" place, or each player gets n tokens to place on the board as desired, or . . . .)
   (b) What is the goal of the game? (Capture the tokens of your opponents or create the greatest number of tokens, or reduce the number of tokens you have to zero as fast as possible, or . . . .)
   (c) Consider the possibilities of colored tokens to distinguish players, and consider appropriate definitions of firing rules.
   (d) Consider assigning points to the different transitions. A player's points are determined by the sum of the transitions fired by that player.

(e) Consider any problems which might arise such as repeated firings of transitions which create new tokens (more outputs than inputs) and the finite number of tokens provided with the game set.

After you have defined your game, try playing it with some friends. Use the Petri net of Figure 2.22 as a playing board.

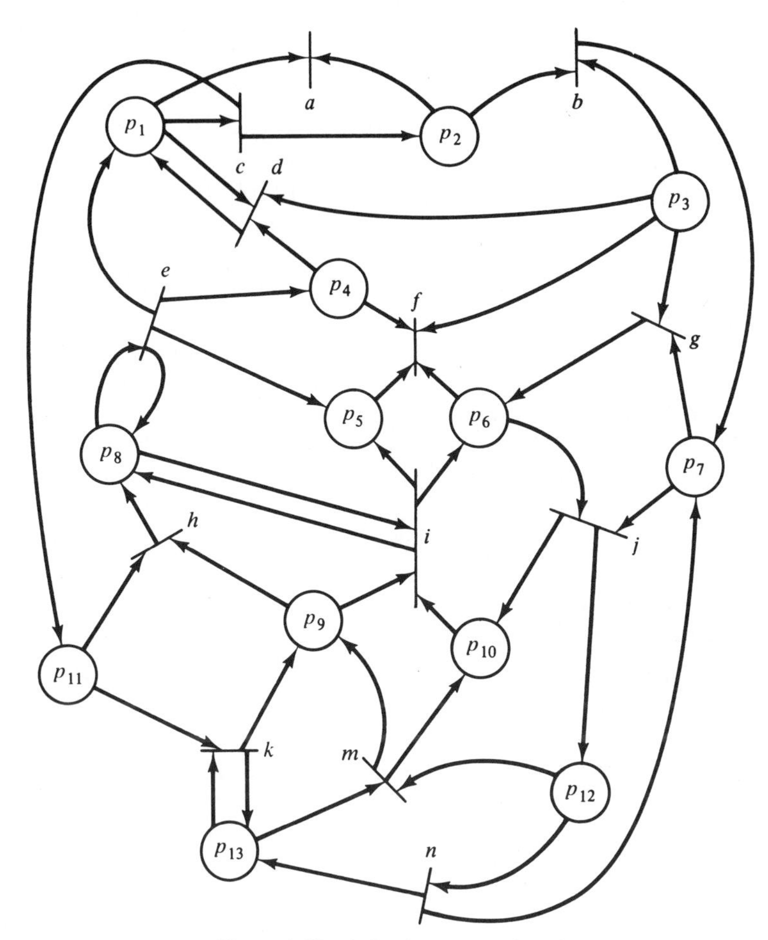

Figure 2.22 A Petri net game board.

2.6 Alternative Forms for Defining Petri Nets

The theory of Petri nets has been developed by a number of people working at different times in different places with different backgrounds and motivations. In part due to this diversity, many of the fundamental concepts have been defined by different researchers in different ways. We present some of these variant definitional forms here to show that there is no substantial difference in the definitions and to prepare the reader for the varying representations which may be encountered in the literature.

Petri's original nets [Petri 1962a], for example, did not allow multiple arcs between places and transitions. This is equivalent to defining the inputs and outputs of a transition to be sets (not bags) of places. Further, the firing rule was limited to requiring that a token reside in each input place to a transition and *no* tokens reside in the output places. A transition fired by removing the tokens from its inputs (which now became empty) and placing tokens in the (previously empty) outputs (which now became full). A transition could not fire if a token already resided in an output place. Thus a marking assigned either zero or one token to each place, $\mu: P \longrightarrow \{0, 1\}$. It should be obvious then that a net with only n places has exactly 2^n possible markings, a finite number of states.

The early work at ADR by Holt and the Information System Theory Project [Holt et al. 1968] continued with these same definitions, but as work progressed, the limitations of this model were recognized. The work of Holt and Commoner presented at the Woods Hole conference [Holt and Commoner 1970] generalized the class of markings and the firing rule to allow arbitrary markings, $\mu: P \longrightarrow \{0, 1, 2, \ldots\}$. This then defined the basic Petri net model as it is defined today (with the exception of the multiple arc feature).

Many of the early researchers did not formally define their models but rather described informally the relevant components, such as places, transitions, tokens, and the firing rule. One of the first formal definitions was by Patil [1970a] in his Ph.D. dissertation where a Petri net was defined as a four-tuple (T,P,A,B) with T the set of transitions, P the set of places, A a set of arcs, and B an initial marking. The arcs in the set A connected either a place with a transition or a transition with a place. Thus $A \subseteq (P \times T) \cup (T \times P)$. Many papers on Petri nets are based on this definition and define a Petri net as a triple (P,T,A) with a separate marking function.

The conversion from the (P,T,A) form of definition to separate input and output functions is relatively straightforward. The set of arcs A is split into a set of input arcs $\{(p_i,t_j) | (p_i,t_j) \in A\}$ and output arcs $\{(t_j,p_i) | (t_j,p_i) \in A\}$. This form leads directly to the generalization allowing multiple inputs and outputs. It is necessary simply to attach a multiplicity to each input or output arc.

Hack [1975c] eventually settled on a definition of Petri nets as a four-tuple (P,T,F,B), with P the set of places and T the set of transitions, as usual. F and B are functions mapping places and transitions onto the number of tokens needed for input (F) or produced for output (B). Thus a transition t_j can fire only if at least $F(t_j,p_i)$ tokens are in each place $p_i \in P$. A transition fires by removing $F(t_j,p_i)$ tokens from each input place and depositing $B(t_j,p_i)$ tokens in each output place. The functions F and B can be represented by matrices.

Peterson in his dissertation [Peterson 1973] tried to combine transitions and their inputs and outputs by defining a transition as an ordered pair of bags of places, $t_j \in P^\infty \times P^\infty$. The first component of the pair is the bag of inputs to the transition; the second component is the bag of outputs of the transition. This reduced the primitive concepts of the theory to places and tokens, since transitions are structures composed of places. It was particularly useful in allowing the easy definition of a transition for a constructed Petri net.

These definitions vary from the one presented here only by notational differences. For most work on Petri nets, this is the case: The differences in definition are strictly notational. However in some cases, the definitions may restrict the class of Petri nets by not allowing multiple input or output arcs or otherwise restricting the form of the transitions; e.g., transitions are required to have a nonnull set of input places and a nonnull set of output places, or the input and output places of a transition must be disjoint (self-loop-free). Even these differences are not important, as is seen in Chapter 5.

Exercises

1. For a Petri net defined by (P,T,F,B), give the equivalent standard (P,T,I,O) definition. For a Petri net defined as (P,T,I,O), give the (P,T,F,B) definition.
2. Why would some researchers prefer a (P,T,A) definition which mixes input and output relationships into the arc set A, while others prefer (P,T,I,O)? Give an advantage and disadvantage of each.

2.7 Further Reading

Few descriptions of Petri nets concentrate simply on the basic definitions. The *Computing Surveys* tutorials by Baer [1973a] and Peterson [1977] are probably the best bets for continued introduction.

Almost any paper on Petri nets will present the basic definitions and notation in the first few sections. Hence you could simply scan the beginning of some of the papers listed in the bibliography to find a variety of definitions. Try Holt et al. [1968], Holt and Commoner [1970], Hack [1974a; 1975b; 1975c], Keller [1975a], Misunas [1973], Murata [1977a], and Thomas [1976].

2.8 Topics for Further Study

1. Develop Petri net theory to allow colored tokens. Consider the changes to the definitions of enabled transitions and transition firings. There are at least three reasonable ways to extend Petri nets for colored tokens; indicate as many as you can think of and evaluate their usefulness.
2. Develop an introduction to Petri net theory for noncomputer scientists. Compare your presentation with those of Meldman and Holt [1971] and Meldman [1977; 1978].
3. Construct a computer simulation system for Petri net execution. To allow a convenient interface for the user of the program, use an erasable graphics display (such as a CRT or plasma panel) to display the Petri net and to show the firing of transitions and consequent movement of tokens. This will involve many subproblems.
   (a) The language used to define the Petri net and its markings and to select options must be defined.
   (b) An internal representation of the Petri net and its marking must be constructed, along with the necessary algorithms for simulation.
   (c) The display must be created. A major problem here will be trying to achieve a planar representation of the net (to the degree reasonable). Also careful thought must be given to the representation of the dynamic properties of the net (token movement).

3

Modeling with Petri Nets

Petri nets were designed for and are used mainly for *modeling*. Many systems, especially those with independent components, can be modeled by a Petri net. The systems may be of many different kinds: computer hardware, computer software, physical systems, social systems, and so on. Petri nets are used to model the occurrence of various events and activities in a system. In particular, Petri nets may model the flow of information or other resources within a system.

In this chapter, we present several examples of the types of systems which have been modeled by Petri nets. From this presentation, you will gain an understanding of the large class of systems which can be modeled by Petri nets, some of the modeling techniques which are used, and some of the properties which are desired for the modeled systems.

3.1 Events and Conditions

The simple Petri net view of a system concentrates on two primitive concepts: *events* and *conditions*. Events are actions which take place in the system. The occurrence of these events is controlled by the state of the system. The state of the system can be described as a set of conditions. A condition is a predicate or logical description of the state of the system. As such, a condition may either *hold* (be true) or *not hold* (be false).

Since events are actions, they may *occur*. For an event to occur, it may be necessary for certain conditions to hold. These are the *preconditions* of the event. The occurrence of the event may cause the preconditions to cease to hold and may cause other conditions, *postconditions*, to become true.

As an example, consider a simple machine shop modeling problem. The machine shop waits until an order appears and then machines the ordered part and sends it out for delivery. The conditions for the system are

a. The machine shop is waiting.
b. An order has arrived and is waiting.
c. The machine shop is working on the order.
d. The order is complete.

The events would be

1. An order arrives.
2. The machine shop starts on the order.
3. The machine shop finishes the order.
4. The order is sent for delivery.

The preconditions of event 2 (the machine shop starts on the order) are obvious: (*a*) the machine shop is waiting, and (*b*) an order has arrived and is waiting. The postcondition of event 2 is (*c*) the machine shop is working on the order. Similarly we can define the preconditions and postconditions of the other events and construct the following chart of events and their preconditions and postconditions:

Event	***Preconditions***	***Postconditions***
1	None	*b*
2	*a*,*b*	*c*
3	*c*	*d*,*a*
4	*d*	None

This view of a system can be easily modeled as a Petri net. Conditions are modeled by places in a Petri net; events are modeled by transitions. The inputs of a transition are the preconditions of the corresponding event; the outputs are the postconditions. The occurrence of an event corresponds to the firing of the corresponding transition. The holding of a condition is represented by a token in the place corresponding to the condition. When the transition fires it removes the enabling tokens representing the holding of the preconditions and creates new tokens which represent the holdings of postconditions.

The Petri net of Figure 3.1 is a Petri net model of the machine shop example given above. We have labeled each transition and place with the corresponding event or condition.

More complicated systems can also be modeled. The machine shop may have three different machines, M_1, M_2, and M_3 and two operators, F_1 and F_2. Operator F_1 can operate machines M_1 and M_2 while operator F_2 can operate machines M_1 and M_3. Orders require two stages of machining. First they must be machined by machine M_1 and then by either machine M_2 or M_3. This more complex system would have the following conditions.

a. An order has arrived and is waiting for machining by M_1.
b. An order has been processed by M_1 and is waiting to be processed by M_2 or M_3.
c. The order is complete.
d. Machine M_1 is idle.
e. Machine M_2 is idle.
f. Machine M_3 is idle.
g. Operator F_1 is idle.
h. Operator F_2 is idle.
i. Machine M_1 is being operated by F_1.
j. Machine M_1 is being operated by F_2.
k. Machine M_2 is being operated by F_1.
l. Machine M_3 is being operated by F_2.

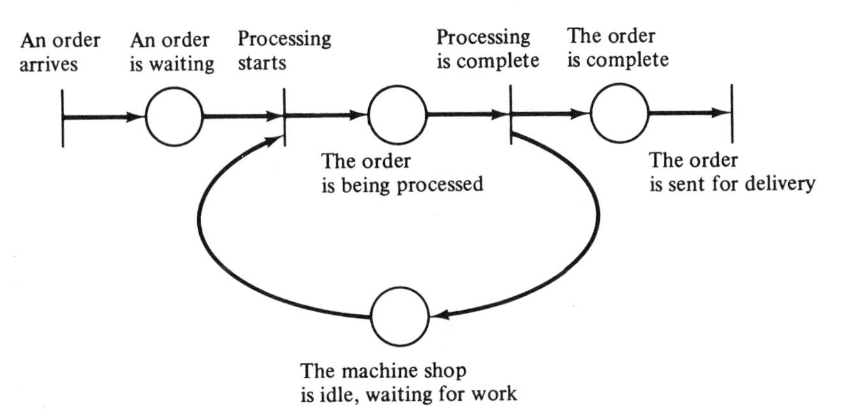

Figure 3.1 A Petri net model of a simple machine shop.

The following events can occur:

1. An order arrives.
2. Operator F_1 starts the order on machine M_1.
3. Operator F_1 finishes the order on machine M_1.
4. Operator F_2 starts the order on machine M_1.
5. Operator F_2 finishes the order on machine M_1.
6. Operator F_1 starts the order on M_2.
7. Operator F_1 finishes the order on M_2.
8. Operator F_2 starts the order on M_3.
9. Operator F_2 finishes the order on M_3.
10. The order is sent for delivery.

The preconditions and postconditions of each event are

Event	***Preconditions***	***Postconditions***
1.	None	a
2.	a,g,d	i
3.	i	g,d,b
4.	a,h,d	j
5.	j	b,h,d
6.	b,g,e	k
7.	k	c,g,e
8.	b,f,h	l
9.	l	c,f,h
10.	c	None

The Petri net for this system is shown in Figure 3.2.

A similar example can be drawn from a computer system which processes jobs from an input device and outputs the results on an output device. Jobs appear on the input device. When the processor is free and there is a job on the input device, the processor starts to process the job. When the job is complete, it is sent to the output device; the processor either continues with another job if one is available or waits until one arrives if there is no job yet on the input device. This system can be modeled by the Petri net of Figure 3.3.

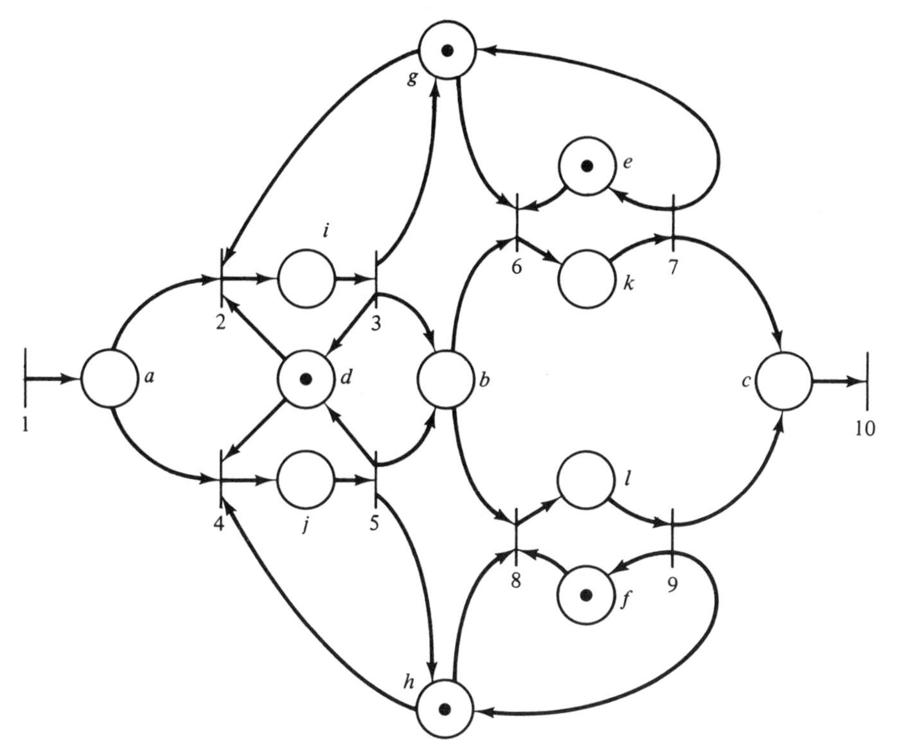

Figure 3.2 An example of a more complex machine shop, modeled by a Petri net.

3.2 Concurrency and Conflict

These examples illustrate several points about Petri nets and the systems which they can model. One is the inherent *parallelism* or *concurrency.* In the Petri net model, two events which are both enabled and do not interact may occur independently. There is no need to synchronize events unless it is required by the underlying system which is being modeled. When synchronization is needed, it is easy to model this also. Thus, Petri nets would seem ideal for modeling systems of distributed control with multiple processes executing concurrently in time.

Another major feature of Petri nets is their *asynchronous* nature. There is no inherent measure of time or the flow of time in a Petri net. This reflects a philosophy of time which states that the only important property of time, from a logical point of view, is in defining a partial ordering of the occurrence of events. Events take variable amounts of

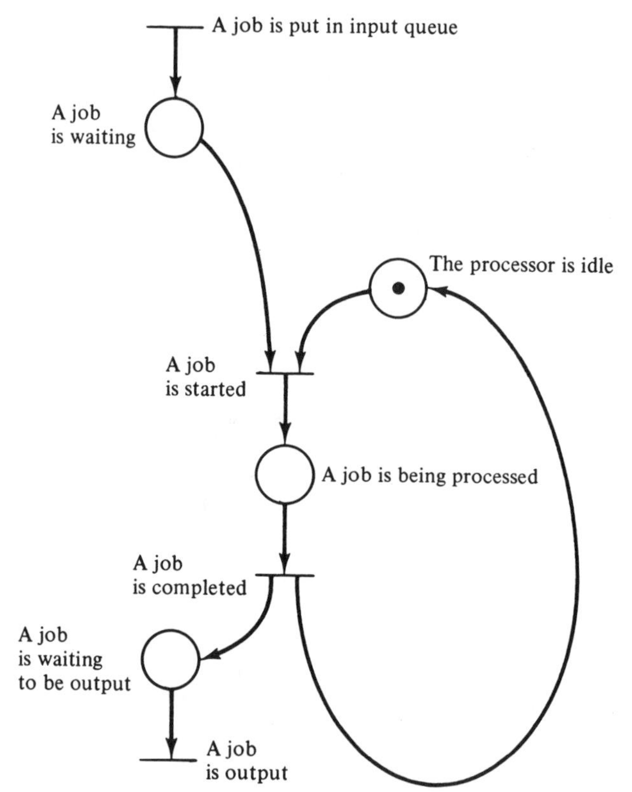

Figure 3.3 Modeling of a simple computer system.

time in real life, and this variability is reflected in the Petri net model by not depending on a notion of time to control the sequence of events. The Petri net structure itself contains all necessary information to define the possible sequences of events. Thus, in Figure 3.3, the event "A job is completed" must follow the corresponding "A job is started" event. However, no information at all is given or needed concerning the amount of time to execute a job.

A Petri net execution (and the system behavior which it models) is viewed here as a *sequence* of discrete events. The order of occurrence of the events is one of possibly many allowed by the basic structure. This leads to an apparent *nondeterminism* in Petri net execution. If, at any time, more than one transition is enabled, then any of the several enabled transitions may be "the next" to fire. From the point of view of the classical execution model, the choice as to which transition fires is made in a nondeterministic manner, i.e., randomly. This feature of Petri nets reflects the fact that in real life situations in which several things are happening concurrently, the apparent order of occurrence of events is not unique, but rather any of a set of sequences of events may occur. However, the partial ordering in which events occur is unique.

The questions involved with these concepts can get quite philosophical in nature. For example, I, personally, tend toward a deterministic view of the universe: All actions are predetermined by the state of the universe, and there is no randomness. Randomness is merely a result of incomplete knowledge of the state of the universe and its individual transitions. In this sense, the selection of one of a set of enabled transitions to fire is determined in the modeled system, but not in the model simply because the model does not represent the complete information about the system.

The theory of relativity should also be considered. One of the basic tenets of relativity theory is that communication is not instantaneous, but rather the information about the occurrence of an event propagates through space at a speed limited by the speed of light, c. The meaning of this is that if two events can occur simultaneously, that is, with no causal relationship, then the order of occurrence may appear different for two separate observers. For two events A and B which occur at essentially the same time, an observer stationed near event A would receive the information concerning event A before the information concerning event B could propagate to the observer. The observer would then deduce that event A occurred before event B. A separate observer stationed near B, on the other hand, could determine that exactly the opposite sequence of events occurred.

These considerations, although necessary for a complete understanding of events, introduce considerable complexity in the description and analysis of the dynamic behavior of a Petri net when viewed as a sequence of transition firings. To help limit this complexity, one limitation in the modeling of systems by Petri nets is generally accepted. The firing of a transition (and the associated event) is considered to be an *instantaneous* event, taking zero time, and the occurrences of two events cannot happen simultaneously. The events modeled are called *primitive* events; primitive events are instantaneous and *nonsimultaneous.* (It is sometimes argued that time is a continuous real variable. Hence if we assign a time of occurrence to each event, the probability of any two separately chosen continuous real variables being identically equal is zero, and hence events are nonsimultaneous.)

A nonprimitive event is an event which does not take zero time. Nonprimitive operations are not nonsimultaneous and hence may overlap in time. Since most events in the real world take time, they are nonprimitive events and hence cannot be properly modeled by transitions in a Petri net. However this need not cause problems in the modeling of a system. A nonprimitive event can be decomposed into two primitive events, "The nonprimitive event starts" and "The

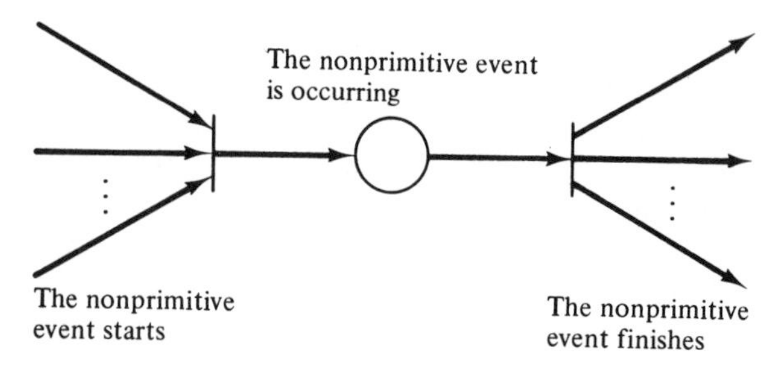

Figure 3.4 Modeling a nonprimitive event.

nonprimitive event finishes," and a condition, "The nonprimitive event is occurring." This can be modeled as shown in Figure 3.4.

Petri and others have suggested that nonprimitive events should be represented by a box in a Petri net [Petri 1975] as shown in Figure 3.5, with primitive events represented by bars, as we have in the past. This would simplify some Petri nets, such as Figure 3.6, which is equivalent to the Petri net of Figure 3.3. However, since the suggested concept can, in principle, be explained in terms of more primitive constructs, we do not use the box notation in this text. The box notation can be of considerable value when modeling a complex system at several hierarchical levels, since it allows entire subnets to be abstracted to a single element of the net. It is in some sense similar to the subroutine or macro concept of programming languages.

The nondeterministic and nonsimultaneous firing of transitions in the modeling of concurrent systems shows up in two ways. One of these is shown in Figure 3.7. In this situation, the two enabled transi-

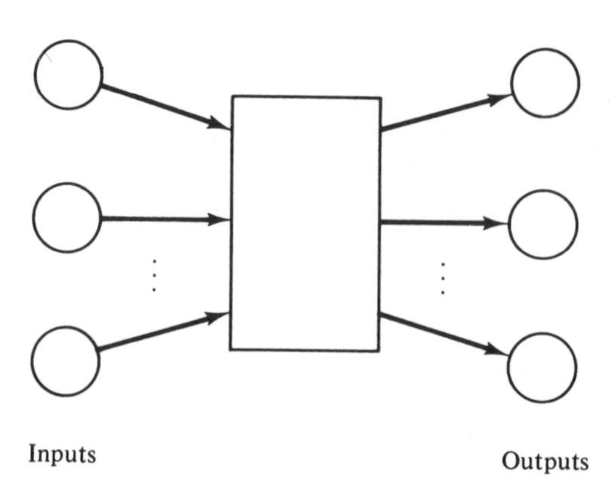

Figure 3.5 A box is sometimes used to represent a nonprimitive event.

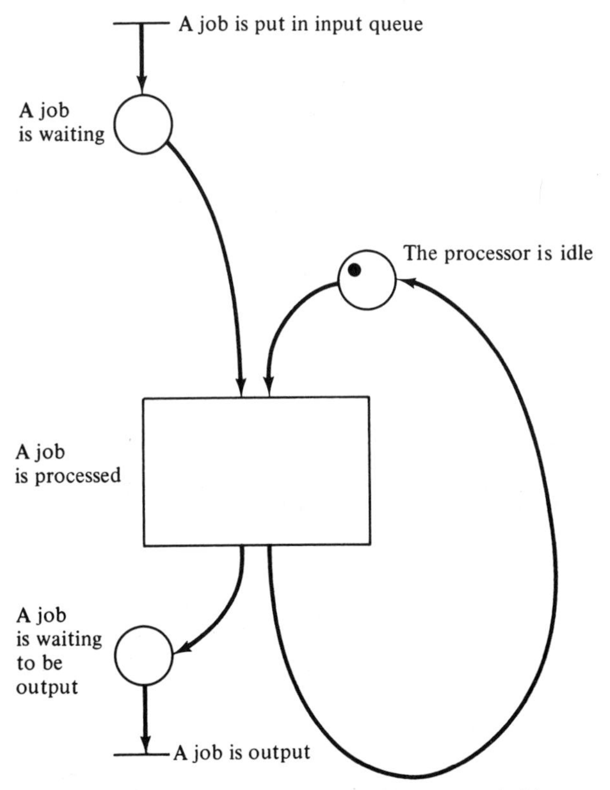

Figure 3.6 Modeling of a computer system using a nonprimitive transition. This net is equivalent to Figure 3.3.

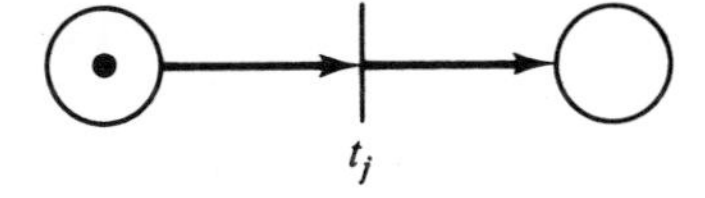

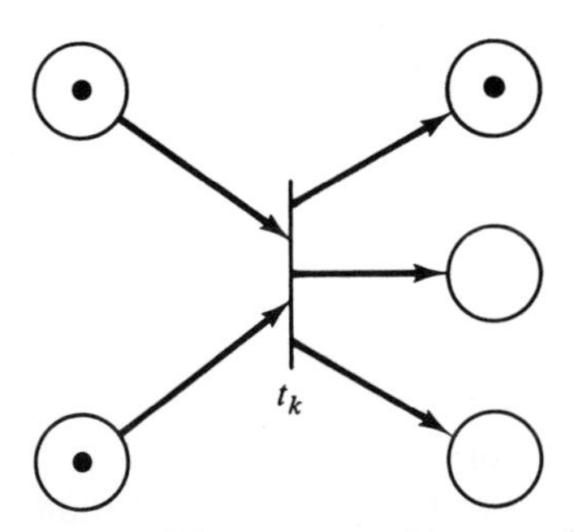

Figure 3.7 Concurrency. These two transitions can fire in any order.

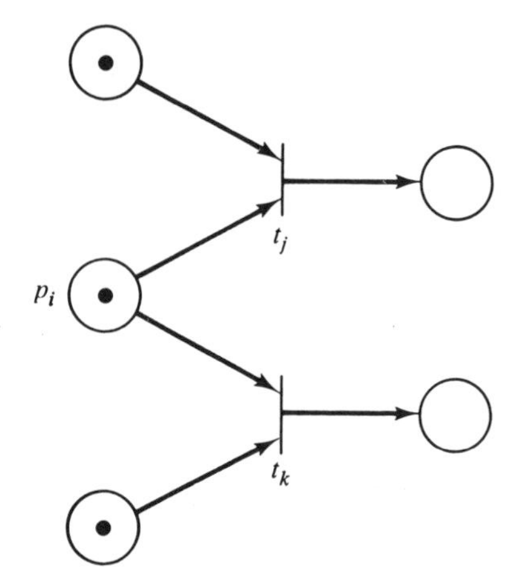

Figure 3.8 Conflict. Transitions t_j and t_k are in conflict since firing either will remove the token from p_i, disabling the other transition.

tions do not affect one another in any way, and the possible sequences of events include some in which one transition occurs first, and some in which the other occurs first. This is termed *concurrency.*

The other situation, where simultaneousness is more difficult to handle and which may be handled by defining events to occur non-simultaneously, is illustrated by Figure 3.8. Here the two enabled transitions are in *conflict.* Only one transition can fire, since, in firing, it removes the token in the shared input and disables the other transition.

These considerations require that systems to be modeled by Petri nets be carefully understood in order to properly model the system behavior. Unfortunately most of the work on Petri nets has been in the investigation of the properties of a given net or class of nets. Little explicit attention has been paid to developing modeling techniques specifically for Petri nets. However, there are certain areas in which Petri nets would seem to be the perfect tool for modeling: those areas in which events occur asynchronously and independently. To give an understanding of Petri net modeling, we show in this chapter how Petri nets can be used to model computer hardware, computer software, and other systems.

3.3 Computer Hardware

Computer hardware can be thought of at several levels, and Petri nets can model each of these levels. At one level, computers are constructed of simple memory devices and gates; at a higher level, functional units and registers are used as the fundamental components of the system. At still a higher level, entire computer systems may be the components in a multicomputer network. One of the powerful features of Petri nets is their ability to model each of these levels. We demonstrate this power by a short discussion and some examples.

3.3.1 Finite State Machines

At the lowest level, computer systems can be described as *state* machines. A state machine is a five-tuple $(Q, \Sigma, \Delta, \delta, \Gamma)$ where

Q is a finite set of states $\{q_1, q_2, \ldots, q_k\}$.

Σ is a finite input alphabet.

Δ is a finite output alphabet.

$\delta: Q \times \Sigma \rightarrow Q$ is the next-state function, mapping the current state and current input into the next state.

$\Gamma: Q \times \Sigma \rightarrow \Delta$ is the output function, mapping the current state and input into the output symbol.

State machines are often represented by a state diagram, such as Figure 3.9. In a state diagram, states are represented by circles which are the nodes of the graph. An arc from state q_i to state q_j labeled a/b means that in state q_i with input a the machine will change to state q_j while outputting the symbol b. Formally, we would have $\delta(q_i, a) = q_j$ and $\Gamma(q_i, a) = b$. The input alphabet defines the inputs to the machine from the outside world, while the output alphabet defines the outputs of the machine to the outside world.

For example, consider the state machine of Figure 3.9. This state machine converts a binary number presented serially low-order bit first to its two's complement negative. Its input alphabet and output alphabet consist of three symbols: 0, 1, and R. The machine starts in state q_1. The reset symbol (R) signals the end (or beginning) of a number

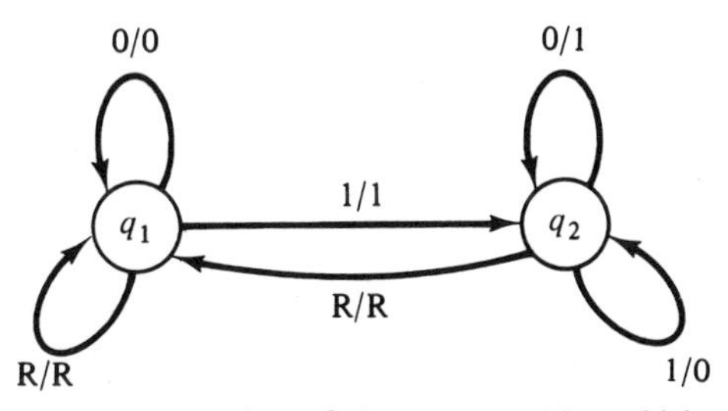

Figure 3.9 A state diagram for a finite state machine which computes the twos' complement of a binary number.

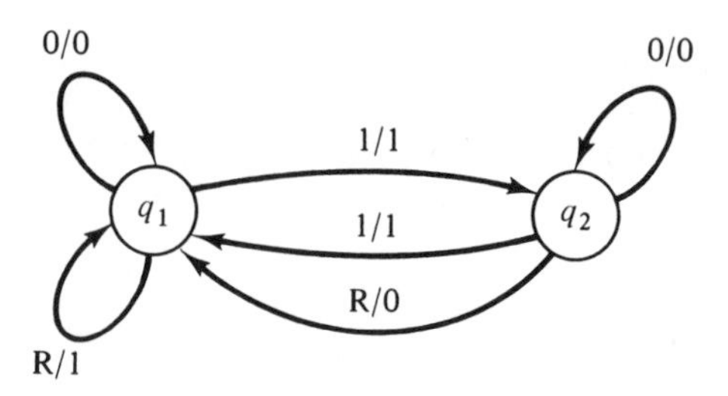

Figure 3.10 A state machine for computing the parity of an input binary number.

and resets the machine to its initial state. The output of the machine for the reset symbol is simply an echo of the reset symbol.

A similar state machine is diagrammed in Figure 3.10. Under the same inputs, this state machine computes the parity of the input number. The machine starts in state q_1. The output merely copies the input until the input symbol is a reset symbol. The output for the reset symbol is 0 for a number with odd parity and 1 for a number with even parity.

Representing a finite state machine as a Petri net requires a little thought since there has been no mention of communication between Petri nets and the outside world. Petri nets are generally studied in isolation. Modeling interactions with the outside world can be done in many ways. For the current problem, we model this interaction with a special set of places. Each input symbol will be represented by a place, and each output symbol will also be represented by a place. We assume that the outside world will deposit a token in the place corresponding to an input symbol and then wait for a token to appear in a place corresponding to an output symbol which will then be removed. This sequence will then repeat as long as desired. Figure 3.11 illustrates the general scheme.

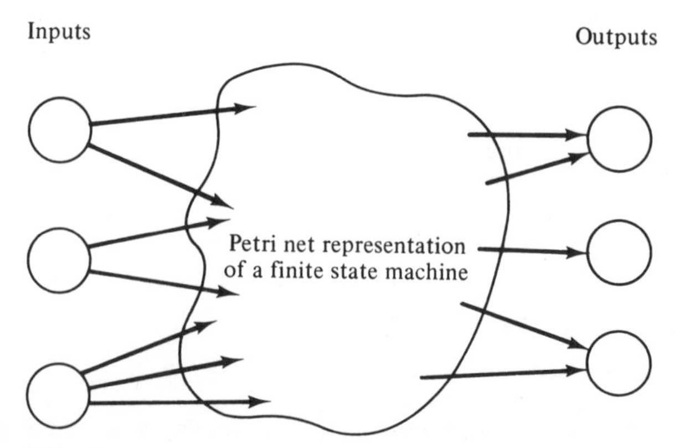

Figure 3.11 A general approach to modeling the communication between a Petri net and the outside world.

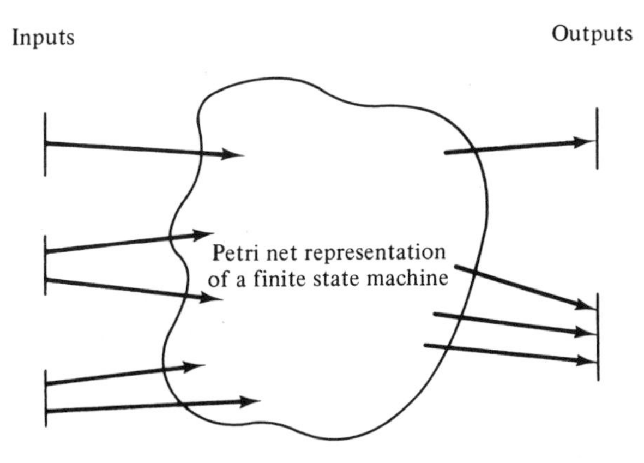

Figure 3.12 An alternative approach to representing communication between a Petri net and the outside world, using transitions instead of places.

Note that there is a potential for notational confusion here since the places associated with the input symbols and output symbols could reasonably be called input places and output places of the net. This should not be confused with the input places of a transition or output places of a transition, however. Despite this potential confusion, the terms are the most natural ones for both concepts.

An alternative approach to modeling the inputs and outputs of the net would be to use transitions. To indicate the next input symbol, the outside world would select an input transition and fire it. The Petri net would respond by (eventually) firing the appropriate one of a set of output transitions corresponding to the appropriate output. The outside world would then fire the next input transition, and so on. This is illustrated in Figure 3.12. These two approaches can easily be shown to be equivalent, so we use the first approach, with places modeling input and output symbols.

Given the place representation of input and output symbols, we can complete the modeling of finite state systems. We represent each state of the state machine by a place in the Petri net. The current state is marked by a token; all other places are empty. Now transitions can be defined to change state and define outputs as follows. For each pair of state and input symbol, we define a transition whose input places are the places corresponding to the state and input symbol and whose output places are the places corresponding to the next state and the output.

For a finite state machine $(Q, \Sigma, \Delta, \delta, \Gamma)$ we define a Petri net (P,T,I,O) by

$$P = Q \cup \Sigma \cup \Delta$$
$$T = \{t_{q,\sigma} \mid q \in Q \text{ and } \sigma \in \Sigma\}$$
$$I(t_{q,\sigma}) = \{q, \sigma\}$$
$$O(t_{q,\sigma}) = \{\delta(q,\sigma), \Gamma(q,\sigma)\}$$

This Petri net is a model of the finite state machine.

Figure 3.13 is the Petri net corresponding to the state machine of Figure 3.9. Figure 3.14 is the Petri net corresponding to Figure 3.10.

Comparing the Petri nets of Figures 3.13 and 3.14 with the equivalent state machines of Figures 3.9 and 3.10, several questions come to mind. The first is: Why would the Petri net model be preferable to the finite state machine description? The state machine description is more understandable than the Petri net description with its 6 transitions, 24 arcs and 7 or 8 places. This is admitted. However, we have shown that Petri nets can represent any system which can be represented by a state machine, thus demonstrating the power of the Petri net model.

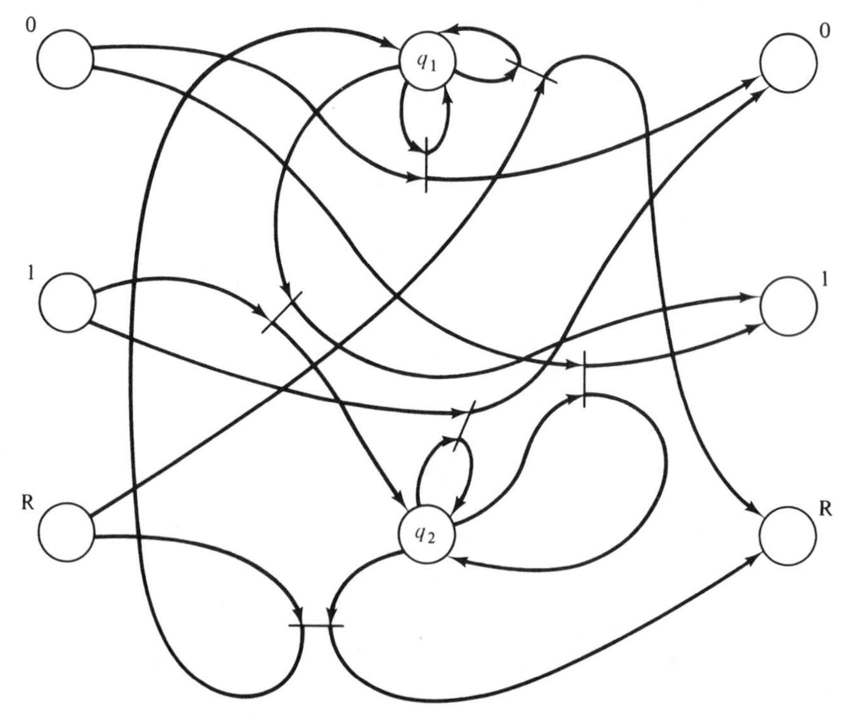

Figure 3.13 A Petri net equivalent to the state machine of Figure 3.9.

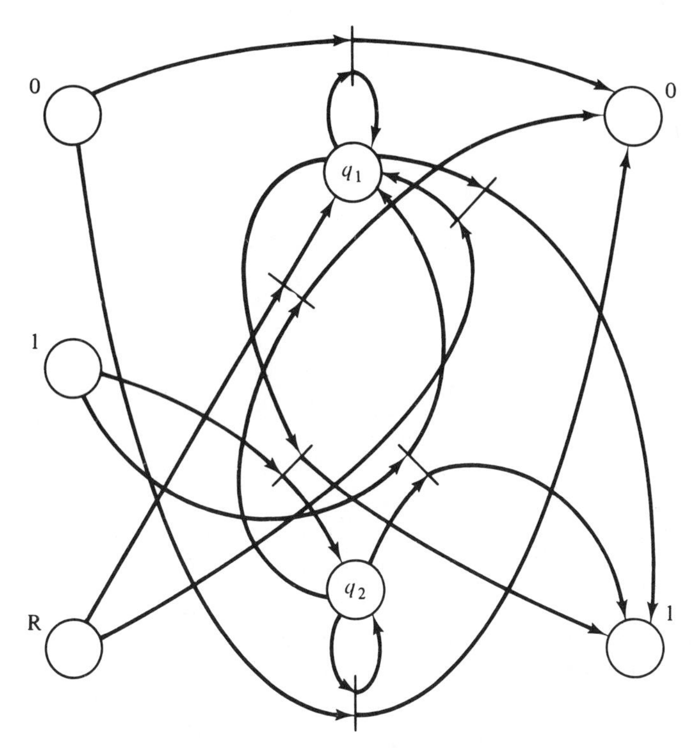

Figure 3.14 A Petri net equivalent to the state machine of Figure 3.10.

In addition the Petri net model has certain advantages in the combination of machines. For example, note that the output alphabet of the machine of Figure 3.13 is identical to the input alphabet of Figure 3.14. By running the output of Figure 3.13 into the input of Figure 3.14, we can construct a composite machine which computes the two's complement negative and its parity. This combination in a state machine is complex, requiring a composite state with components of both submachines, a *cross-product* machine. For a Petri net machine, on the other hand, the composition is simply the overlapping of the output places of the first net with the input places of the second net. Figure 3.15 shows the cross-product machine, while Figure 3.16 shows the composite Petri net machine.

Another advantage of the Petri net representation is with other forms of composition. For example, a parallel composition allows the component machines to execute simultaneously. For a state machine, this again involves a cross-product machine, while for a Petri net, it involves simply duplicating the input tokens which represent the input symbols, feeding these into each component Petri net machine. Finally, on output we simply select the appropriate output places. For

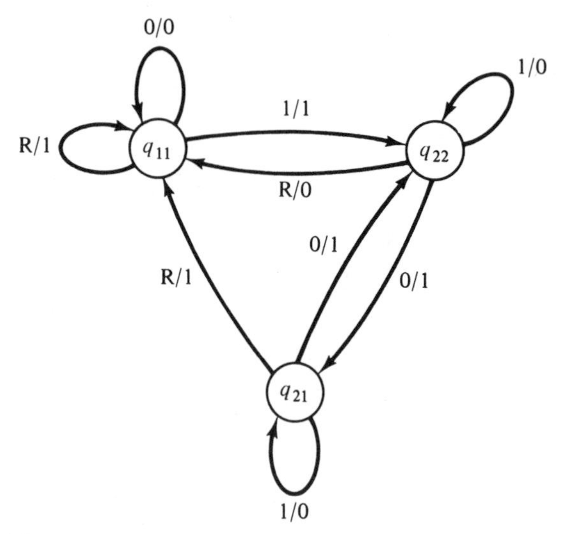

Figure 3.15 The composite machine representing the serial composition of the state machines of Figures 3.9 and 3.10.

example, if we wish to combine the two Petri net machines of Figures 3.13 and 3.14 in parallel, this would result in a Petri net like Figure 3.17, which computes the two's complement negative of a number and its parity. The parity is output when the reset symbol is input.

Exercises

1. Show that the two approaches to modeling interactions between a Petri net and its environment (using transitions or using places) are equivalent.

3.3.2 Pipelined Computers

The ability to model parallelism and to easily combine subsystems modeled as Petri nets makes the Petri net model very useful for modeling more complex computer hardware. Computer systems are constructed of many components, and many designs attempt to increase throughput by executing several functions in parallel. This makes the Petri net a particularly appropriate representation of the system.

An example of this approach to the construction of a high-performance computer is the use of *pipelines* [Chen 1971]. This technique is similar to the operation of an assembly line and is especially useful for vector and array processing. A pipeline is composed of a

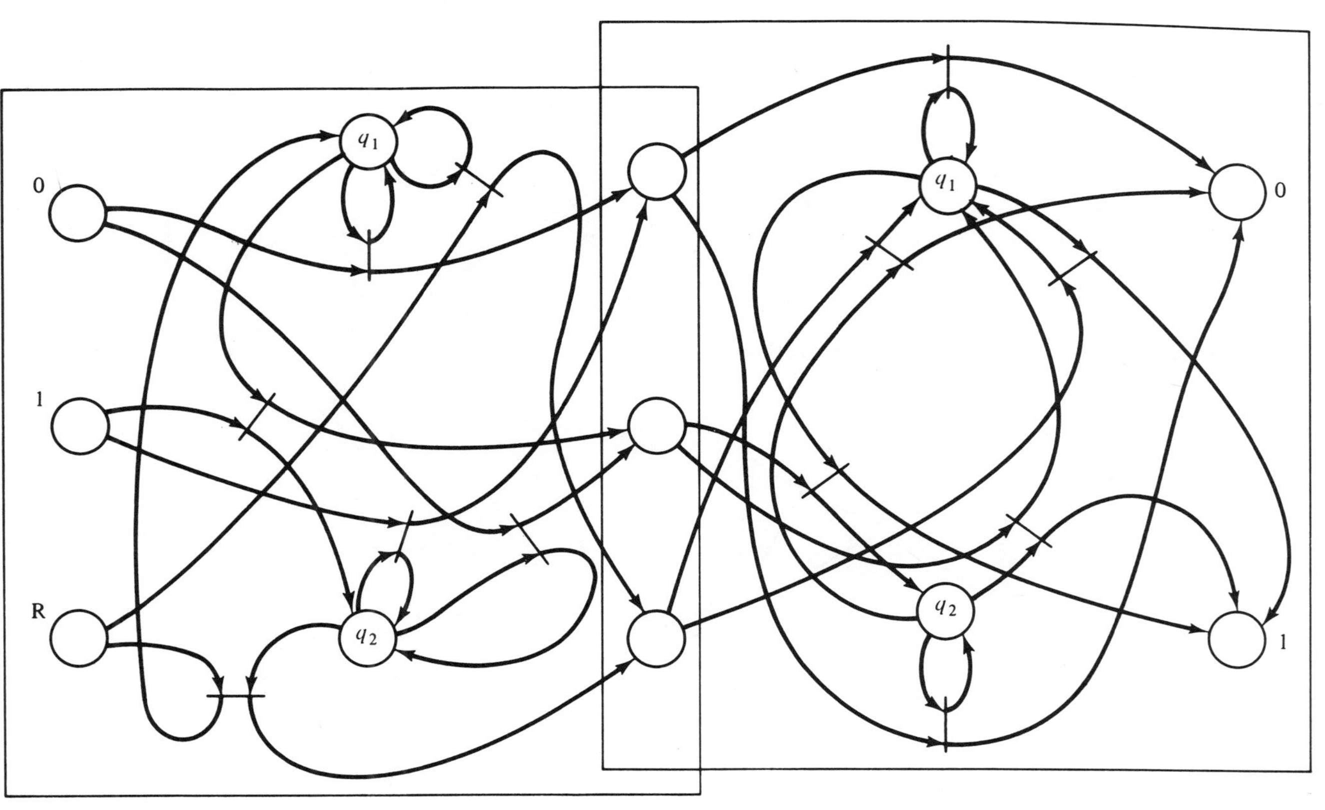

Figure 3.16 The composite Petri net machine which is a serial composition of the Petri nets of Figures 3.13 and 3.14.

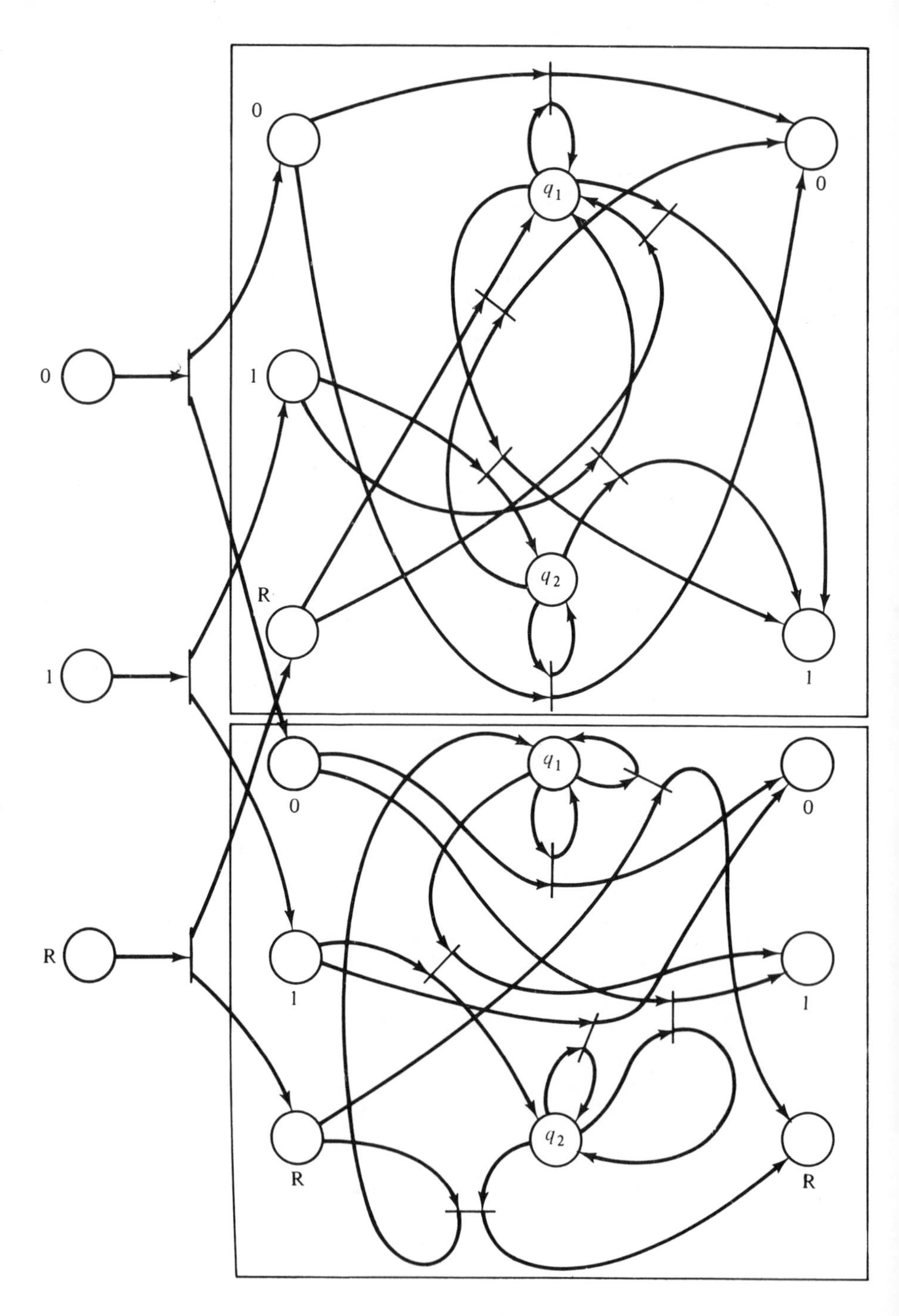

Figure 3.17 A parallel composition of the Petri net machines of Figures 3.13 and 3.14. A duplicator subnet is needed to provide inputs for both component Petri nets.

number of stages, which may be in execution simultaneously. When stage k finishes, it passes on its results to stage $(k+1)$ and looks to stage $(k-1)$ for new work. If each stage takes t time units and there are n stages, then the complete operation for one operand takes nt time units. However, if the pipeline is kept supplied with new operands, it can turn out results at the rate of one every t time units.

As an example, consider the addition of two floating point numbers. The gross steps involved are

1. Extract the exponents of the two numbers.
2. Compare the exponents, and interchange if necessary to properly order the larger and smaller of the exponents.
3. Shift the smaller fraction to equalize exponents.
4. Add fractions.
5. Postnormalize.
6. Consider exponent overflow or underflow and pack the exponent and fraction of the result.

Each of these steps can be performed by a separate computational unit, with a particular operand being passed from unit to unit for the complete addition operation. This would allow as many as six additions to be underway simultaneously.

The coordination of the different units can be handled in several ways. Typically, the pipeline control is synchronous; the time allowed for each step of the pipe is a fixed constant time t. Every t time units, the result of each unit is shifted down the pipeline to become the input for the next unit. The synchronous approach can unnecessarily hold up processing, however, since the time needed may vary from unit to unit and may also vary within a given unit for different inputs. For example, the postnormalization step in the floating point addition above may take different amounts of time depending on how long the normalization shift should be and whether it should be to the left or to the right. In this case, since the time t must be selected as the maximum time which could be needed by the slowest unit of the pipeline, it could well be the case that all units sit idle most of the time waiting for the remainder of the t time units.

An asynchronous pipeline can speed this up, on average, by signaling when each stage of the pipeline is complete and ready to pass its operand on and receive new operands. The results of stage k of the pipeline can be sent on to stage $(k+1)$ as soon as stage k is done and stage $(k+1)$ is free. Consider an arbitrary stage in the pipeline. Obviously, there must be a place to put the inputs and outputs while they are being used or produced. Typically, this involves registers: the unit uses the values in its input (buffer) register to produce values in its output (buffer) register. It must then wait until (1) its output register has been emptied by being copied into the input register of the next stage, and (2) a new input is available in its input register. Thus, the control for stage k of the pipeline needs to know when the following conditions hold:

- Input register full
- Input register empty
- Output register full
- Output register empty
- Unit busy
- Unit idle
- Copying taking place

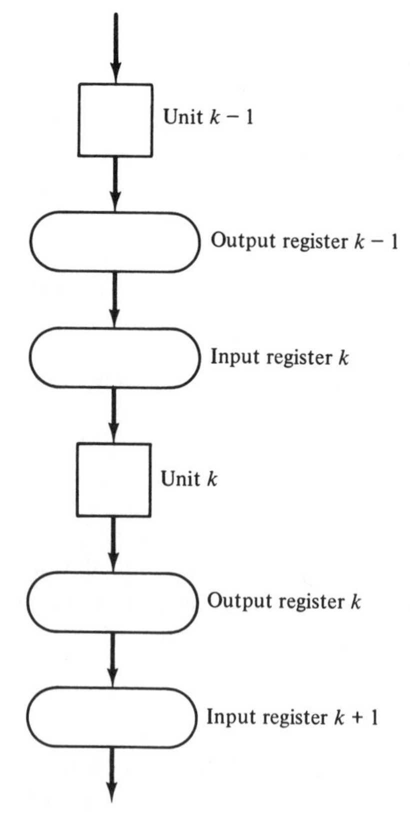

Figure 3.18 Block diagram of an asynchronous control unit for a pipelined computer.

Figures 3.18 and 3.19 show how an asynchronous pipeline of this sort can be modeled. Figure 3.18 is a block diagram of the pipeline that is modeled as a Petri net in Figure 3.19.

Notice that in this model we have modeled the actual execution of the units of the pipeline as nonprimitive events. This allows us to ignore, at this level, the specific details of what each unit does and concentrate on their proper interaction. Each operation could also be modeled as a Petri net. The Petri nets for each unit could then be substituted into the Petri net of Figure 3.19 to obtain a more detailed Petri net. This ability to model a system at several different levels of abstraction, in a hierarchical manner, can be very useful.

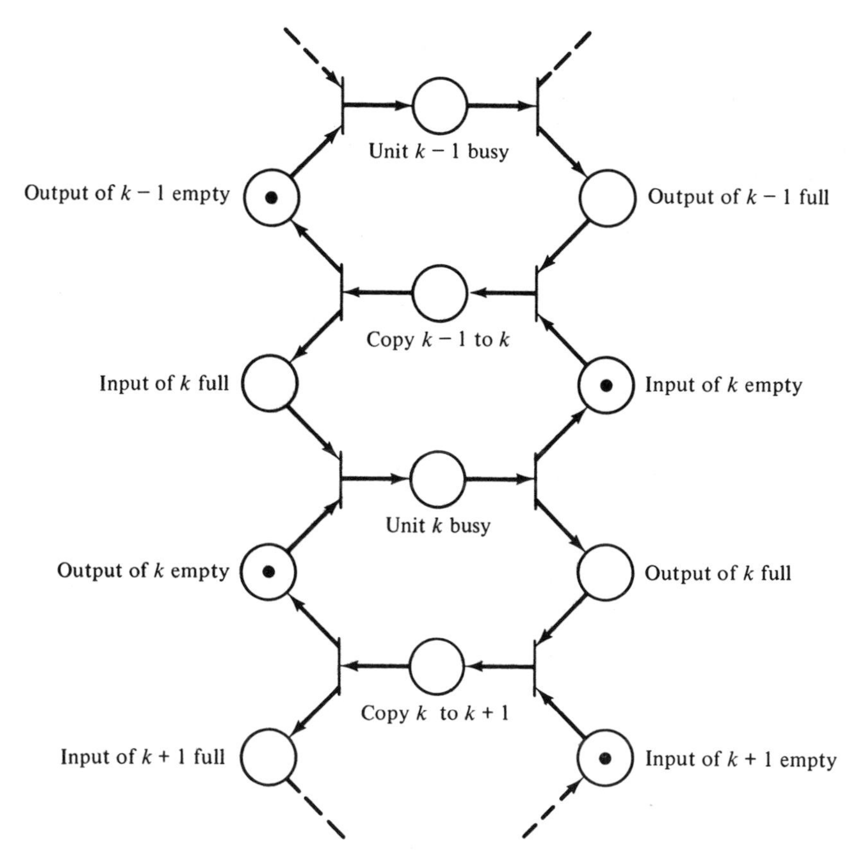

Figure 3.19 A Petri net model of the control unit of an asynchronous pipelined computer.

3.3.3 Multiple Functional Units

The pipelined control structure of the previous section is one approach used to build very large fast computer systems. Another approach, used in the CDC 6600 [Thornton 1970] and IBM 360/91 [Anderson et al. 1967], for example, is to provide *multiple functional units.* On the 6600, 10 functional units are available: 1 branch unit (for conditional jumps), 1 boolean unit (for Boolean operations), 1 shift unit, 1 floating point add unit, 1 fixed point add unit, 2 multiply units, 1 divide unit, and 2 increment units (for indexing). In addition multiple registers are provided to hold the inputs and outputs of the functional units. The control unit of the computer attempts to keep several of the independent units in operation simultaneously.

For example, consider the following sequence of instructions based on the CDC 6600 computer system.

1. Multiply X1 by X1, giving X0.
2. Multiply X3 by X1, giving X3.
3. Add X2 to X4, giving X4.
4. Add X0 to X3, giving X3.
5. Divide X0 by X4, giving X6.

When these instructions are executed, the control unit issues the first instruction to the multiply unit. Then, since there are two multiply units, the second instruction can also be issued. Notice that both units can read the contents of X1 with no problem. Instruction 3 can be issued to the add unit. Now to issue instruction 4, we must wait until instructions 1, 2, and 3 are complete since instruction 4 uses the add unit (which is in use by instruction 3) to process X0 (being computed by instruction 1) and X3 (being computed by instruction 2). Instruction 5 must wait for instruction 1 (to finish computing X0) and instruction 3 (to finish computing X4).

The introduction of this type of parallelism, executing several instructions of a program simultaneously, must be controlled so that the results of executing the program with and without parallelism are the same. Certain instructions in the program will require that the results of previous instructions have been successfully computed before the following instructions can proceed. A system which introduces parallelism into a sequential program in such a way as to maintain correct results is *determinate.* The conditions for maintaining determinacy have been considered by Bernstein [1966]. They are the following: For two operations a and b such that a precedes b in the linear precedence of the program, b can be started before a is done if and only if b does not need the results of a as inputs and the results of b do not change either the inputs or results of a.

A *reservation table* is one method of applying these constraints to the construction of a control unit that is to issue instructions to separate functional units. An instruction for functional unit u using registers i, j, and k can be issued only if all four of these components are not reserved; when the instruction is issued, all four of them become reserved. If the instruction cannot be issued at this time because either the functional unit or one of the registers is in use, the control unit waits until the instruction can be issued before continuing on to the next instruction.

This sort of scheme can be modeled as a Petri net. To each functional unit and each register, we associate a place. If the unit or register is free, a token will be in the place; if it is not, no token will be in the place. Multiple identical functional units can be indicated by multiple tokens in the places. Figure 3.20 shows a portion of a Petri net which could be used to model the execution of an instruction using unit u and registers i, j, and k. Modeling the entire control unit would of course require a much larger Petri net.

The scheme described above is a very simple method of introducing parallelism and does not consider, for example, the fact that multiple functional units can use the same register as an input simultaneously. Thus, this scheme may not produce schedules with maximal parallelism [Keller 1975b]. However, there are other schemes which

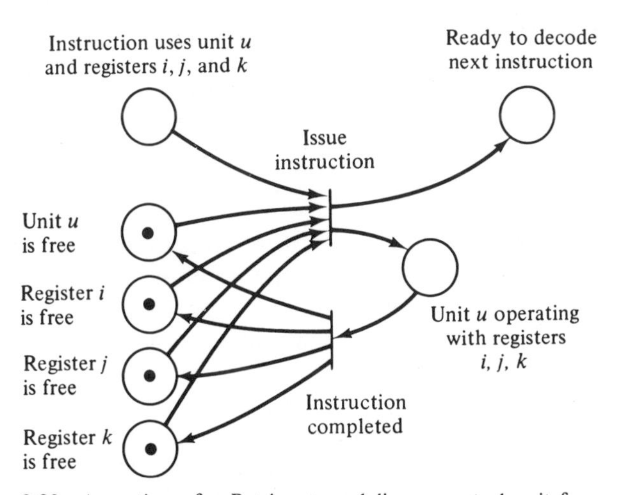

Figure 3.20 A portion of a Petri net modeling a control unit for a computer with multiple registers and multiple functional units.

can do so. These (more complicated) schemes can also be modeled by (more complicated) Petri nets. These nets may be very large. Consider that the CDC 6600 has 24 different registers and 64 different instructions. If each instruction and triple of registers needed a place corresponding to "unit *u* is operating with registers *i*, *j*, and *k*," then over a half million places and transitions would be needed. The main problem here is the difficulty of modeling the fact that the contents of an internal register may specify which registers and units are to be used (i.e., indexing).

(Any given program will not use all possible combinations of registers and units, however. This allowed Shapiro and Saint [1970] to model the 6600 computer system with a Petri net. This Petri net model was then used to optimize code generation for a FORTRAN compiler, as we shall see in Section 3.5.)

3.4 Computer Software

In addition to computer hardware, computer software can be modeled by Petri nets. This is perhaps the most common use of Petri nets and has the greatest potential for useful results. Many systems have been developed over the years for the description and modeling of computer hardware, but it is only in the past few years that efforts have been made to formally model computer software. Much of this recent effort is concerned with the analysis, specification and description of sequential programs; systems of concurrent processes are still major research topics. In this section we show how Petri nets can faithfully model many systems of concurrently executing cooperating processes.

3.4.1 Flowcharts

The degenerate case of a system of concurrent processes is a system with exactly one process. We first examine how a single process can be represented by a Petri net, and then by combining Petri nets representing several processes we would have a system of concurrent processes.

A single process is described by a program. This program can be written in many languages, but for convenience, let us assume a general-purpose language such as ALGOL, FORTRAN, PL/I, COBOL, Pascal, BASIC, or even assembly language. The program represents two separate aspects of the process: computation and control. Computation is concerned with the actual arithmetic and logical operations, the input and output, and general manipulation of memory locations and their values. Control, on the other hand, is not concerned with the values or computations being performed but only with the order of their performance.

Petri nets can best represent the control structure of programs. Petri nets are meant to model the sequencing of instructions and the flow of information and computation but not the actual information values themselves. A model of a system, by its nature, is an *abstraction* of the modeled system. As such it ignores the specific details as much as possible. If all the details were modeled, then the model would be a duplicate of the modeled system, not an abstraction.

One standard means of representing the control structure of a program is with a *flowchart.* A flowchart represents the flow of control in a program. For example, the program of Figure 3.21 is represented by the flowchart of Figure 3.22. Notice that the flowchart of Figure 3.22 does not specify the computations to be done, only the structure of the program. This flowchart is *uninterpreted.* Figure 3.23 shows how an interpretation can be applied to the actions of the flowchart to represent the program of Figure 3.21.

Every sequential program can be represented by a flowchart.

```
begin
    Input( y1);
    Input( y2);
    y3 := 1;
    while y1 > 0
        do begin
            if odd ( y1)
                then begin
                    y3 := y3 * y2;
                    y1 := y1 - 1;
                end;
            y2 := y2 * y2;
            y1 := y1 - 2;
        end;
    Output ( y3);
end;
```

Figure 3.21 A simple program. This program is modeled as a flowchart in Figure 3.22 and as a Petri net in Figure 3.25.

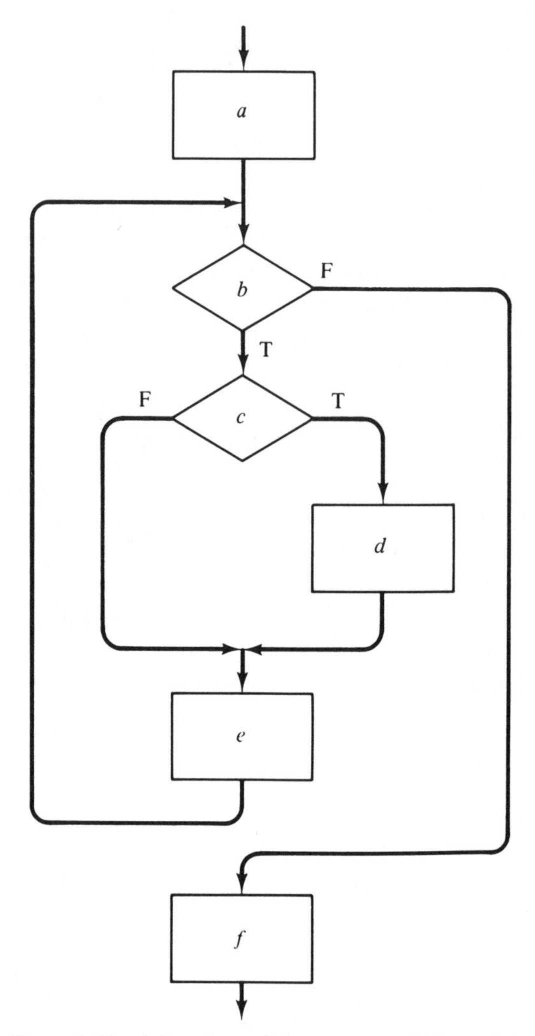

Figure 3.22 A flowchart of the program of Figure 3.21.

Thus, by showing how a flowchart can be represented by a Petri net, we have shown how to represent an uninterpreted program by a Petri net.

A flowchart would appear to be very similar to a Petri net: A flowchart is composed of nodes (of two types: decisions represented by the diamond shapes and computations represented by the rectangles) and arcs between them. A convenient way to execute a flowchart is to introduce a token which represents the current instruction. As the

Action	***Interpretation***
a	Input(y_1); Input (y_2); $y_3 := 1$;
b	$y_1 > 0$?
c	odd(y_1) ?
d	$y_3 := y_3 * y_2$; $y_1 := y_1 - 1$;
e	$y_2 := y_2 * y_2$; $y_1 := y_1 / 2$;
f	Output (y_3);

Figure 3.23 An interpretation of the actions of the flowchart of Figure 3.22 to represent the program of Figure 3.21.

instructions execute, the token moves around the flowchart. The similarity between these graphical representations of a program and a Petri net would seem to indicate that we can replace the nodes of the flowchart by places and the arcs by transitions to create an equivalent Petri net. This is the approach taken to converting a finite state machine into a Petri net (see Section 3.3.1).

However, consider that in the Petri net model, the transitions model actions, while in the flowchart model, the nodes model actions. Also, if our current-instruction token in a flowchart were to want to "rest," it would pause between nodes, on an arc, not within a box.

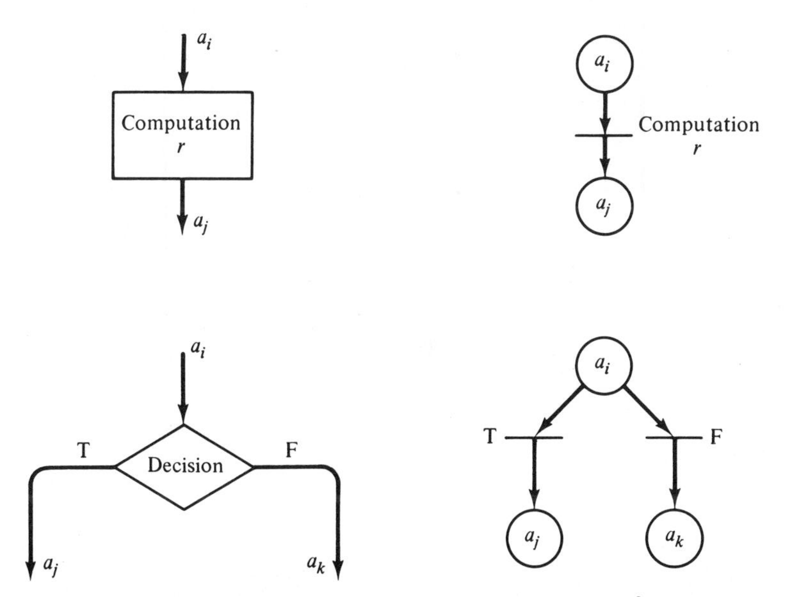

Figure 3.24 Translating computation and decision nodes in a flowchart to transitions in a Petri net.

Thus the appropriate translation from a flowchart to a Petri net replaces the nodes of the flowchart with transitions in the Petri net and the arcs of the flowchart with places in the Petri net. Each arc of the flowchart is represented by exactly one place in the corresponding Petri net. The nodes of the flowchart are represented in different ways, depending on the type of the node: computation or decision. Figure 3.24 illustrates the two methods of translation. Figure 3.25 applies this translation to the flowchart of Figure 3.22 to produce an equivalent Petri net.

A point or two to notice about the Petri net of Figure 3.25 con-

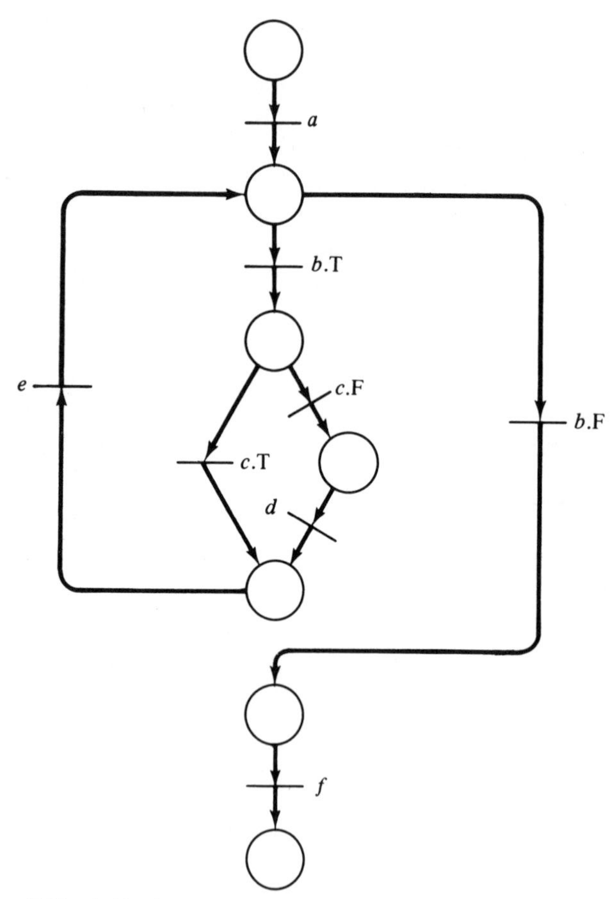

Figure 3.25 A Petri net representation of Figure 3.21, derived from the flowchart of Figure 3.22.

cerns the meaning of the Petri net components. What is the meaning of the places? The easiest answer is to consider the program counter interpretation of the flowchart token. In this sense, a token residing in a place means that the program counter is positioned ready to execute the next instruction. Every place has a unique output transition, except for places which precede decisions; these places have two output transitions corresponding to true and false outcomes of the decision predicate.

The transitions are obviously associated with the actions of the program: the computations and the decisions. If we wish to interpret the Petri net, we must provide an interpretation for each transition. Notice also that transitions for computational actions have a unique input and unique output, and that no conflict can exist for a transition representing a computation, since its input place is not an input place for any other transition. Decision actions do introduce conflict into the net, but in a very constrained way: either choice can be freely made. This choice can either be made nondeterministically (i.e., randomly) or may be controlled by some external force (i.e., by an agent which computes the truth or falsity of the decision and forces the correct transition to fire). The distinction between these two interpretations of conflict resolution is a matter of philosophy.

3.4.2 Parallelism

Parallelism or concurrency can now be introduced in several ways. Consider the case of two concurrent processes. Each process can be represented by a Petri net. Thus the composite Petri net which is simply the union of the Petri nets for each of the two processes can represent the concurrent execution of the two processes. The initial marking of the composite Petri net has two tokens, one in each place representing the initial program counter of a process. This introduces a parallelism which cannot be represented in a flowchart, but still not a very useful one.

Another approach is to consider how parallelism would normally be introduced into a process in a computer system. Several proposals have been made. One of the simplest involves the FORK and JOIN operations originally proposed by Dennis and Van Horn [1966]. A FORK j operation executed at location i results in the current process continuing at location $i+1$, and a new process being created with execution started at location j. A JOIN operation will recombine two processes into one (or equivalently will destroy one of the two and let the other proceed). These operations can be modeled in a Petri net as shown in Figure 3.26.

Another suggestion for introducing parallelism is the *parbegin* and *parend* control structure [Dijkstra 1968]. This control structure was suggested by Dijkstra and is of the general form *parbegin* $S_1;S_2;\ldots;S_n$ *parend*, where the S_i are statements. The meaning of the *parbegin/parend* structure is to execute each of the statements, $S_1,S_2, \ldots,S_n$ in parallel. This construct can be represented in a Petri net as shown in Figure 3.27.

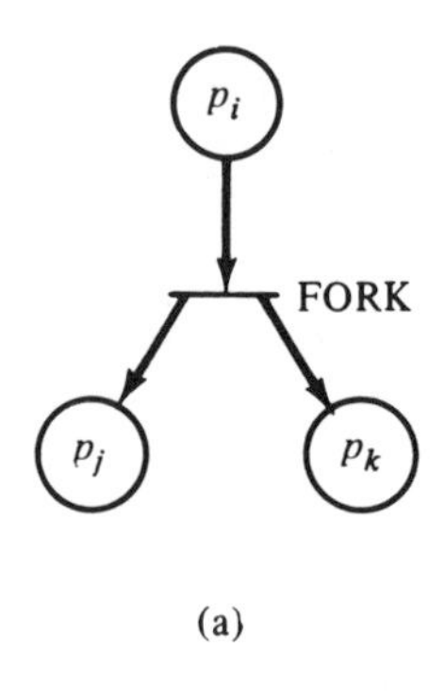

(a)

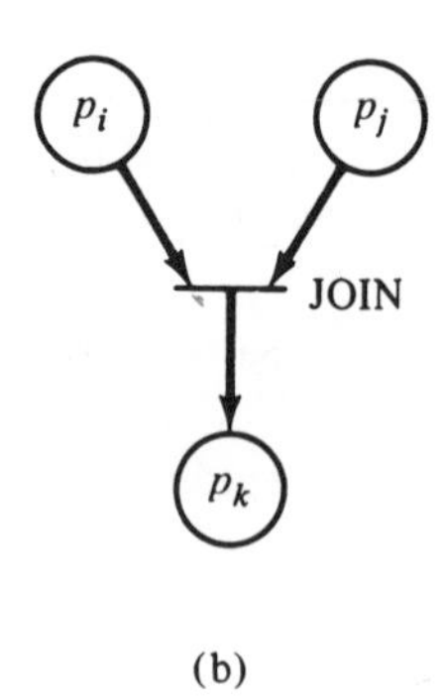

(b)

Figure 3.26 Modeling FORK and JOIN operations with a Petri net. (a) FORK (executed at location i, creating two new processes at locations j and k). (b) JOIN (join the two processes which end at locations i and j into one process which continues at location k).

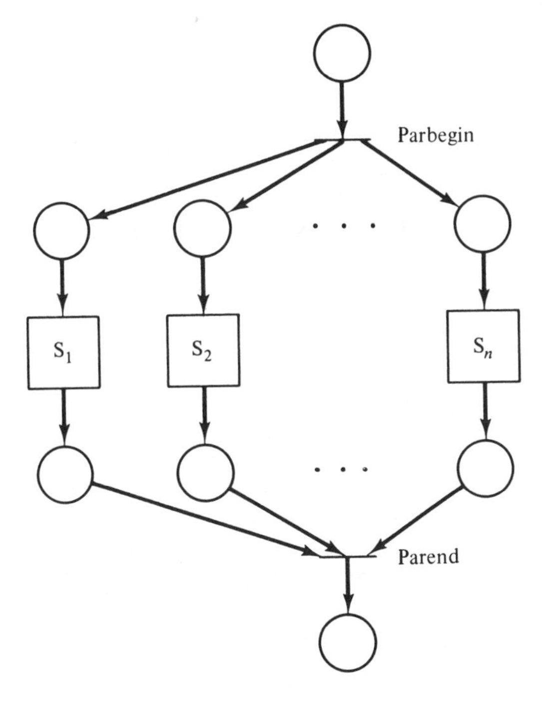

Figure 3.27 The modeling of *parbegin* $S_1;S_2; \cdots ;S_n$ *parend* in a Petri net. Each of the square boxes represents the Petri net representation of the statements S_1, S_2, and so on. This figure also illustrates the hierarchical nature of Petri net modeling.

3.4.3 Synchronization

Parallelism is usefully introduced into the solution of a problem only if the component processes can cooperate in the solution of the problem. Such cooperation requires the sharing of information and resources between the processes. This sharing must be *controlled* to ensure correct operation of the overall system. A variety of synchronization problems have been proposed in the literature to illustrate the types of problems which can arise between cooperating processes. Among these are the mutual exclusion problem [Dijkstra 1965], the producer/consumer problem [Dijkstra 1968], the dining philosophers problem [Dijkstra 1968], and the readers/writers problem [Courtois et al. 1971].

These problems are classics in the field of synchronization problems; every new suggestion for a synchronization mechanism must be able to handle these problems. Although Petri nets are a modeling scheme, and not a synchronization mechanism, Petri nets must certainly be able to model synchronization mechanisms which solve these problems. Thus, we present here some solutions to these problems, as Petri nets. This presentation is based in part on the work of Cooprider [1976].

3.4.4 The Mutual Exclusion Problem

Assume that several processes share a common variable, record, file, or other data item. This shared data item may be used in several ways by the processes but these uses can be grossly classified as needing to either *read* the value of the data item or *write* a new value. These two operations are often the only primitive operations. This means that to update the shared data item, a process must first read the old value, then compute the new value, and finally write the new value in place. A problem may occur if two processes attempt to execute this sequence of instructions at the same time. The following sequence may occur.

1. The first process reads the value x from the shared object.
2. The second process reads the value x from the shared object.
3. The first process computes an updated value $x' = f(x)$.
4. The second process computes an updated value $x'' = g(x)$.
5. The first process writes x' into the shared object.
6. The second process writes x'' into the shared object, destroying the x' value.

The effect of the computation of the first process has been lost, since now the value of the shared object is $g(x)$, while it should be either $g(f(x))$ or $f(g(x))$. [Consider the effect if $g(x)$ is "withdraw $1000 from account x" and $f(x)$ is "deposit $1000 into account x" and processes 1 and 2 are bank tellers.]

To prevent these sorts of problems, it is necessary to provide a mechanism for mutual exclusion. *Mutual exclusion* is a technique of defining entry and exit code so that at most one process is accessing a shared data object at a time. The code which accesses the shared object and needs protection from interference by other processes is called a *critical section.* The idea is that as a process is about to execute its critical section, it first waits until no other process is executing its own critical section. Then it "locks" access to the critical section, preventing any other process from entering its critical section. It enters the critical section, executes it, and as it leaves the critical section "unlocks" it to allow other processes to access it.

This problem can be solved by a Petri net such as Figure 3.28. The place m represents the permission to enter the critical section. For a process to enter the critical section, it must have a token in p_1 or p_2, as appropriate, signaling that it wishes to enter the critical section, *and* there must be a token in m signaling permission to enter. If both processes wish to enter simultaneously, then transitions t_1 and t_2 are in conflict, and only one of them can fire. Firing t_1 will disable transition t_2, requiring process 2 to wait until the first process exits its critical section and puts a token back in place m.

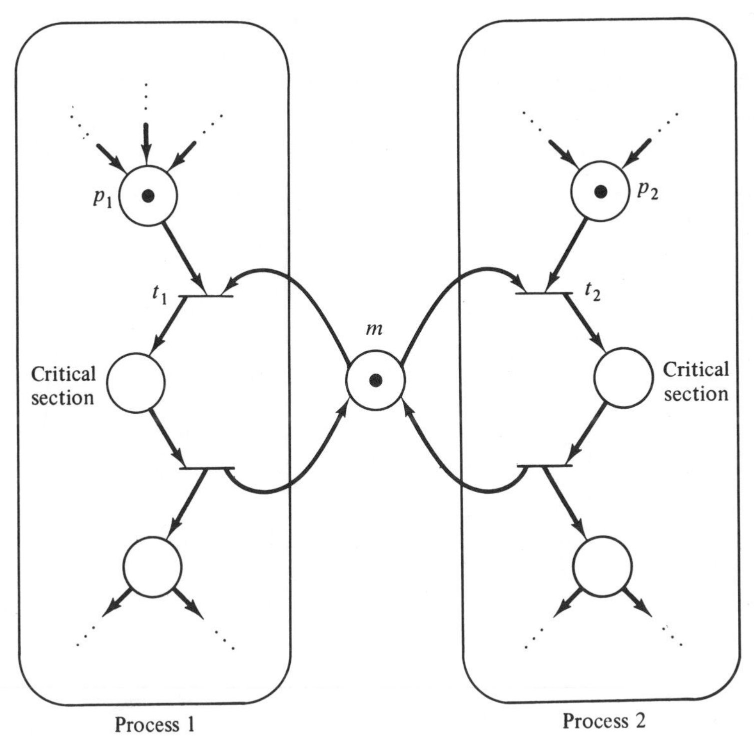

Figure 3.28 Mutual exclusion. Access to the critical sections of the two processes is controlled so that both processes cannot simultaneously execute their critical sections.

3.4.5 The Producer/Consumer Problem

The producer/consumer problem also involves a shared data object, but in this case the shared object is specified to be a buffer. The producer process creates objects which are put in the buffer; the consumer waits until an object has been put in the buffer, removes it, and consumes it. This can be modeled as shown in Figure 3.29. The place *B* represents the buffer; each token represents an item which has been produced but not yet consumed.

A variant on this problem is the multiple-producer/multiple-consumer problem. In this variant, multiple producers produce items which are placed in a common buffer for the multiple consumers. Figure 3.30 is a Petri net solution to this problem. It is the same as Figure 3.29, except that to represent s producers and t consumers, we start the system with s tokens in the initial place of the producer process and t tokens in the initial place of the consumer process. This represents s producers and t consumers executing reentrant, shared bodies of code. An alternative would be to duplicate the code for the producer and consumer processes, but this results in the same behavior with a much larger net.

Another variant is the bounded buffer producer/consumer prob-

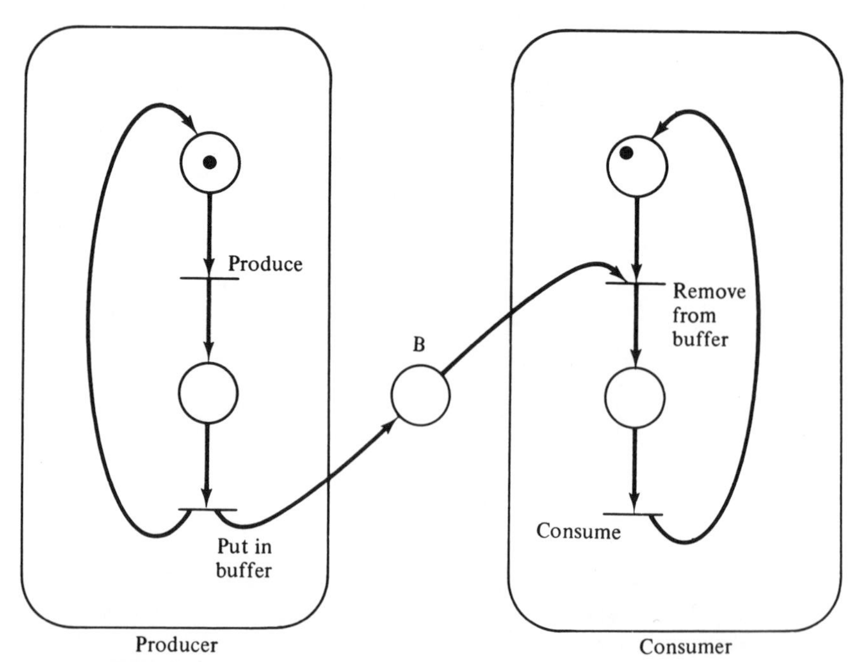

Figure 3.29 The producer/consumer problem modeled as a Petri net.

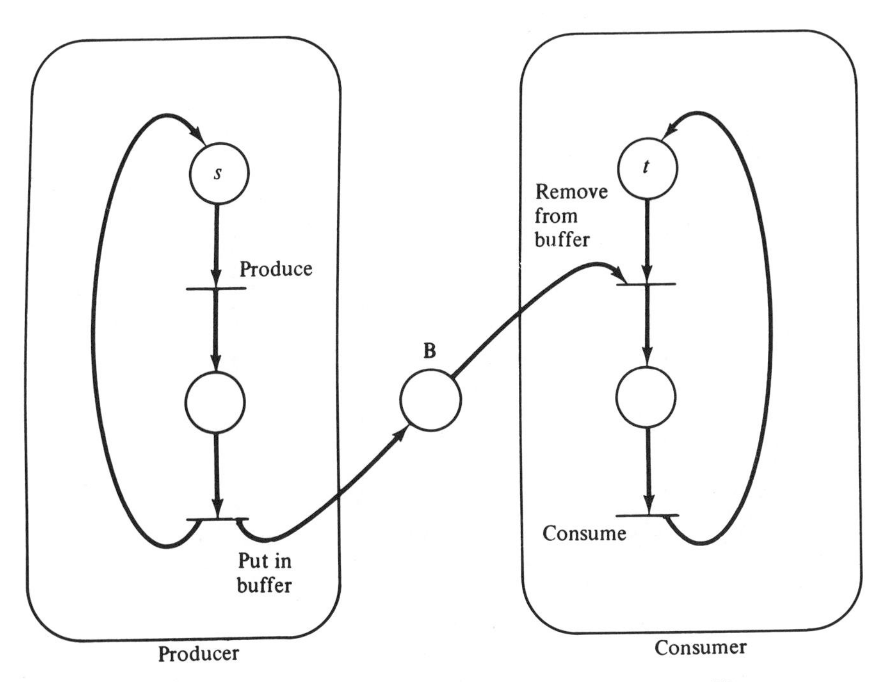

Figure 3.30 The multiple-producer/multiple-consumer problem. There are s producers and t consumers for fixed s and t.

lem. In this version of the producer/consumer problem, it is recognized that the buffer between the producer and the consumer is likely to be bounded, that is, have only n positions for items. Thus the producer cannot always produce as fast as desired but may have to wait if the consumer is slow and the buffer has filled. Figure 3.31 is a solution to this problem. The bounded buffer is represented by two places: B represents the number of items which have been produced but not yet consumed (the number of full positions); B' represents the number of empty positions in the buffer. Originally B' has n tokens and B has zero. If the buffer gets full, then B' will have zero tokens and B will have n. At this point if the producer tries to put another item in the buffer, it will be stopped because there is no token in B' to enable that transition.

3.4.6 The Dining Philosophers Problem

The dining philosophers problem was suggested by Dijkstra [1968] and concerns five philosophers who alternatively think and eat. The philosophers are seated at a large round table on which are a large number of Chinese foods. Between each philosopher is one chopstick.

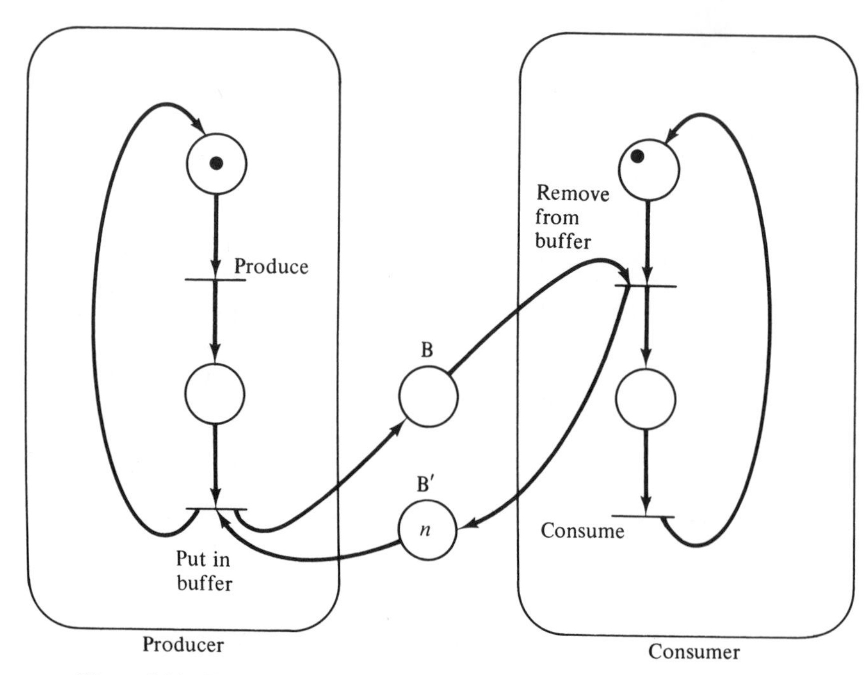

Figure 3.31 The bounded buffer producer/consumer problem. The buffer, represented by the places B and B', is limited to at most n items.

However, to eat Chinese food, you need two chopsticks; hence each philosopher must pick up both the chopstick on the left and the chopstick on the right. The problem, of course, is that if all philosophers pick up the chopstick on their left and then wait for the chopstick on their right, they will all wait forever and starve (a *deadlock* condition).

Figure 3.32 illustrates the Petri net solution to this problem. The places $C_1, \ldots, C_5$ represent the chopsticks, and since each is initially free, a token resides in each in the initial marking. Each philosopher is represented by two places M_i and E_i representing the meditating and eating states, respectively. For a philosopher to move from the meditating state to the eating state, both chopsticks (the one to the left and the one to the right) must be available. This is easily modeled in the Petri net.

3.4.7 The Readers/Writers Problem

There are several variants of the readers/writers problem [Courtois et al. 1971], but the basic structure is the same. Processes are of

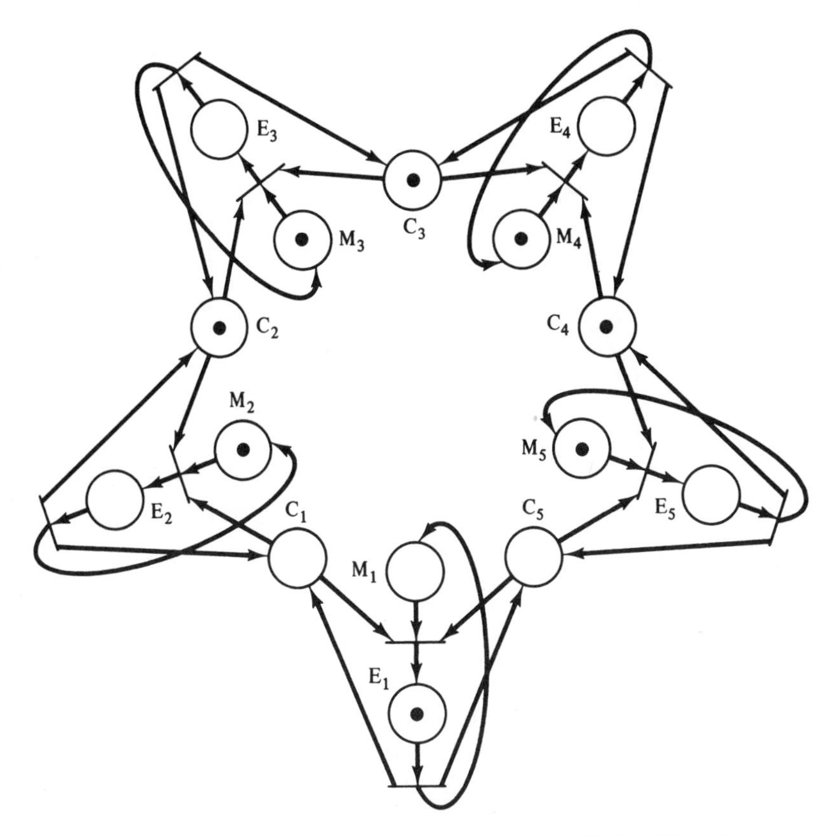

Figure 3.32 The dining philosophers problem. Each philosopher is modeled by two places, meditating (M_i) and eating (E_i).

two types: reader processes and writer processes. All processes share a common file, variable, or data object. Reader processes never modify the object, while writer processes do modify it. Thus writer processes must mutually exclude all other reader and writer processes, but multiple reader processes can access the shared data simultaneously. The problem is to define a control structure which does not deadlock or allow violations of the mutual exclusion criteria.

Figure 3.33 illustrates a solution when the number of reader processes is bounded by n. In a system where the number of reader processes is not bounded, then only n readers may read at a time.

A problem occurs, however, if the number of readers is unbounded and we wish to allow an unbounded number of readers to simultaneously read. In this case, it can be argued that it will be necessary for the readers to keep a count of the number of readers reading. Each reader adds one to this counter when it starts reading and sub-

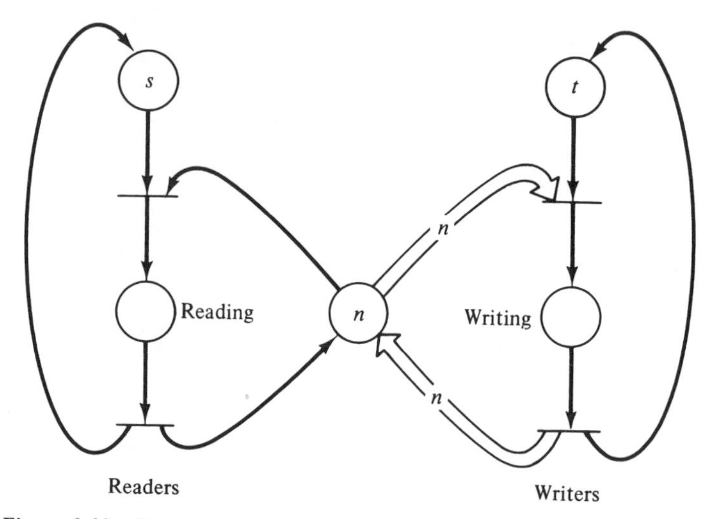

Figure 3.33 The readers/writers problem when the number of readers is bounded by *n*. There are initially *s* readers and *t* writers.

tracts one when it finishes reading. This can be easily modeled by a place with a number of tokens equal to the number of readers. However, now in order to allow a writer to begin writing, it is necessary that this counter be zero, i.e., the corresponding place be empty. There is no mechanism in Petri nets which allows an unbounded place to be tested for zero. Thus, it would appear that the readers/writers problem with unbounded readers cannot be solved with Petri nets. This is our first indication that Petri nets may not be able to model all systems and is a topic that deserves more study (Chapter 7).

3.4.8 P and V Systems

Most synchronization problems will not be solved directly by Petri nets but rather in terms of an established synchronization mechanism. In particular, one of the most popular synchronization mechanisms has been the P and V operations on semaphores originally defined by Dijkstra [1968]. A *semaphore* is a data item which can take on only nonnegative integer values. The V operation increments the value by 1, while the P operation decrements the value by 1. The P operation can occur only when the semaphore value will remain nonnegative; if the semaphore value is zero, the P operation must wait until some other process executes a V operation. Both the P and V operations are defined to be primitive; no other operation may simultaneously modify the semaphore value.

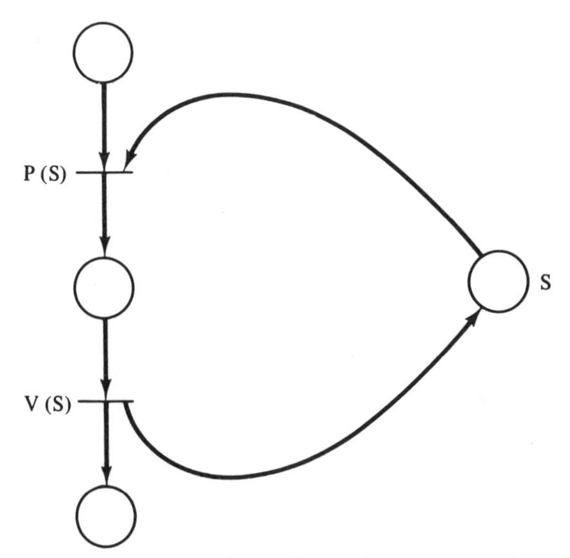

Figure 3.34 Modeling *P* and *V* operations on a semaphore *S*.

primitive; no other operation may simultaneously modify the semaphore value.

These operations can be easily modeled by Petri nets as shown in Figure 3.34. Each semaphore is modeled by a place; the number of tokens in that place indicate the value of the semaphore. A P operation uses the semaphore place as an input; a V operation uses the semaphore as an output.

The advantage of this ability to model P and V operations is that many systems are written or designed using P and V operations. For example, the Venus operating system [Liskov 1972] provides P and V operations as the basic interprocess communication mechanism. These systems can thus be modeled as Petri nets.

3.5 Other Systems

The systems which have been described so far are those which are the most obvious types of systems for Petri net modeling: computer hardware and software. But in large part this "obviousness" is a result of the fact that Petri nets have been defined and developed mainly for this purpose. Petri nets can also be directly applied to the modeling of a large number of other systems, some quite distinct from computer systems. In this section we briefly survey some of the systems to which Petri nets have been or could be applied.

PERT charts have long been used in the planning and scheduling of large projects. A PERT chart is a graphical representation of the relationships between the various activities which make up a large project. A project consists of a number of activities; some activities must be completed before other activities can start. In addition a time is associated with each activity indicating the amount of time it will take. (Sometimes three times — worst case, average, and best case — are associated with each activity.) The activities are represented graphically by a node; arcs are used to connect activity nodes to show precedence requirements.

PERT charts show the same type of scheduling constraints as Petri nets. We can easily convert a PERT chart to a Petri net. Each activity in a PERT chart is represented by a place, while the precedence constraints are represented by the transitions. The PERT chart of Figure 3.35 can be converted to the equivalent Petri net shown in Figure 3.36.

The Petri net is an excellent vehicle for representing the concurrency and precedence constraints of the PERT chart, but PERT charts also provide timing information which is useful for determining minimum project completion time, latest starting time for an activity which will not delay the project, and so on. A Petri net does not provide any of this type of information. The addition of timing information might provide a powerful new feature for Petri nets but may not be possible in a manner consistent with the basic philosophy of Petri nets. Research is needed on this extension to basic Petri nets [Sifakis 1977; Han 1978].

The pipeline of Section 3.3.2 is a special case of a *production system* [Hack 1972]. An assembly line is another example of a production system. Production systems and assembly lines can be modeled as Petri nets.

One of the early applications of Petri nets was as a tool for the *generation of optimal code* for a CDC 6600 FORTRAN compiler. The approach suggested by Shapiro and Saint [1970] was to model the FORTRAN program as a Petri net, in a manner similar to the flowchart modeling of Section 3.4.1. Then the individual statements of the program are examined to determine the minimal precedence constraints between statements, allowing the Petri net to drop some of the artificial sequencing constraints of the program. These artificial constraints are introduced because the FORTRAN programmer must express the program as a total ordering of statements, even though only a partial order-

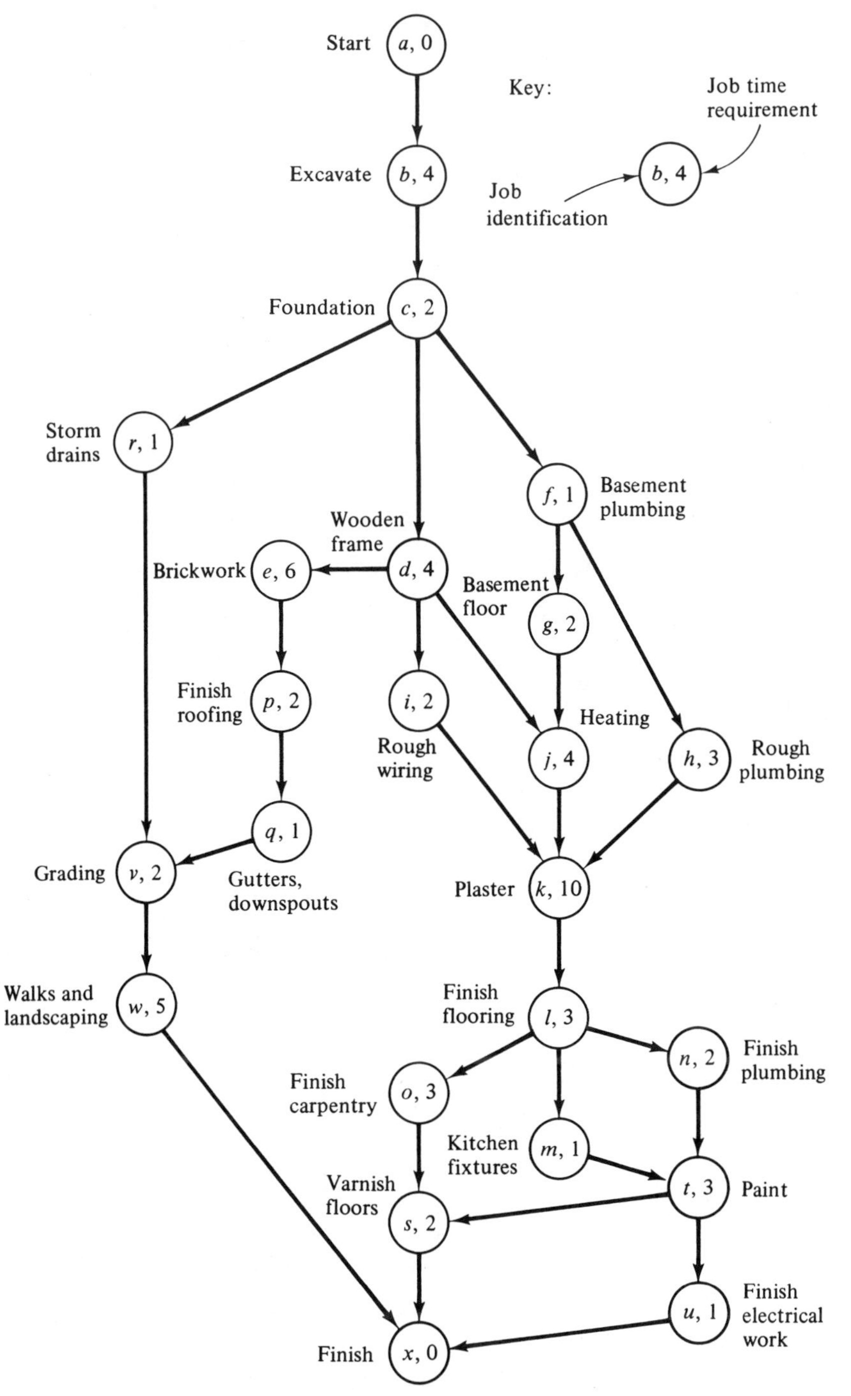

Figure 3.35 A PERT chart of the construction of a house. (From F. Levy, G. Thompson, and J. Wiest, "Introduction to the Critical-Path Method," in *Industrial Scheduling*, edited by John F. Muth and Gerald L. Thompson, copyright 1963, p. 335. Adapted by permission of Prentice-Hall, Inc., Englewood Cliffs, New Jersey.)

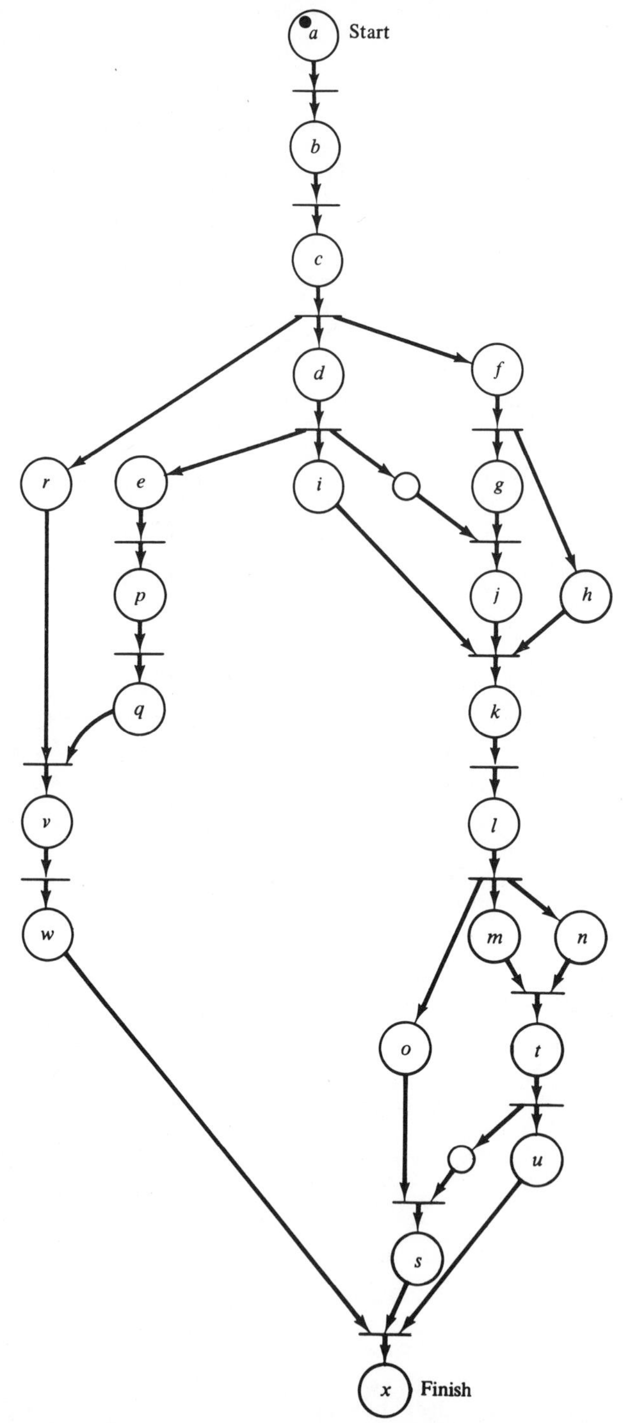

Figure 3.36 A Petri net representation of the PERT chart of Figure 3.35. Notice that some extra nodes have been added. These are necessary to properly reflect the precedence constraints and explicitly represent situations where waiting must occur.

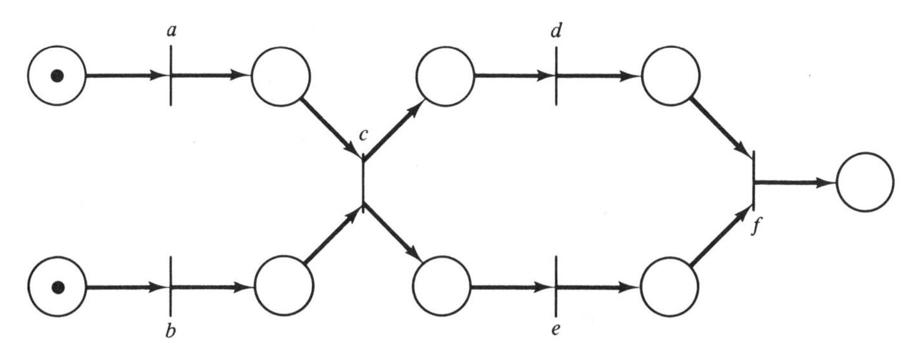

Figure 3.37 A Petri net which represents several sequences of instruction executions.

ing is necessary. For example, consider the following three statements.

$$10\, x = x + 1$$

$$20\, y = y + 1$$

$$30\, z = x + y$$

Statements 10 and 20 are written as statement 10 before statement 20, but this constraint is unnecessary. Statements 10 and 20 can be executed in any order (or concurrently) with no effect on the program. Statement 30, however, is constrained to follow both statements 10 and 20. Control flow must also be considered in this restatement of the sequencing requirements. This analysis is the application of Bernstein's conditions to assure determinacy.

The result of this analysis is a Petri net which represents the program with only the minimal sequencing constraints, i.e., allowing maximal parallelism. Now the problem is to compile this program. This requires mapping variables into registers and ordering the instructions to produce a totally ordered sequence of machine language instructions. The 6600 is a multiple register, multiple functional unit computer, as described in Section 3.3.3. Since the functional units can execute in parallel on separate instructions, it is very important to generate instructions in an order which maximizes the parallelism in the execution of the functional units. This is also affected by the assignment of variables to registers. The Petri net model of the program representing the constraints of the program is combined with a Petri net model of the CDC 6600 control unit, representing the constraints imposed by the hardware. This composite net then represents all possible sequences of instructions which can execute on the hardware and perform the algorithm of the program. This net is then executed to produce all such sequences of instructions. Two (or more) sequences are created whenever two (or more) transitions are simultaneously enabled. Firing one transition will produce one sequence; firing the other produces another sequence. For example, the Petri net of Figure 3.37 represents the

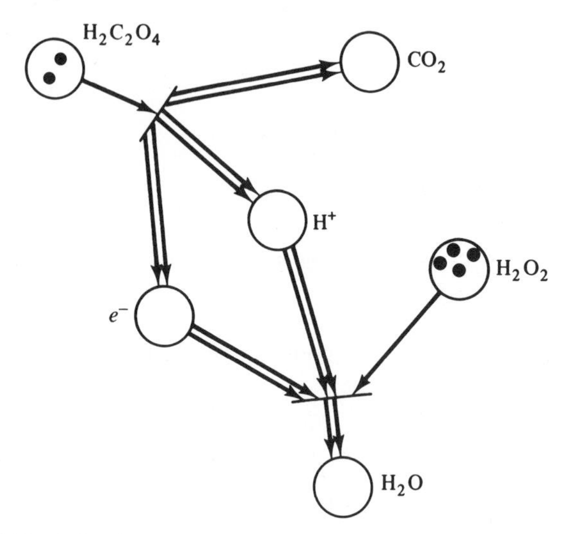

Figure 3.38 A Petri net representing the oxidation-reduction of oxalic acid and hydrogen peroxide into carbon dioxide and water.

sequences *abcdef*, *bacdef*, *abcedf*, and *bacedf*. As these sequences are produced, the amount of time to execute each is computed, and the fastest sequence is generated by the compiler for actual execution later.

Chemical systems are another example of a system which can be modeled by Petri nets. Chemical equations are modeled by transitions; reactants are modeled by places. The number of tokens in a place indicate the amount of that reactant in the system. For example, the Petri net in Figure 3.38 represents the two chemical equations

$$H_2C_2O_4 \longrightarrow 2CO_2 + 2H^+ + 2e^-$$

$$2e^- + 2H^+ + H_2O_2 \longrightarrow 2H_2O$$

Catalytic reactions can also be represented. The combination of hydrogen and ethylene to form ethane ($H_2 + C_2H_4 \rightarrow C_2H_6$) only occurs in the presence of platinum. This is diagrammed in Figure 3.39.

Meldman and Holt [1971] have suggested that *legal systems* may be modeled by Petri nets. In these systems, several actors (judges, lawyers, defendants, clerks, and so on) may concurrently perform activities which relate to a particular legal matter. The activities and their relationships can be represented by a Petri net. For example, Figure 3.40 is a model of the initial stages of a civil action.

Another suggested use for Petri nets has been in the modeling and analysis of *communication protocols* [Merlin 1975]. Computer net-

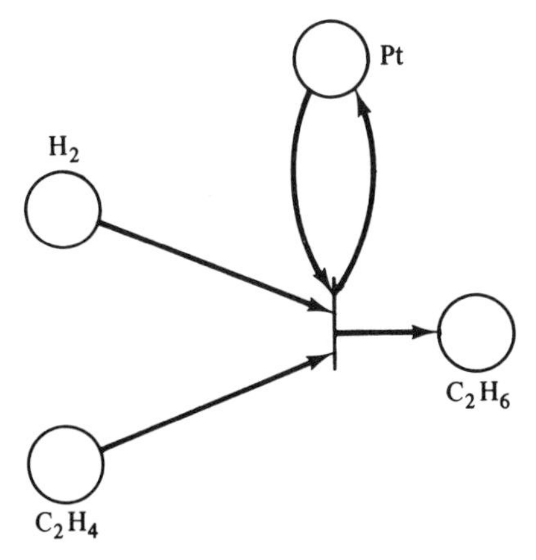

Figure 3.39 The production of ethane from hydrogen and ethylane in the presence of the catalyst platinum.

works and systems of distributed processes require the ability to transmit information from computer to computer. This intrinsically involves parallelism and therefore falls into the class of problems for which Petri nets have been defined. Farber and his students [Merlin 1974; Postel 1974; Merlin and Farber 1976; Postel and Farber 1976] have developed methodologies for the specification, design, and analysis of simple communications protocols using Petri nets and similar models.

Other systems which could be modeled by Petri nets include queueing networks (where the queues would be represented by places and the jobs by tokens), brain models (neuron firings would be modeled by transition firings), propositional calculus [Genrich 1975; Genrich and Lautenbach 1978] (the places represent literals, and the transitions combine them to define clauses in conjunctive normal form), and many others. The list is limited in the main by the time and imagination of the modeler, and not by the properties of the Petri net model.

However, we have seen at least one example (the readers/writers problem) which may not be able to be modeled by a Petri net. Also, although modeling as a Petri net may aid the description of a system, we need to develop analytic tools which will allow us to examine a Petri net and determine properties of the Petri net. This leads us to the next chapter, in which we present analysis methods for Petri nets.

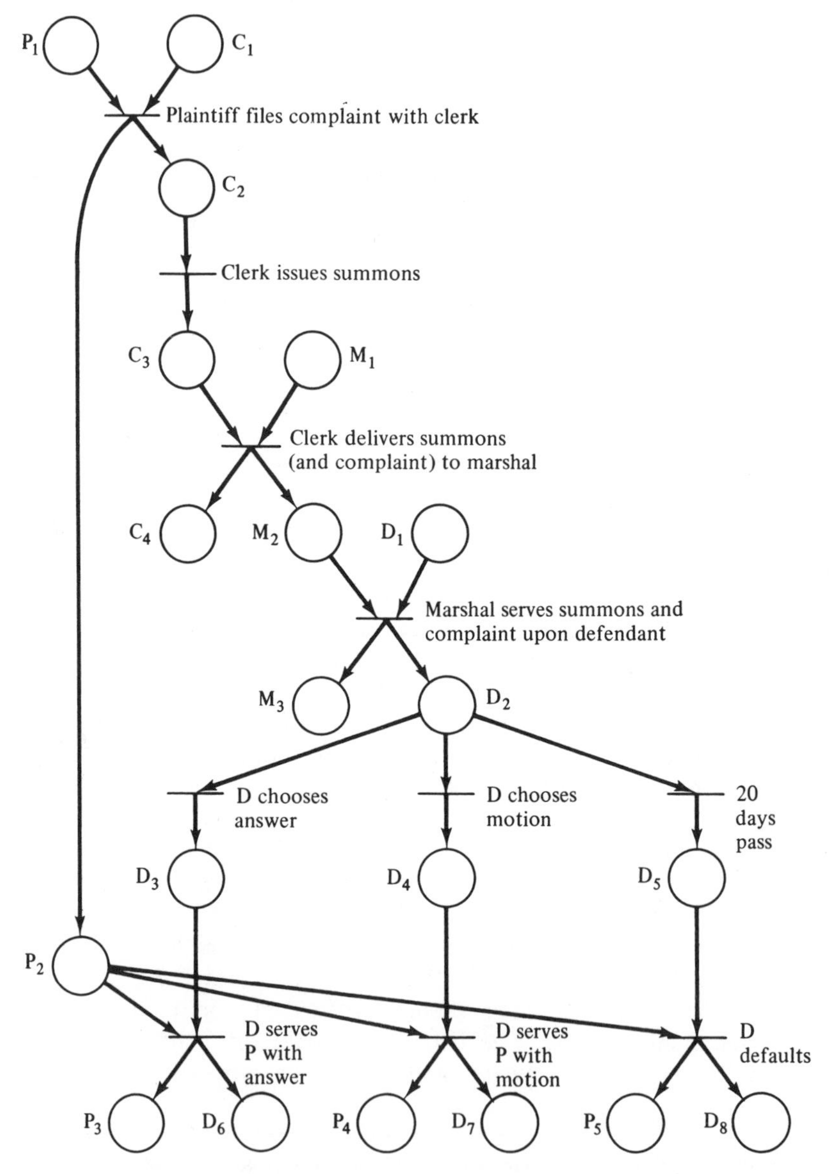

Figure 3.40 A Petri net representing the first stages of a civil legal action. (From [Meldman and Holt 1971], copyright American Bar Association, Section of Science and Technology. Redrawn with permission.)

Exercises

1. Model a computer system with three processes and four resources: card reader, line printer, disk, and two memory partitions. Any process can run in either partition. The resource usage of the three processes is
   (a) Process 1 requests the card reader and line printer and then later releases both of these resources.
   (b) Process 2 requests the card reader and the disk and then later releases the card reader, requests the line printer, and finally releases both the line printer and the disk.
   (c) Process 3 wants all three resources at once and then releases them all later.

3.6 Further Reading

Most of the research on Petri nets has been on analysis, not modeling. Surveys of applications of Petri nets to modeling have appeared in [Peterson 1977; Agerwala 1978]. Hardware modeling was considered in [Dennis 1970a; Huen and Sieworek 1975]. The paper by Shapiro and Saint [1970] combines hardware and software modeling to implement a compiler. The notes by Cooprider [1976] are concerned with modeling software systems by Petri nets. Hack's Master's thesis [Hack 1972] considered the modeling of production schemata which includes assembly line-type systems.

Baer and Ellis [1977] used Petri nets to model a compiler, while Noe [1971] and Best [1976] used Petri nets to model operating systems. Noe and Kehl [1975] have modeled the hardware of a computer system. Azema et al. [1975], Azema et al. [1976], and Foo and Musgrave [1975] have suggested the use of Petri nets for design automation.

The work of Noe and Nutt is specifically directed at modeling systems to determine performance properties. Their work [Noe 1971; Nutt 1972a; Nutt 1972b; Noe and Nutt 1973] eventually led to the development of a model, E-nets, which is related to Petri nets.

3.7 Topics for Further Study

1. Apply Petri net modeling to describe the interactions of subatomic particles in high-energy physics.
2. PERT charts normally have timing information associated with them. Investigate how to add timing information to a Petri net.

How does this affect the firing rules? Devise algorithms for computing minimal (and maximal) finishing times.

3. Take any one of the following topics and investigate the use of Petri nets for modeling:
   (a) Queueing theory
   (b) Brain modeling
   (c) Chemical reactions
   (d) Military conflict
   (e) Political systems
   (f) Economics (particularly macroeconomic theory)
   (g) Traffic flow on roads and highways
   (h) Biological populations
   (i) Semantic nets for natural language representation

4

Analysis of Petri Nets

In the last chapter we demonstrated the modeling power of Petri nets. Petri nets are capable of modeling a large variety of systems, properly representing the interactions between the various actions which can occur. The major strength of Petri nets is, of course, in the modeling of systems which may exhibit concurrency; concurrency is modeled in a natural and convenient way. A Petri net model can be used to represent and communicate the design of a concurrent system.

However, modeling by itself is of little use. It is necessary to *analyze* the modeled system. This analysis will hopefully lead to important insights into the behavior of the modeled system. Thus, we turn now to presenting analysis techniques for Petri nets. Several techniques have been developed for the analysis of Petri nets, but many problems in the analysis of Petri nets are still open. To better evaluate the usefulness of the analysis techniques which have been developed, we first consider what types of problems may need to be solved for Petri nets. The objective of the analysis of a Petri net is to determine the answer to a question about the Petri net. What types of questions might be asked about Petri nets?

4.1 Analysis Problems for Petri Nets

The following properties and questions have been considered in the literature about Petri nets. We define and illustrate these properties here; we show the appropriate analysis techniques in the second portion of this chapter.

4.1.1 Safeness

For a Petri net which is to model a real hardware device, one of the more important properties is *safeness.* A place in a Petri net is safe if the number of tokens in that place never exceeds one. A Petri net is *safe* if all places in the net are safe.

DEFINITION 4-1 A place $p_i \in P$ of a Petri net $C = (P,T,I,O)$ with initial marking μ is *safe* if for all $\mu' \in R(C,\mu)$, $\mu'(p_i) \leqslant 1$. A Petri net is safe if each place in that net is safe.

Safeness is a very important property for hardware devices. If a place is safe, then the number of tokens in that place is either 0 or 1. Thus the place can be implemented by a single flip-flop.

The original Petri nets were safe by definition, since a transition could not fire unless all of its output places were empty (and multiple arcs were not allowed). This was motivated by the interpretation of a place as representing a condition. A condition, being a logical statement, is either true (represented by a token in the place) or false (represented by an absence of a token); multiple tokens have no interpretation. Thus, the marking of each place should be safe under an interpretation as conditions and events.

As long as a place is not a multiple input or multiple output of a transition, it is possible to force that place to be safe. A place p_i which is to be forced to be safe is supplemented by another place p_i'. Transitions which use p_i as an input or output are modified as follows:

If $p_i \in I(t_j)$ and $p_i \notin O(t_j)$, then add p_i' to $O(t_j)$.
If $p_i \in O(t_j)$ and $p_i \notin I(t_j)$, then add p_i' to $I(t_j)$.

The object of this new place p_i' is to represent the condition "p_i is empty." Thus p_i and p_i' are complementary; p_i has a token only if p_i' has no token and vice versa. Any transition which removes a token from p_i must deposit one in p_i', and any transition which removes a token from p_i' must deposit one in p_i. The initial marking must also be modified to provide exactly one token in either p_i or p_i'. (We assume that the initial marking is safe.) Notice that this forced safeness is only possible for places which are safe in the initial marking and whose input and output multiplicity is either 0 or 1 for all transitions. A place which has an output multiplicity of two for a transition will receive two tokens when that transition fires and hence cannot be safe. Figure 4.1 is a simple Petri net which has been forced to be safe in Figure 4.2.

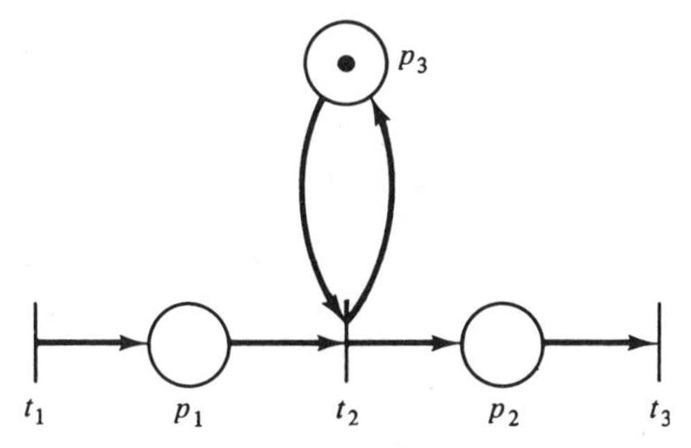

Figure 4.1 A Petri net. This net is not safe.

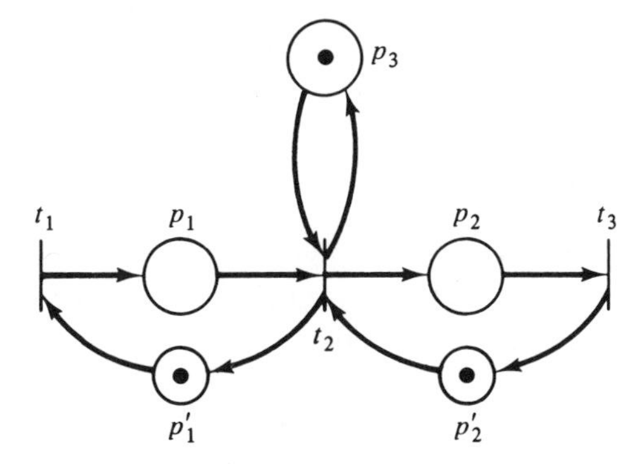

Figure 4.2 The Petri net of Figure 4.1 can be forced to be safe as shown here.

4.1.2 Boundedness

Safeness is a special case of the more general *boundedness* property. Some thought about the real limitation to implementing places in hardware shows that it is not necessary to require safeness. Safeness allows a place to be implemented with a flip-flop, but, more generally, a counter could be used. However, any such hardware counter would be limited in the maximum number which could be represented. A place is *k-safe* or *k-bounded* if the number of tokens in that place cannot exceed an integer k.

DEFINITION 4-2 A place $p_i \in P$ of a Petri net $C = (P,T,I,O)$ with an initial marking μ is *k-safe* if for all $\mu' \in R(C,\mu)$, $\mu'(p_i) \leqslant k$.

A place which is 1-safe is simply called safe. Notice that the bound k on the number of tokens which can be in a place may be a function of the place (e.g., place p_1 may be 3-safe while place p_2 is 8-safe). However, if a place p_i is k-safe, then it is also k'-safe for all $k' \geqslant k$. Since there are only a finite number of places, we can pick k to be the maximum of the bounds of each place and define a Petri net to be k-safe if every place of the net is k-safe.

We may sometimes be concerned only with whether or not the number of tokens in a place is bounded or not, but we are not concerned with the specific value of the bound. A place is *bounded* if it is k-safe for some k; a Petri net is bounded if all places are bounded. A bounded Petri net could be realized in hardware, while a Petri net with an unbounded place could not in general be implemented in hardware. (Remember that these definitions are independent of interpretation. In implementation a place might represent some entity which is bounded, although the net structure itself does not reflect this fact.)

4.1.3 Conservation

Petri nets can be used to model resource allocation systems. For example, a Petri net can model the requests, allocations, and releases of input/output devices in a computer system. In these systems some tokens may represent the resources. A set of three line printers is represented by a place with an initial marking of three tokens. A request for a line printer is a transition which has this place as an input; the line printer is later released by a transition with an output to the line printer place.

For these types of Petri nets, among others, *conservation* is an important property. We would like to show that tokens which represent resources are neither created nor destroyed. The simplest way to do this would be to require that the total number of tokens in the net remain constant.

DEFINITION 4-3 A Petri net $C = (P, T, I, O)$ with initial marking μ is *strictly conservative* if for all $\mu' \in R(C, \mu)$,

$$\sum_{p_i \in P} \mu'(p_i) = \sum_{p_i \in P} \mu(p_i).$$

Strict conservation is a very strong relationship. For example, one can immediately show that the number of inputs to each transition must equal the number of outputs, $|I(t_j)| = |O(t_j)|$. If this were not the case, firing transition t_j would change the number of tokens in the net.

For a broader view, however, consider Figure 4.3. Figure 4.3 is not strictly conservative since the firing of either transition t_1 or t_3 will decrease the number of tokens by one, while firing transition t_2 or t_4 will add a token to the marking. We could, however, convert the Petri net of Figure 4.3 to the net of Figure 4.4, which is strictly conservative.

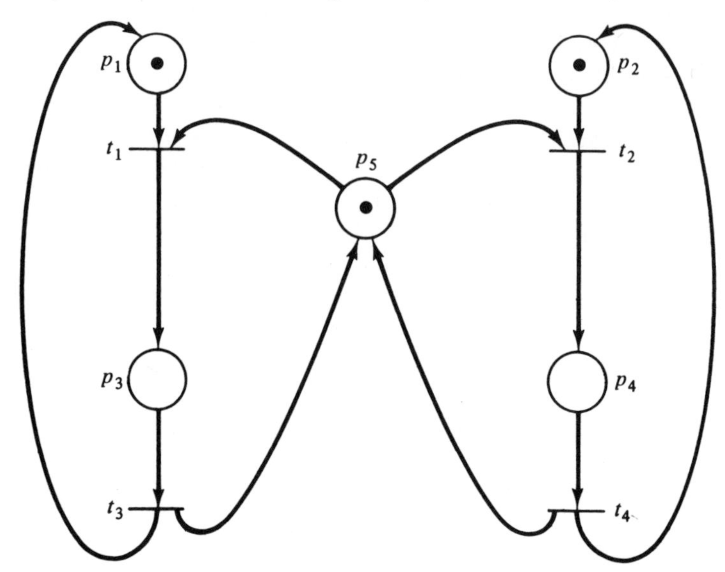

Figure 4.3 A Petri net which is not strictly conservative.

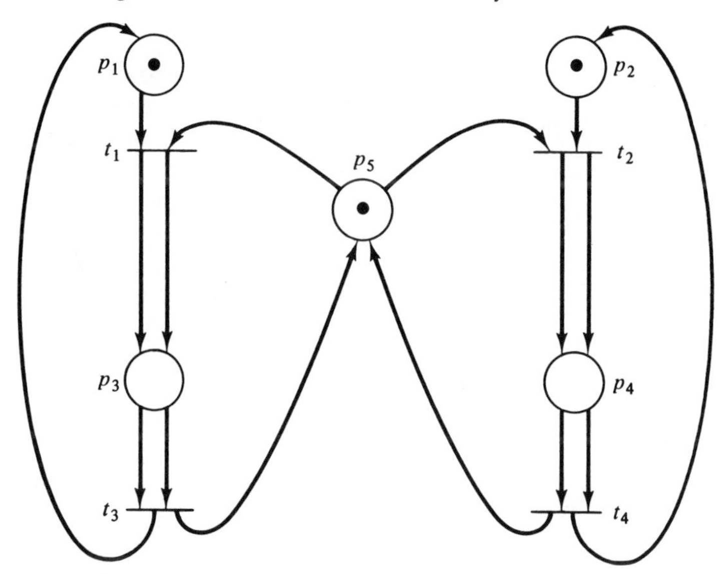

Figure 4.4 A strictly conservative Petri net which is equivalent to the net of Figure 4.3.

A Petri net should conserve the resources which it is modeling. However, there is no one-to-one mapping between tokens and resources. Some tokens represent program counters or other items; other tokens may represent several resources with one token. This token is later used to create multiple tokens (one per resource) by firing a transition with more outputs than inputs. In general, we would like to define a *weighting* of tokens. The *weighted* sum for all reachable markings should be constant. Tokens which are not important can be assigned a weight of 0; other tokens can be assigned weights of 1, 2, 3, or any other integer. (Rational numbers would be acceptable since the weightings could be multiplied by a common denominator to define an integer weighting. Irrational weights would not seem to be needed.)

A token is defined by its place in the net, and all tokens in a place are indistinguishable. Thus the weights are associated with each place of the Petri net. A weighting vector $w = (w_1, w_2, \ldots, w_n)$ defines a weight w_i for each place $p_i \in P$.

DEFINITION 4-4 A Petri net $C = (P, T, I, O)$ with initial marking μ, is *conservative with respect to a weighting vector* w, $w = (w_1, w_2, \ldots, w_n)$, $n = |P|$, $w_i \geqslant 0$, if for all $\mu' \in R(C, \mu)$,

$$\sum_i w_i \cdot \mu'(p_i) = \sum_i w_i \cdot \mu(p_i).$$

A strictly conservative Petri net is conservative with respect to the weighting vector (1, 1, ..., 1). All Petri nets are conservative with respect to the weighting vector (0, 0, ..., 0). The latter observation is disturbing since we would like to define a Petri net as conservative if it is conservative with respect to some weighting vector. However, since every Petri net is conservative with respect to the zero vector, this would not be satisfactory. Thus, a Petri net is *conservative* if it is conservative with respect to some positive nonzero weighting vector, $w > 0$ (with positive nonzero weights, $w_i > 0$).

The Petri net of Figure 4.3 is therefore conservative since it is conservative with respect to (1, 1, 2, 2, 1). The Petri net of Figure 4.5 is not conservative.

4.1.4 Liveness

The motivation for considering conservation in a Petri net was resource allocation in a computer operating system. Another problem

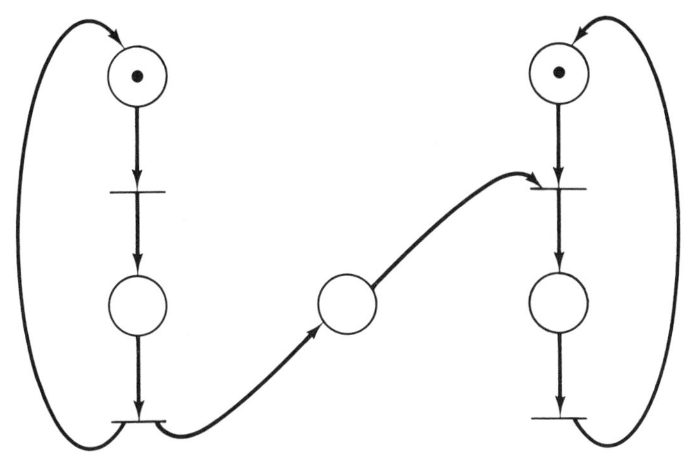

Figure 4.5 A nonconservative Petri net.

which may arise in resource allocation for a computer system is *deadlock*. Deadlock has been the subject of a number of studies in computer science [Hebalkar 1970]. A simple example can best illustrate the problem: Consider a system with two different resources q and r and two processes a and b. If both processes need both resources, it will be necessary to share them. To accomplish this we require each process to request a resource and later release it. Now suppose process a first requests resource q and then resource r and finally releases both q and r. Process b is similar but first requests r and then q. Figure 4.6 illustrates these two processes and the resource allocation with a Petri net.

The initial marking indicates resources $q\,(p_4)$ and $r\,(p_5)$ are available and processes a and b are ready. One execution of this net is $t_1t_2t_3t_4t_5t_6$; another is $t_4t_5t_6t_1t_2t_3$. Neither of these executions results in deadlock. However, consider the sequence which starts t_1t_4: Process a has q and wants r; process b has r and wants q. The system is deadlocked; neither process can proceed.

A deadlock in a Petri net is a transition (or a set of transitions) which cannot fire. In Figure 4.6, deadlock occurs if transitions t_2 and t_5 cannot fire. A transition is *live* if it is not deadlocked. This does not mean that the transition is enabled but rather that it can be enabled. A transition t_j of a Petri net C is *potentially fireable* in a marking μ if there exists a marking $\mu' \in R(C,\mu)$ and t_j is enabled in μ'. A transition is *live* in a marking μ if it is potentially fireable in every marking in $R(C,\mu)$. Thus if a transition is live, it is always possible to maneuver the Petri net from its current marking to a marking which would allow the transition to fire.

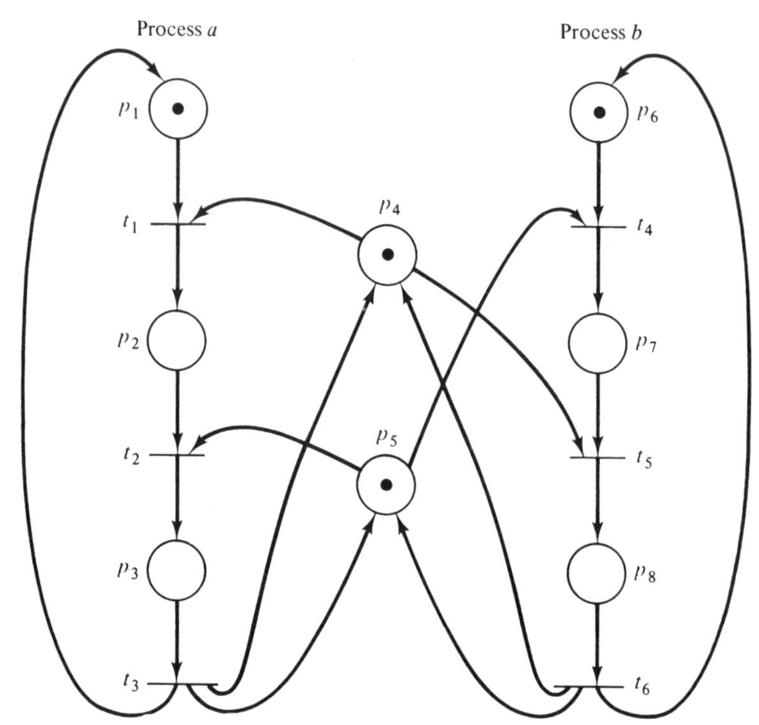

Figure 4.6 Resource allocation for two processes (*a* and *b*) and two resources (q [modeled by p_4] and r [modeled by p_5]).

There are other concepts, related to liveness, which have been considered in studies of deadlock [Commoner 1972]. These can be categorized as *levels* of liveness and can be defined for a Petri net C with marking μ as

Level 0: A transition t_j is *live at level 0* if it can never be fired.

Level 1: A transition t_j is *live at level 1* if it is potentially fireable; that is, if there exists a $\mu' \in R(C,\mu)$ such that t_j is enabled in μ'.

Level 2: A transition t_j is *live at level 2* if for every integer n there exists a firing sequence in which t_j occurs at least n times.

Level 3: A transition t_j is *live at level 3* if there is an infinite firing sequence in which t_j occurs infinitely often.

Level 4: A transition t_j is *live at level 4* if for each $\mu' \in R(C,\mu)$ there exists a firing sequence σ such that t_j is enabled in $\delta(\mu',\sigma)$.

A transition which is live at level 0 is *dead.* A transition which is live at level 4 is *live*. A Petri net is live at level i if every transition is live at level i.

As an example of these levels of liveness, consider Figure 4.7. Transition t_0 can never fire; it is dead. Transition t_1 can fire exactly once; it is live at level 1. Transition t_2 can be made to fire an arbitrary number of times, but the number of times is dependent on the number of times that t_3 fires. If we want to fire t_2 five times, we fire t_3 five times, then t_1, and then t_2 five times. However, once t_1 fires (and t_1 must fire before t_2 can fire), the number of times t_2 can fire is fixed. Thus, t_2 is live at level 2, but not at level 3. Transition t_3, on the other hand, can be fired an infinite number of times and so is live at level 3, but not at level 4, since once t_1 fires t_3 can no longer fire.

4.1.5 Reachability and Coverability

Most of the problems which have been mentioned so far are concerned with reachable markings. Perhaps the simplest problem (to state) is the *reachability* problem.

DEFINITION 4-5 *The Reachability Problem* Given a Petri net C with marking μ and a marking μ', is $\mu' \in R(C,\mu)$?

The reachability problem is perhaps the most basic Petri net

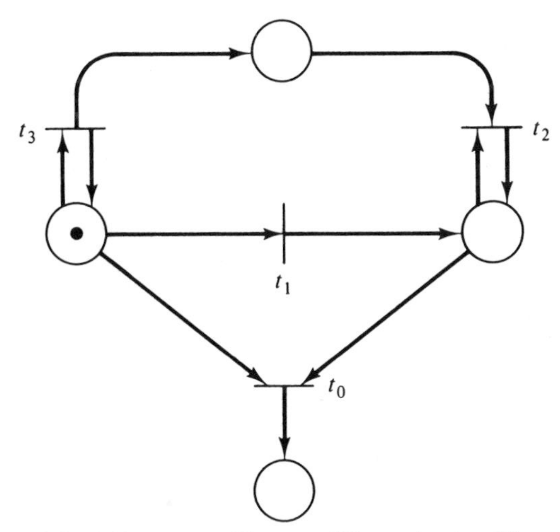

Figure 4.7 A Petri net to illustrate different levels of liveness.

analysis problem; many other analysis problems can be stated in terms of the reachability problem. For example, for the Petri net of Figure 4.6, deadlock can occur if the state (0, 1, 0, 0, 0, 0, 1, 0) is reachable.

Figure 4.8 shows a Petri net which purports to solve the mutual exclusion problem — places p_4 and p_9 are expected to be mutually exclusive. We wish to know if any state is reachable with $p_4 \geqslant 1$ and $p_9 \geqslant 1$. This problem is similar to reachability but is slightly different; it is called the *coverability* problem. A marking μ'' *covers* a marking μ' if $\mu'' \geqslant \mu'$.

DEFINITION 4-6 *The Coverability Problem* Given a Petri net C with initial marking μ and a marking μ', is there a reachable marking $\mu'' \in R(C,\mu)$ such that $\mu'' \geqslant \mu'$?

Another possible use of reachability-type problems would be to ignore the contents of some places, concentrating only on matching or covering the contents of a few important places. For example, in the Petri net of Figure 4.8, our interest is confined to places p_4 and p_9; the markings of the remaining places are not important. Thus, we can consider reachability or coverability modulo a set of places. These problems are called the *submarking reachability* and *submarking coverability* problems.

These problems can be further complicated by wanting to know reachability or coverability for a set of markings, the *set reachability* and *set coverability* problems. However, if the set is finite, the set problems can obviously be solved by repeated solutions of the reachability and coverability problems for one marking.

4.1.6 Firing Sequences

Another approach to analysis which has been suggested concentrates on *sequences* of transition firings rather than on states. This is related to liveness, since we may ask: Can transition t_j be fired (i.e., is it dead)? But more generally we may want to determine if a specific sequence of transition firings is possible or if any of a set of firing sequences is possible. In Figure 4.8, for example, mutual exclusion would be violated if the sequence $t_3 t_9$ can occur, or if $t_4 t_{10}$ can occur, or, more generally, if $t_3\, \sigma\, t_9$ can occur, where σ is any sequence of firings not including t_4. These analysis questions introduce the concept

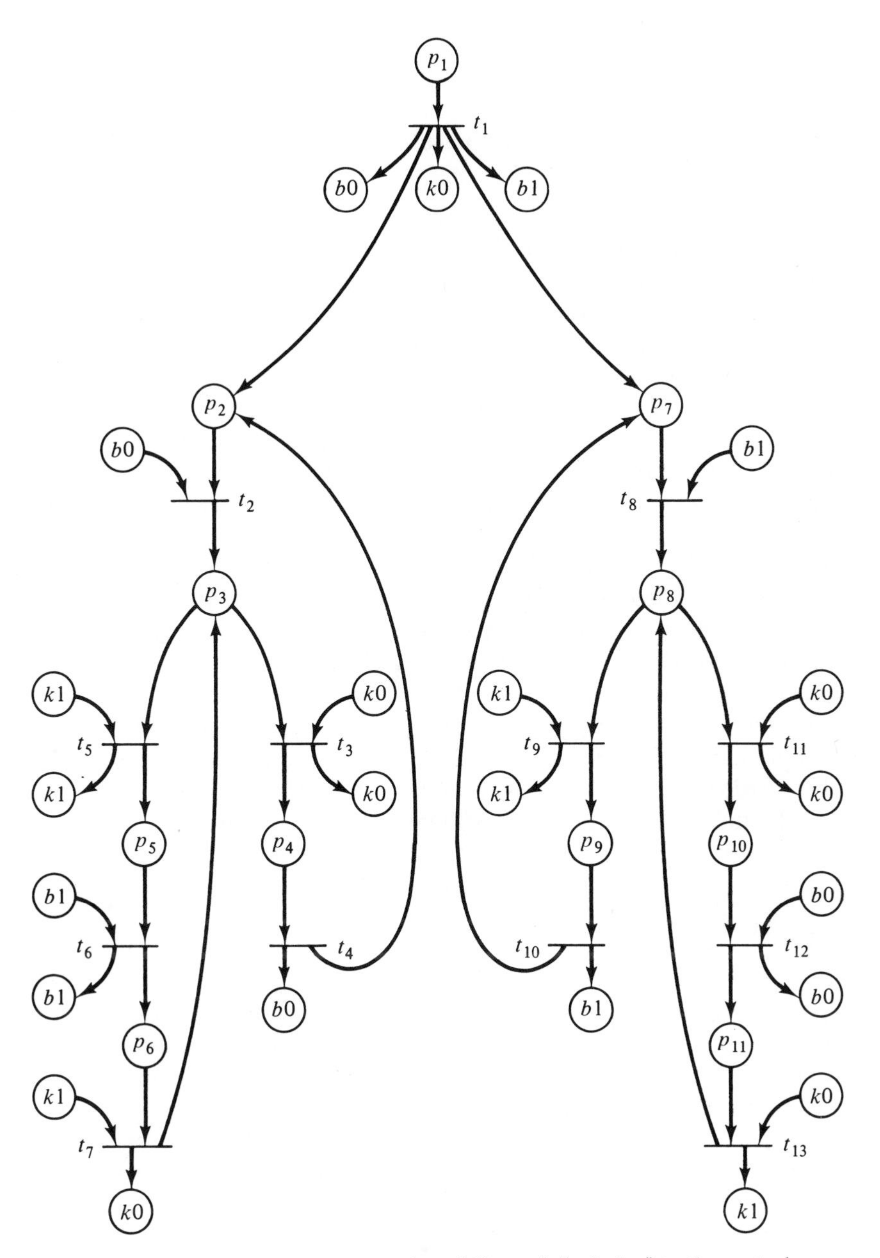

Figure 4.8 A Petri net representation of Hyman's "solution" to the mutual exclusion problem [Hyman 1966]. Due to the constant use of places $k0$, $k1$, $b1$, and $b0$, these are repeated in the graph to make it easier to read. All places labeled $k0$ are the same place.

of *languages* of Petri nets and will be investigated in more detail in Chapter 6.

4.1.7 Equivalence and Subset Problems

A final class of problems arises from optimization considerations. If a Petri net exhibits a certain behavior, as indicated by its set of transition firing sequences and its reachability set, can the Petri net be changed (optimized) without affecting its behavior? This may involve deleting dead transitions (which can never be fired) and dead places (which can never be marked) or perhaps the redefinition of some transitions. Can we show that two different marked Petri nets with the same number of transitions (but perhaps different numbers of places) will generate the same sequence of transition firings or that two different marked Petri nets with the same number of places (but perhaps different numbers of transitions) will generate the same reachability set? This might allow us to modify Petri nets to increase parallelism, decrease the cost of implementation, or other optimizations.

In these cases, we are concerned with determining if two Petri nets are *equivalent* or if one is a subset of the other. We must be careful with these problems to define the notion of equivalence or containment carefully. If we define equivalence as equal reachability sets, then we cannot change the number of places, while if we require equality of sets of transition firing sequences, we may not be able to change transitions. Our definition of the problem is therefore quite important.

4.2 Analysis Techniques

There are other problems that can be considered, but those presented here are the most common problems mentioned in the literature; we mention others as it becomes necessary to introduce them. You can see that there are a number of problems which may need solution for Petri nets. Can we develop analysis techniques to solve these problems? In particular, of course, we want to develop techniques which can be easily implemented on a computer, to allow automatic analysis of modeled systems.

Two major Petri net analysis techniques have been suggested, and we present these in this section. These techniques provide solution mechanisms for some of the above problems. The major analysis technique which has been used with Petri nets is the *reachability tree*; the other technique involves *matrix equations.* We discuss each of these in turn.

4.2.1 The Reachability Tree

The reachability tree represents the reachability set of a Petri net. As an example, consider the marked Petri net of Figure 4.9. The initial marking is (1, 0, 0). In this initial marking two transitions are enabled: t_1 and t_2. Since we wish to consider the entire reachability set, we define new nodes in the reachability tree for the (reachable) markings which result from firing both transitions. An arc, labeled by the transition fired, leads from the initial marking to each of the new markings (Figure 4.10). This (partial) tree shows all markings which are immediately reachable from the initial marking.

Now we must consider all markings reachable from these new markings. From marking (1, 1, 0), we can again fire t_1 [giving (1, 2, 0)] and t_2 [giving (0, 2, 1)]; from (0, 1, 1) we can fire t_3 [giving (0, 0,

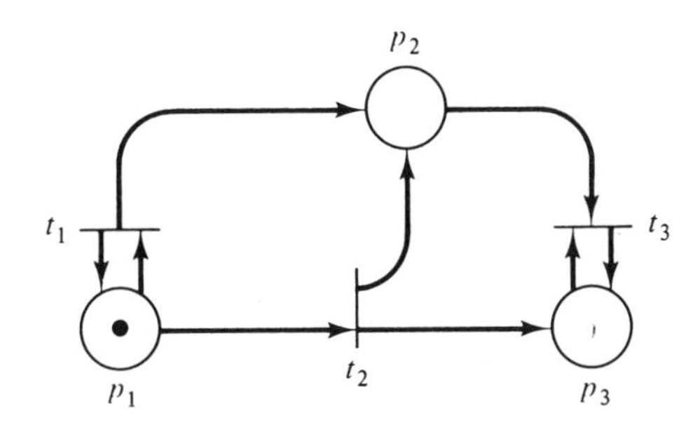

Figure 4.9 A marked Petri net for illustrating the construction of a reachability tree.

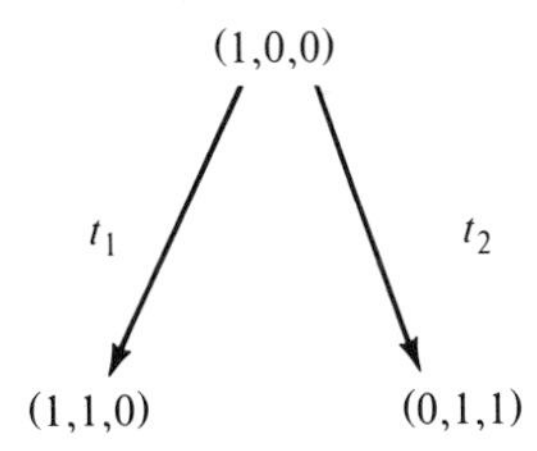

Figure 4.10 The first steps in building a reachability tree.

1)]. This produces the tree of Figure 4.11. With the three new markings, we must repeat this process, producing new markings to add to the tree as shown in Figure 4.12. Notice that the marking (0, 0, 1) is dead; no transitions are enabled, and so no new markings are produced in the tree by this dead marking. Also notice that the marking produced by firing t_3 in (0, 2, 1) is (0, 1, 1); the marking (0, 1, 1) was also produced directly from the initial marking by firing t_2.

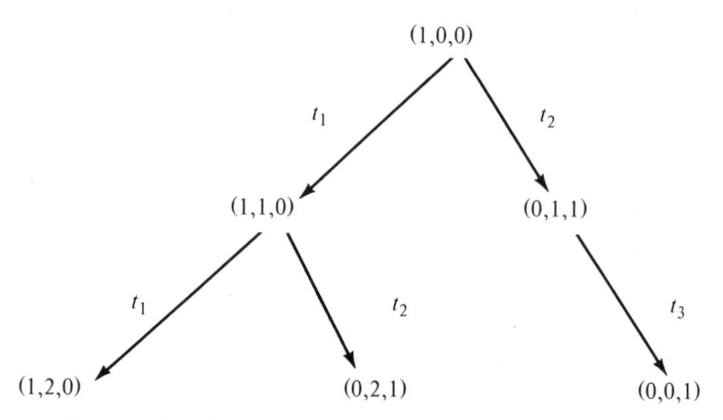

Figure 4.11 The second step in building a reachability tree.

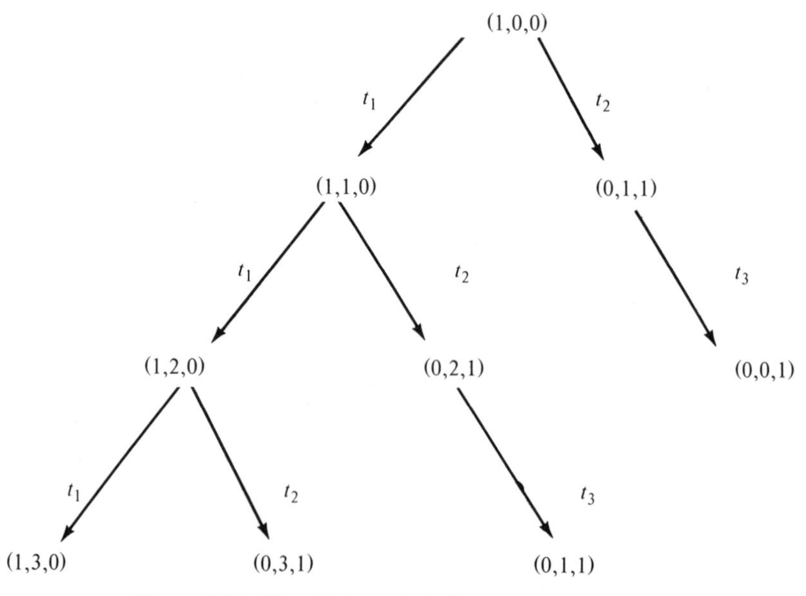

Figure 4.12 The third step in building a reachability tree.

If this procedure is repeated over and over, every reachable marking will eventually be produced. However, the resulting reachability tree might well be infinite. Every marking in the reachability set will be produced, and so for any Petri net with an infinite reachability set, the corresponding tree would also be infinite. Even a Petri net with a finite reachability set can have an infinite tree (Figure 4.13). The tree represents all possible sequences of transition firings. Every path in the tree, starting at the root, corresponds to a legal transition sequence. If the tree is going to be a useful analysis tool, we must find a means to limit it to a finite size. (Notice that if the representation of an infinite set is finite, then an infinite number of markings must be mapped onto the same representation. This will, in general, result in a loss of information, which may mean that some properties of Petri nets cannot be determined, but this depends on how the representation is done.)

The reduction to a finite representation is accomplished by several means. We must find a means of limiting the new markings (called *frontier* nodes) introduced at each step. This is helped by dead markings — markings in which no transition is enabled. These dead markings are known as *terminal* nodes. Another class of markings are those markings which have previously appeared in the tree. These duplicate markings are known as *duplicate* nodes, and no successors of a duplicate node need be considered; all these successors will be produced from the first occurrence of the marking in the tree. Thus in the tree of Figure 4.12, the marking (0, 1, 1) which results from the sequence $t_1 t_2 t_3$ does not produce any further nodes in the tree, since it occurred earlier in

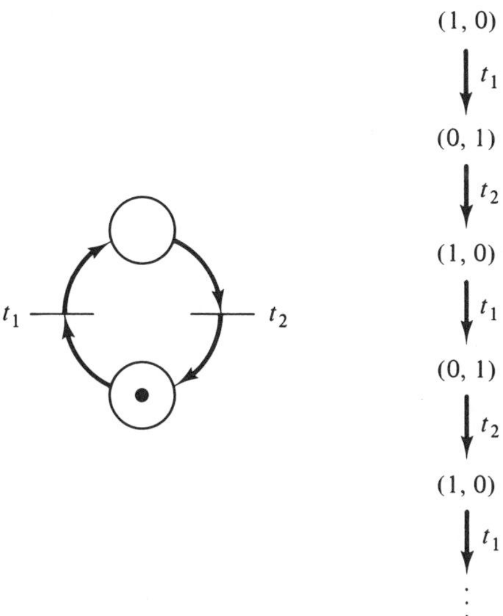

Figure 4.13 A simple Petri net with an infinite reachability tree.

the tree as a result of the sequence t_2 from the initial marking.

One final means is used to reduce the reachability tree to a finite representation. Consider a sequence of transition firings σ which starts at a marking μ and ends at a marking μ' with $\mu' > \mu$. The marking μ' is the same as the marking μ except that it has some "extra" tokens in some places, that is, $\mu' = \mu + (\mu' - \mu)$ and $(\mu' - \mu) > 0$. Now, since transition firings are not affected by extra tokens, the sequence σ can be fired again, starting in μ', leading to a marking μ''. Since the effect of the sequence of transitions σ was to add $\mu' - \mu$ tokens to the marking μ, it will also add $\mu' - \mu$ tokens to the marking μ', so $\mu'' = \mu' + (\mu' - \mu)$ or $\mu'' = \mu + 2(\mu' - \mu)$. In general, we can fire the sequence σ n times to produce a marking $\mu + n(\mu' - \mu)$. Thus, for those places which gained tokens from the sequence σ, we can create an arbitrarily large number of tokens simply by repeating the sequence σ as often as desired. In the Petri net of Figure 4.9, for example, we can fire transition t_1 as many times as we want to build up an arbitrary number of tokens in p_2.

We represent the infinite number of markings which result from these types of loops by using a special symbol, ω, which can be thought of as "infinity" and which represents a number of tokens which can be made arbitrarily large. For any constant a, we define

$$\omega + a = \omega$$

$$\omega - a = \omega$$

$$a < \omega$$

$$\omega \leqslant \omega$$

These are the only operations on ω which are necessary for the construction of the reachability tree.

The actual algorithm to construct the reachability tree can now be precisely stated. Each node i in the tree is associated with an extended marking $\mu[i]$; the marking is extended to allow the number of tokens in a place to be either a nonnegative integer or the ω symbol. Each node is also classified as either a frontier node, a terminal node, a duplicate node, or an internal node. Frontier nodes are nodes which have not yet been processed by the algorithm; they are converted by the algorithm to terminal, duplicate, or interior nodes.

The algorithm begins by defining the initial marking to be the root of the tree and, initially, a frontier node. As long as frontier nodes remain, they are processed by the algorithm.

Let x be a frontier node to be processed.

1. If there exists another node y in the tree which is not a frontier node, and has the same marking associated with it, $\mu[x] = \mu[y]$, then node x is a *duplicate* node.
2. If no transitions are enabled for the marking $\mu[x]$, [i.e., $\delta(\mu[x], t_j)$ is undefined for all $t_j \in T$], then x is a *terminal* node.
3. For all transitions $t_j \in T$ which are enabled in $\mu[x]$, [i.e., $\delta(\mu[x], t_j)$ is defined], create a new node z in the reachability tree. The marking $\mu[z]$ associated with this new node is, for each place p_i,
   (a) If $\mu[x]_i = \omega$, then $\mu[z]_i = \omega$.
   (b) If there exists a node y on the path from the root node to x with $\mu[y] < \delta(\mu[x], t_j)$ and $\mu[y]_i < \delta(\mu[x], t_j)_i$, then $\mu[z]_i = \omega$.
   (c) Otherwise, $\mu[z]_i = \delta(\mu[x], t_j)_i$.

   An arc, labeled t_j, is directed from node x to node z. Node x is redefined as an interior node; node z becomes a frontier node.

When all nodes have been classified as terminal, duplicate, or interior, the algorithm halts.

Figure 4.14 is the reachability tree of the Petri net of Figure 4.9. The reachability tree of the Petri net of Figure 4.15 is shown in Figure 4.16.

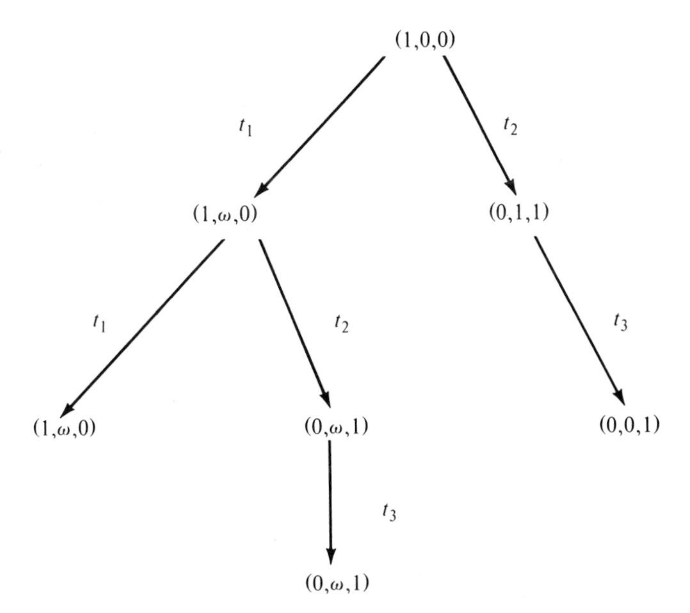

Figure 4.14 The reachability tree of the Petri net of Figure 4.9.

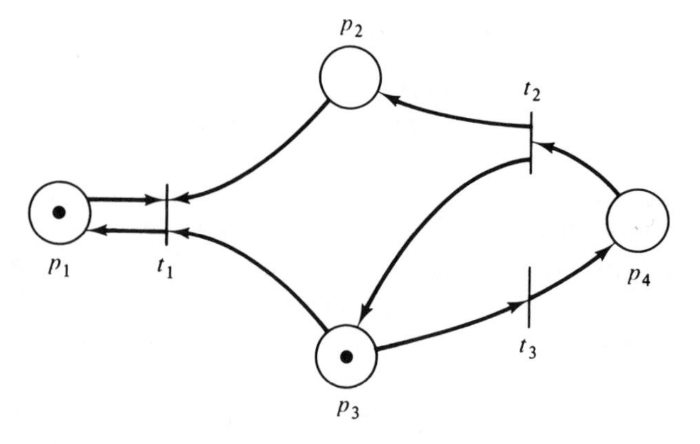

Figure 4.15 A Petri net to illustrate the construction of a reachability tree.

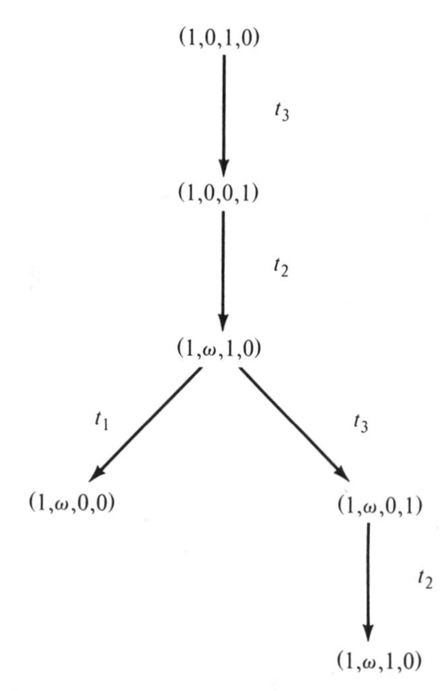

Figure 4.16 The reachability tree of the Petri net of Figure 4.15.

A very important property of the algorithm to construct a reachability tree is the fact that it terminates. To prove this we must show that the algorithm cannot continue to create new frontier nodes forever. The proof of this property requires three lemmas.

LEMMA 4-1 In any infinite directed tree in which each node has only a finite number of direct successors, there is an infinite path leading from the root.

***Proof*:** Start at the root with node x_0. Since there are only a finite number of direct successors to x_0, but the total number of nodes in the tree is infinite, at least one of the direct successors of x_0 must be the root of an infinite subtree. (If all the subtrees of direct successors of x_0 were finite, then the tree of x_0 would be finite.) Pick a node x_1 which is a direct successor of x_0 with an infinite subtree. Now one of its direct successors is also the root of an infinite subtree; pick x_2 to be such a direct successor. Continuing in this manner, we produce the infinite path $x_0, x_1, x_2, \ldots$ in the tree. □

LEMMA 4-2 Every infinite sequence of nonnegative integers contains an infinite nondecreasing subsequence.

***Proof*:** There are two cases:

1. If any element of the sequence occurs infinitely often, then let x_0 be such an element. The infinite subsequence, $x_0, x_0, x_0, \ldots$ is an infinite nondecreasing subsequence.
2. If no element occurs infinitely often, then each element occurs only a finite number of times. Let x_0 be an arbitrary element of the sequence. There are at most x_0 integers which are nonnegative and less than x_0 $[0, \ldots, x_0 - 1]$, and each of these occurs only a finite number of times in the sequence. Thus, by continuing far enough in the sequence, we must encounter an element x_1 with $x_1 \geqslant x_0$. Similarly, there must exist an x_2 farther on in the sequence with $x_2 \geqslant x_1$, and so on. This defines an infinite nondecreasing subsequence, $x_0, x_1, x_2, \ldots$.

In either case, an infinite nondecreasing sequence exists. □

LEMMA 4-3 Every infinite sequence of n-vectors over the extended nonnegative integers (the nonnegative integers plus the symbol ω) contains an infinite nondecreasing subsequence.

***Proof*:** By induction on n, the dimension of the vector space.

1. Base case ($n = 1$). If there are an infinite number of (ω) vectors in the sequence, then they form an infinite nondecreasing sequence. If not, then the infinite sequence formed by deleting the finite number of (ω) vectors has an infinite nondecreasing subsequence by Lemma 4-2.
2. Induction hypothesis. (Assume the lemma is true for n, prove for $n+1$.) Consider the first coordinate. If there are infinitely many vectors with ω as their first coordinate, then select this infinite subsequence which is nondecreasing (constant) in the first coordinate. If there are only a finite number of vectors with a first coordinate of ω, then consider the infinite sequence of integers which are the first coordinates. By Lemma 4-2, this sequence has an infinite nondecreasing subsequence. This defines an infinite subsequence of vectors which are nondecreasing in their first coordinate.

   In either case, we have a sequence of vectors that are nondecreasing in their first coordinates. Apply the induction hypothesis on the sequence of n-vectors which result from ignoring the first component of the $n+1$-vectors. The infinite subsequence which is selected in this manner is nondecreasing in each coordinate. □

Now we can prove the following theorem.

THEOREM 4-1 The reachability tree of a Petri net is finite.

***Proof*:** The proof is by contradiction. Assume there exists an infinite reachability tree. Then by Lemma 4-1 there is an infinite path $x_0, x_1, x_2, \ldots$ from the root (x_0). (The number of successors for each node in the tree is limited by m, the number of transitions.) Then $\mu[x_0], \mu[x_1], \mu[x_2], \ldots$ is an infinite sequence of n-vectors over $N \cup \{\omega\}$ and by Lemma 4-3 has an infinite nondecreasing subsequence $\mu[x_{i_0}] \leqslant \mu[x_{i_1}] \leqslant \mu[x_{i_2}] \leqslant \cdots$. But by construction, we cannot have $\mu[x_i] = \mu[x_j]$, since then one would be a duplicate node and would have no successors. Thus, we must have an infinite strictly increasing sequence $\mu[x_{i_0}] < \mu[x_{i_1}] < \cdots$. But, again by construction, since $\mu[x_i] < \mu[x_j]$, we would replace at least one component of $\mu[x_j]$

which is not ω by an ω in $\mu[x_j]$. Thus, $\mu[x_{i_1}]$ has at least one component which is ω, $\mu[x_{i_2}]$ has at least two ω-components, and $\mu[x_{i_n}]$ has at least n ω-components. Since the markings are n-dimensional, $\mu[x_{i_n}]$ has all components ω. But then $\mu[x_{i_{n+1}}]$ cannot be greater than $\mu[x_{i_n}]$. This is a contradiction, proving that our assumption that an infinite reachability tree existed was incorrect. □

The construction of the reachability tree was first described by Karp and Miller [1968]. A variant has been given by Keller [1972]. The proof of the finiteness of the tree given here is taken from Hack [1974a], who based his proof on the proof of Karp and Miller [1968].

The reachability tree is an extremely useful tool for the analysis of Petri nets. In the following sections, we show how it can be used to solve several of the problems presented in Section 4.1.

4.2.1.1 Safeness and Boundedness. A Petri net is safe if the number of tokens in each place cannot exceed one; a Petri net is bounded if there exists an integer k such that the number of tokens in any place cannot exceed k. Both of these properties can be tested using the reachability tree. *A Petri net is bounded if and only if the symbol ω never appears in its reachability tree.* The appearance of the symbol ω as part of a reachability tree means that the number of tokens is potentially unbounded; there exists a sequence of transition firings which can be repeated arbitrarily many times to increase the number of tokens to an arbitrary, unbounded number. Thus, if the symbol ω appears, the net is unbounded. In addition, the ω symbol indicates by its position which places are unbounded.

Conversely, if the Petri net is unbounded, then the number of reachable markings is infinite. Since the reachability tree is finite, the symbol ω must occur to represent the infinite number of reachable markings.

If the Petri net is bounded, and the ω symbol is not in the reachability tree, then the Petri net represents a *finite state* system. The reachability tree then is essentially a state graph and will contain a node corresponding to every reachable marking. This allows any, and all, other analysis questions to be solved by simply the exhaustive examination of the finite set of all reachable markings. For example, to determine the bound on a particular place, generate the reachability tree and scan the tree for the largest value of the component of the markings corresponding to that place. This is the bound on the number of tokens for this place. If the bound for all places is 1, then the net is safe.

Figure 4.17 demonstrates using the reachability tree to determine boundedness.

Notice that even for Petri nets which are not bounded (because some place is unbounded) it is possible to determine the bounds for those places which are bounded from the reachability tree. Thus the reachability tree effectively solves the analysis of Petri nets to determine boundedness or safeness for individual places or entire nets.

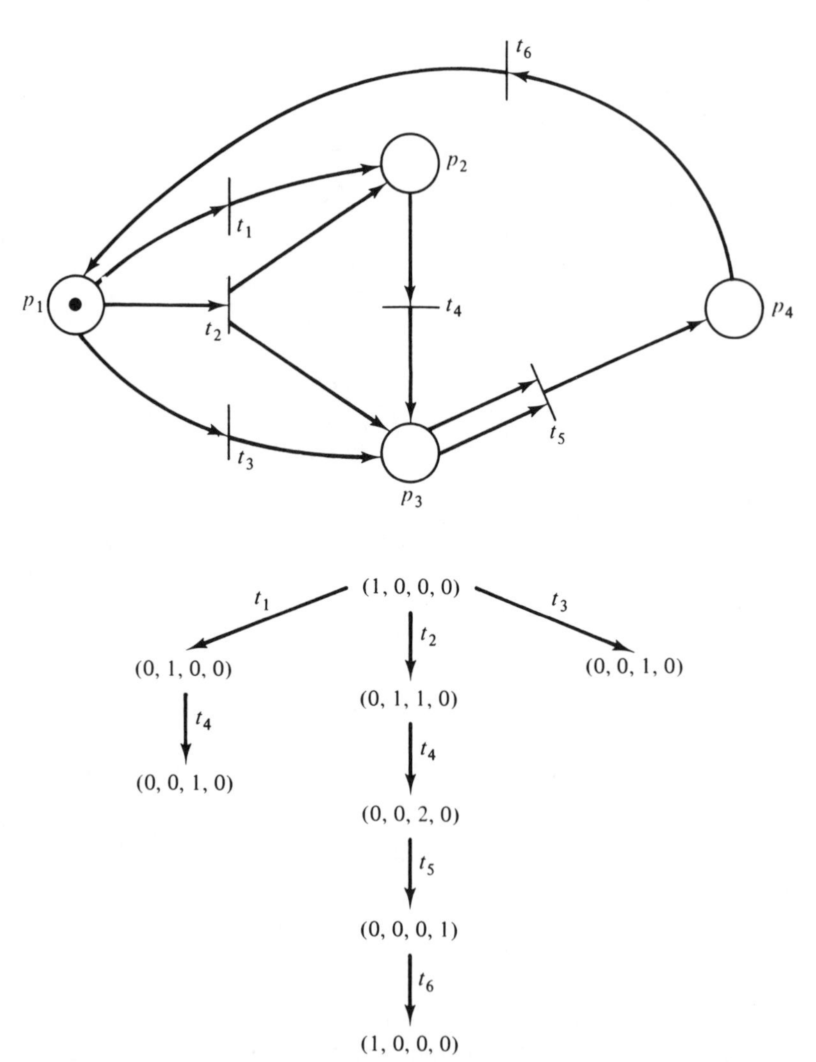

Figure 4.17 Determining boundedness for a Petri net using the reachability tree.

4.2.1.2 Conservation. A Petri net is *conservative* if it does not lose or gain tokens but merely moves them around. Since two tokens may be encoded as one token which later causes a transition to fire, creating two tokens, a vector of weights defines the value of a token in each place; the weights are nonnegative. A Petri net is conservative with respect to a weighting vector if the weighted sum of tokens is constant over all reachable markings.

Conservation can be effectively tested using the reachability tree. Since the reachability tree is finite, the weighted sum can be computed for each marking. If the sums are the same for each reachable marking, the net is conservative with respect to the given weight. If the sums are not equal. the net is not conservative.

The ω symbol must be carefully considered in the evaluation of conservation. If a marking has ω as the marking for place p_i, then the weight of that place must be 0 for the net to be conservative. Remember that the ω symbol represents an infinite set of values. Since all weights are nonnegative, either the weight must be zero (indicating that the value of the number of tokens in this place is unimportant) or positive. If the weight is positive, then the sum will vary for two markings which differ in their ω-component. Hence, if any marking with nonzero weight is ω, the net is not conservative.

The above considerations refer to conservation with respect to a defined weighting. A Petri net is conservative if it is conservative with respect to some weight vector w, with $w_i > 0$. The reachability tree can be used to determine if a Petri net is conservative by finding a positive weight vector w, if one exists. To determine a positive weight vector with respect to which a Petri net is conservative, first note that the net must be bounded. As pointed out above, an unbounded place must have a weight of zero, which is not possible in a net with positive weight vector. (If we wish to allow zero components, we simply set the weights of all unbounded places to zero and consider further only the remaining components.) Now if the net is conservative, a weighted sum, call it s, and a weight vector, $w = (w_1, w_2, \ldots, w_n)$, exist. For each marking $\mu[x]$ of the reachability tree we must have

$$w_1 \cdot \mu[x]_1 + w_2 \cdot \mu[x]_2 + \cdots + w_n \cdot \mu[x]_n = s$$

This defines, for k nodes in the reachability tree, a set of k linear equations in $n+1$ unknowns. Add to this the constraints

$$w_i > 0, \ i = 1, \ldots, n$$

and we have defined the constraints on the weight vector.

Solution of this system of linear equations is a well-known problem with many algorithms for solution. It can be considered a linear programming problem or simply a system of linear equations. In either case, if a solution exists, it can be computed. (The solutions from these techniques will in general be rational, not integer, but the weights can be multiplied by a common denominator to produce an integer solution.)

If the weights are overly constrained and hence no weighting vector exists, this will be determined. In either case, it can be determined whether or not the Petri net is conservative, and if so, a weight vector is produced.

4.2.1.3 Coverability. A final problem which can be solved with the aid of the reachability tree is the coverability problem. For the coverability problem, we wish to determine, for a given marking μ', if a marking $\mu'' \geqslant \mu'$ is reachable. This problem can be solved by inspection of the reachability tree. Given an initial marking μ, we construct the reachability tree. Then we search for any node x, with $\mu[x] \geqslant \mu'$. If no node is found, the marking μ' is not covered by any reachable marking; if such a node is found, $\mu[x]$ gives a reachable marking which covers μ'.

The path from the root to the covering marking defines the sequence of transitions which leads from the initial marking to the covering marking, and the marking associated with that node defines the covering marking. Again, of course, the symbol ω should be treated as representing an infinite set of values. If a component of the covering marking is ω, then a "loop" will exist in the path from the root to the covering marking. It will be necessary to repeat this loop enough times to raise the corresponding components to not less than the given marking.

Note that if there are several components in the covering marking which are ω, there may be an interaction between the marking changes which result from the loops. Consider the Petri net of Figure 4.18 and

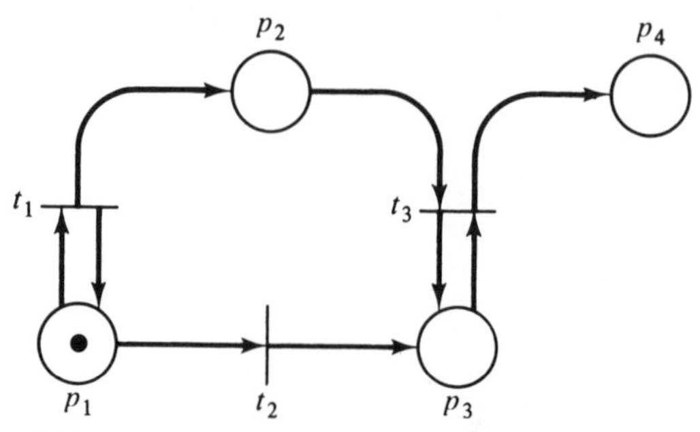

Figure 4.18 A Petri net with the reachability tree of Figure 4.19.

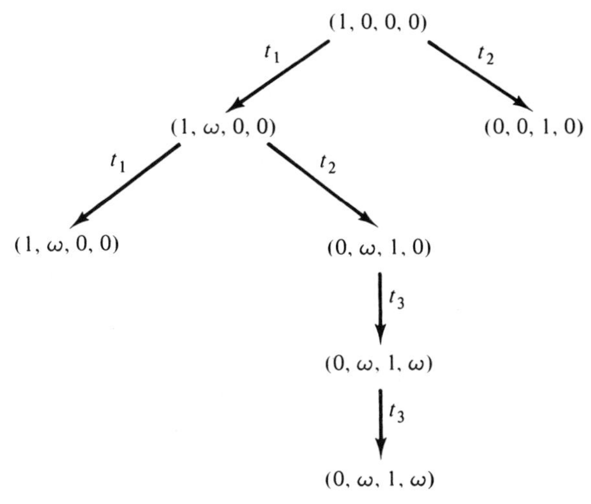

Figure 4.19 The reachability tree of the Petri net of Figure 4.18.

its reachability tree given in Figure 4.19. According to the analysis given, the marking (0, 14, 1, 7) is covered in the reachability set. The path to generate a covering marking consists of some number of t_1 followed by t_2 followed by some number of t_3. The problem is to determine how many t_1 and t_3. Since we want 14 tokens in p_2 and t_1 puts a token in p_2, we might try 14 t_1. However, we need 7 t_3, and each t_3 removes a token from p_2, so we actually need at least 21 t_1, then t_2, and then at least 7 t_3(but not so many t_3 as to empty p_2 too much). Karp and Miller [1968] give an algorithm which will determine the minimal number of transition firings needed to cover a given marking.

4.2.1.4 Limitations of the Reachability Tree. As we have seen, the reachability tree can be used to solve the safeness, boundedness, conservation and coverability problems. Unfortunately, it cannot, in general, be used to solve the reachability or liveness problems or to define or determine which firing sequences are possible. These problems are limited by the existence of the ω symbol. The ω symbol is a loss of information; the individual numbers are discarded, with only the existence of the large number of them being remembered.

Consider, for example, the Petri nets of Figures 4.20 and 4.21 whose reachability tree is given in Figure 4.22. The same reachability tree represents these two similar (but different) Petri nets. The reachability *sets* are not the same, however. In the Petri net of Figure 4.20, the number of tokens in place p_2 is always an even number (until t_1

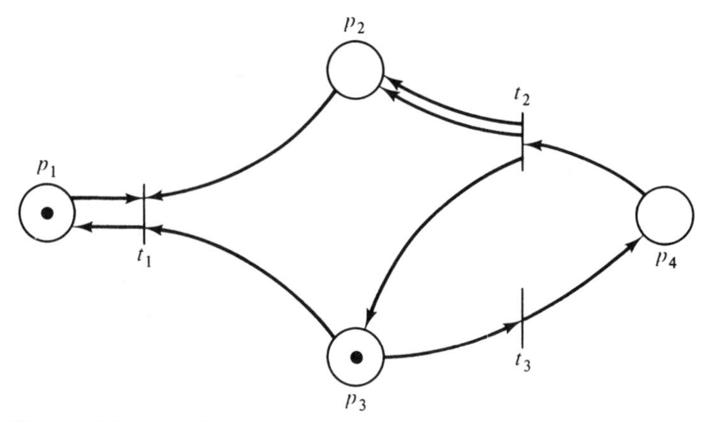

Figure 4.20 A Petri net whose reachability tree is shown in Figure 4.22.

fires), whereas in Figure 4.21 it may be an arbitrary integer. The ω symbol does not allow this kind of information to be detected, preventing the use of the reachability tree to solve the reachability problem.

A similar problem exists for the liveness problem. Figures 4.23 and 4.24 are two Petri nets whose reachability tree is given in Figure 4.25. However, the net in Figure 4.23 can deadlock (the sequence $t_1t_2t_3$, for example), while the net in Figure 4.24 cannot. Again, however, the reachability tree cannot distinguish between these two cases.

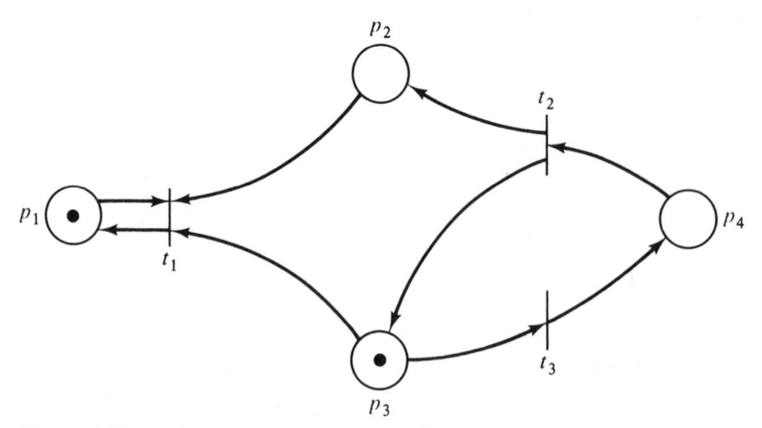

Figure 4.21 A second Petri net whose reachability tree is shown in Figure 4.22. The Petri net of Figure 4.20 whose reachability tree is in Figure 4.22 has only even numbers of tokens in place p_2 while this Petri net has an arbitrary integer marking for place p_2.

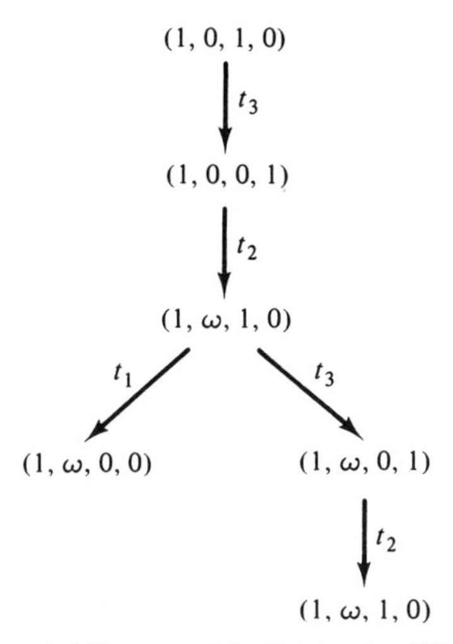

Figure 4.22 The reachability tree of the Petri nets of Figures 4.21 and 4.20.

Note that although the reachability tree does not necessarily contain enough information to always solve the reachability or liveness problems, it is the case that the tree *may* have sufficient information to solve many such problems. A net whose reachability tree has a terminal node (one with no successors) is not live (since some reachable marking has no successors). Similarly, a marking μ' of a reachability problem may appear in the reachability tree, and if so, it is reachable. Also, if a marking is not covered by some node of the reachability tree, then it is not reachable.

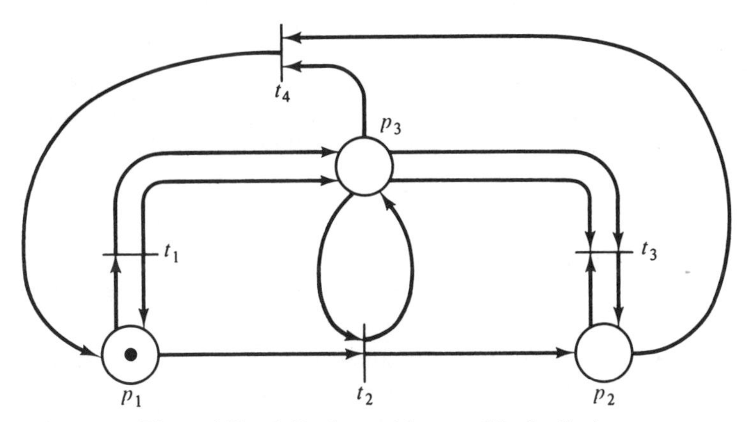

Figure 4.23 A Petri net with a possible deadlock.

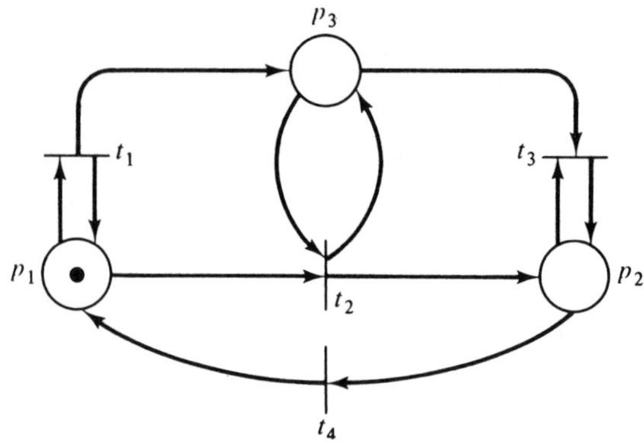

Figure 4.24 A Petri net which cannot deadlock. This net is live, yet its reachability tree (Figure 4.25) is identical to the reachability tree of the non-live Petri net of Figure 4.23.

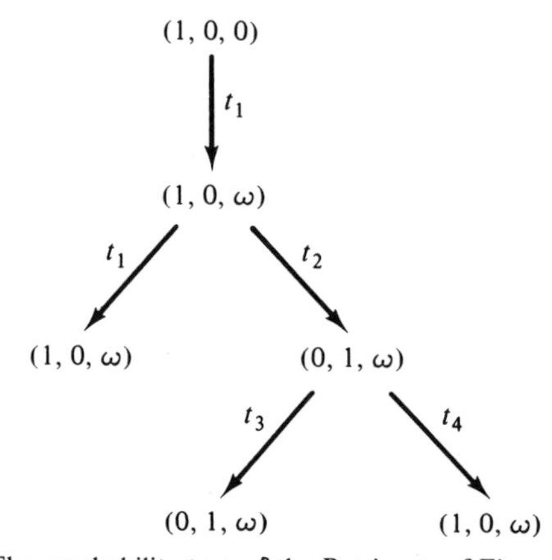

Figure 4.25 The reachability tree of the Petri nets of Figures 4.23 and 4.24.

These conditions are sufficient to solve some reachability and liveness problems, but they do not solve these problems in general. Thus, to solve these two problems, other approaches are needed.

Exercises

1. Construct the reachability trees for the marked Petri nets of Figures 4.1 and 4.2.
2. Write a computer program to construct the reachability tree from a Petri net description and an initial marking.
3. Most statements of the algorithm for constructing the reachability

tree allow a node to be classified as a duplicate node only if the first node occurs on the path from the root to the duplicate node. Thus nodes in separate subtrees which represent the same markings continue to be processed by the algorithm. This change increases the size of the subtree, but it will not affect the (eventual) termination of the algorithm. Prove that the algorithm will terminate even if duplicate markings must be in the same subtree as the first marking, and prove that allowing duplicate markings to be in separate subtrees will not result in any reachable markings being missed from the tree.

4.2.2 Matrix Equations

A second approach to the analysis of Petri nets is based on a matrix view of Petri nets. An alternative to the (P,T,I,O) definition of Petri nets is to define two matrices D^- and D^+ to represent the input and output functions. (These are equivalent to the F and B functions of Hack's Petri net definition; see Section 2.6.) Each matrix is m rows (one for each transition) by n columns (one for each place). We define $D^-[j,i] = \#(p_i, I(t_j))$ and $D^+[j,i] = \#(p_i, O(t_j))$. D^- defines the inputs to the transitions and, D^+ defines the outputs.

The matrix definitional form of a Petri net (P,T,D^-,D^+) is equivalent to the standard form we have used but allows the definitions to be recast in vector and matrix terms. Let $e[j]$ be the unit m-vector which is zero everywhere except in the jth component. The transition t_j is represented by the unit m-vector $e[j]$.

Now a transition t_j is enabled in a marking μ if

$$\mu \geqslant e[j] \cdot D^-$$

and the result of firing transition t_j in marking μ, if it is enabled, is

$$\begin{aligned}\delta(\mu,t_j) &= \mu - e[j] \cdot D^- + e[j] \cdot D^+ \\ &= \mu + e[j] \cdot (-D^- + D^+) \\ &= \mu + e[j] \cdot D\end{aligned}$$

where we have defined the composite change matrix $D = D^+ - D^-$.

Now for a sequence of transition firings $\sigma = t_{j_1} t_{j_2} \cdots t_{j_k}$, we have

$$\begin{aligned}\delta(\mu,\sigma) &= \delta(\mu, t_{j_1} t_{j_2} \cdots t_{j_k}) \\ &= \mu + e[j_1] \cdot D + e[j_2] \cdot D + \cdots + e[j_k] \cdot D \\ &= \mu + (e[j_1] + e[j_2] + \cdots + e[j_k]) \cdot D \\ &= \mu + f(\sigma) \cdot D\end{aligned}$$

The vector $f(\sigma) = e[j_1] + e[j_2] + \cdots + e[j_k]$ is called the *firing vector* of the sequence $t_{j_1}t_{j_2}\cdots t_{j_k}$. The *i*th element of $f(\sigma)$, $f(\sigma)_i$, is the number of times that transition t_i fires in the sequence $t_{j_1}t_{j_2}\cdots t_{j_k}$. The firing vector is thus a vector of nonnegative integers. (The vector $f(\sigma)$ is the *Parikh mapping* of the sequence σ [Parikh 1966]).

To show an example of the usefulness of this matrix approach to Petri nets, consider the conservation problem: Given a marked Petri net, is it conservative? To show conservation, it is necessary to find a (nonzero) weighting vector for which the weighted sum over all reachable markings is constant. Let w be an $n \times 1$ column vector. Then if μ is the initial marking and μ' is an arbitrary reachable marking, we need

$$\mu \cdot w = \mu' \cdot w$$

Now since μ' is reachable, there exists a sequence σ of transition firings which takes the net from μ to μ'. So,

$$\begin{aligned} \mu' &= \delta(\mu, \sigma) \\ &= \mu + f(\sigma) \cdot D \end{aligned}$$

Thus,

$$\begin{aligned} \mu \cdot w &= \mu' \cdot w \\ &= (\mu + f(\sigma) \cdot D) \cdot w \\ &= \mu \cdot w + f(\sigma) \cdot D \cdot w \end{aligned}$$

and so

$$f(\sigma) \cdot D \cdot w = 0$$

Since this must be true for all $f(\sigma)$, we must have

$$D \cdot w = 0$$

Thus a Petri net is conservative if and only if there exists a positive vector w such that $D \cdot w = 0$. This provides a simple test for conservation and also produces the weighting vector, w.

The development of this matrix Petri net theory provides a useful tool for attacking the reachability problem. Assume that a marking μ' is reachable from a marking μ. Then there exists a sequence σ (possi-

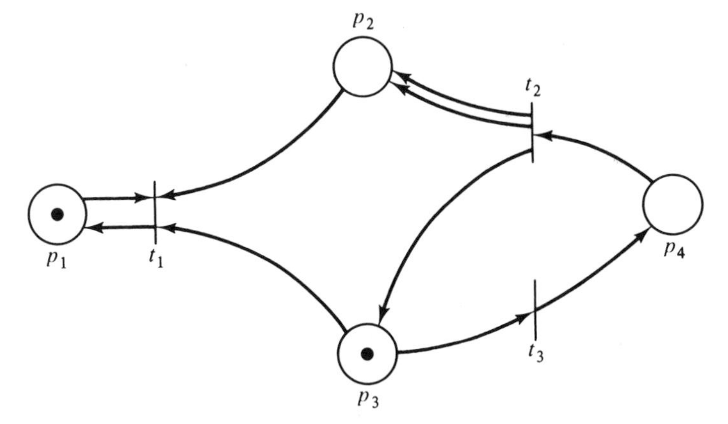

Figure 4.26 A Petri net for illustrating matrix equation analysis.

bly null) of transition firings which will lead from μ to μ'. This means that $f(\sigma)$ is a solution, in nonnegative integers, for x in the following matrix equation.

$$\mu' = \mu + x \cdot D \tag{4-1}$$

Thus, if μ' is reachable from μ, then Equation (4-1) has a solution in nonnegative integers; if Equation (4-1) has no solution, then μ' is not reachable from μ.

As an example, consider the marked Petri net of Figure 4.26. The matrices D^- and D^+ are

$$D^- = \begin{bmatrix} 1 & 1 & 1 & 0 \\ 0 & 0 & 0 & 1 \\ 0 & 0 & 1 & 0 \end{bmatrix}$$

$$D^+ = \begin{bmatrix} 1 & 0 & 0 & 0 \\ 0 & 2 & 1 & 0 \\ 0 & 0 & 0 & 1 \end{bmatrix}$$

and the D matrix is

$$D = \begin{bmatrix} 0 & -1 & -1 & 0 \\ -1 & +2 & +1 & -1 \\ +1 & 0 & -1 & +1 \end{bmatrix}$$

With an initial marking $\mu = (1, 0, 1, 0)$, transition t_3 is enabled and leads to marking μ', where

$$\mu' = (1, 0, 1, 0) + (0, 0, 1) \cdot \begin{bmatrix} 0 & -1 & -1 & 0 \\ 0 & +2 & +1 & -1 \\ 0 & 0 & -1 & +1 \end{bmatrix}$$

$$= (1, 0, 1, 0) + (0, 0, -1, +1)$$

$$= (1, 0, 0, 1)$$

The sequence $\sigma = t_3 t_2 t_3 t_2 t_1$ is represented by the firing vector $f(\sigma) = (1, 2, 2)$ and produces the marking μ',

$$\mu' = (1, 0, 1, 0) + (1, 2, 2) \cdot \begin{bmatrix} 0 & -1 & -1 & \\ 0 & +2 & +1 & 0{-}1{+}1 \\ 0 & 0 & -1 & \end{bmatrix}$$

$$= (1, 0, 1, 0) + (0, 3, -1, 0)$$

$$= (1, 3, 0, 0)$$

To determine if the marking (1, 8, 0, 1) is reachable from the marking (1, 0, 1, 0), we have the equation

$$(1, 8, 0, 1) = (1, 0, 1, 0) + x \cdot \begin{bmatrix} 0 & -1 & -1 & 0 \\ 0 & +2 & +1 & -1 \\ 0 & 0 & -1 & +1 \end{bmatrix}$$

$$(0, 8, -1, 1) = x \cdot \begin{bmatrix} 0 & -1 & -1 & 0 \\ 0 & +2 & +1 & -1 \\ 0 & 0 & -1 & +1 \end{bmatrix}$$

which has a solution $x = (0, 4, 5)$. This corresponds to the sequence $\sigma = t_3 t_2 t_3 t_2 t_3 t_2 t_3 t_2 t_3$.

We can further show that marking (1, 7, 0, 1) is not reachable from the marking (1, 0, 1, 0), since the matrix equation

$$(1, 7, 0, 1) = (1, 0, 1, 0) + x \cdot \begin{bmatrix} 0 & -1 & -1 & 0 \\ 0 & +2 & +1 & -1 \\ 0 & 0 & -1 & +1 \end{bmatrix}$$

$$(0, 7, -1, 1) = x \cdot \begin{bmatrix} 0 & -1 & -1 & 0 \\ 0 & +2 & +1 & -1 \\ 0 & 0 & -1 & +1 \end{bmatrix}$$

has no solution.

The matrix approach to the analysis of Petri nets has great promise but also has some severe problems. First notice that the matrix D

by itself does not properly reflect the structure of the Petri net. Transitions which have both inputs and outputs from the same place (self-loops) will be represented in the same position of the D^- and D^+ matrices and so will cancel out in the $D = D^+ - D^-$ matrix. This was reflected in the previous example by place p_1 and transition t_1.

Another problem is the lack of sequencing information in the firing vector. Consider the Petri net of Figure 4.27. Assume we wish to determine if the marking (0, 0, 0, 0, 1) is reachable from (1, 0, 0, 0, 0). Then one equation is

$$(0,0,0,0,1) = (1,0,0,0,0) + x \cdot \begin{bmatrix} -1 & 2 & 1 & 0 & 0 \\ 0 & 0 & 0 & 0 & 0 \\ 0 & 0 & 1 & 0 & 0 \\ 0 & -1 & 0 & 1 & 0 \\ 0 & 2 & 0 & 0 & -1 \\ 0 & 0 & -1 & -1 & 1 \end{bmatrix}$$

This equation does not have a unique solution but reduces to the set of solutions $\{\sigma \mid f(\sigma) = (1, x_2, x_6 - 1, 2x_6, x_6 - 1, x_6)\}$. This defines the relationships between the firings of the transitions. If we let $x_6 = 1$ and $x_2 = 1$, we have $f(\sigma) = (1,1,0,2,0,1)$, but both the sequence $t_1 t_2 t_4 t_4 t_6$ and the sequence $t_1 t_4 t_2 t_4 t_6$ correspond to this firing vector. Thus, although we know the number of transition firings, we do not know the order of transition firings.

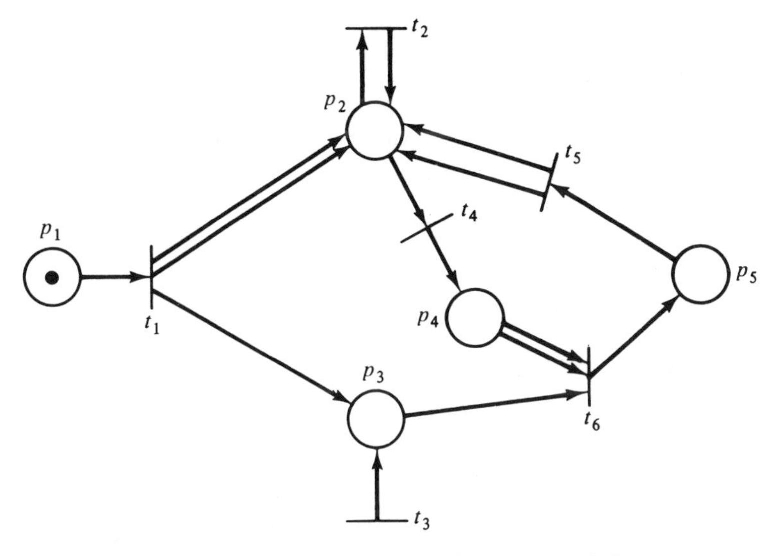

Figure 4.27 Another Petri net for matrix analysis.

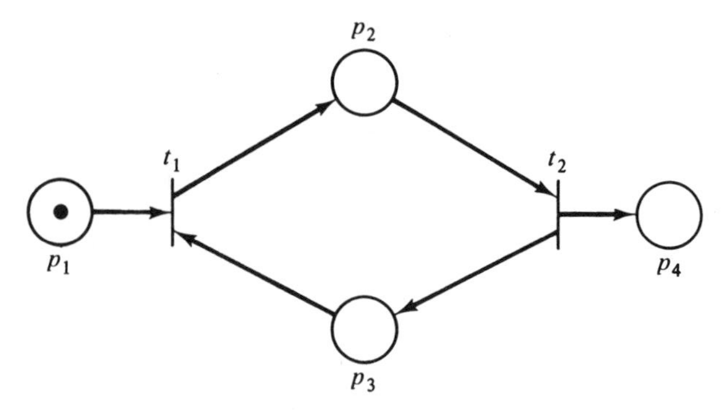

Figure 4.28 A Petri net showing that a solution to the matrix equation is a necessary but not a sufficient condition for reachability.

Another problem is that although a solution to Equation (4-1) is *necessary* for reachability, it is not *sufficient*. Consider the simple Petri net of Figure 4.28. If we wish to determine if (0, 0, 0, 1) is reachable from (1, 0, 0, 0) we must solve the equation

$$(0,0,0,1) = (1,0,0,0) + f(\sigma) \cdot \begin{bmatrix} -1 & +1 & -1 & 0 \\ 0 & -1 & +1 & +1 \end{bmatrix}$$

This equation has the solution $f(\sigma) = (1,1)$, corresponding to the two sequences $t_1 t_2$ or $t_2 t_1$. But neither of these two transition sequences are possible, since neither t_1 nor t_2 is enabled in (1, 0, 0, 0). Thus a solution to Equation (4-1) is not sufficient to prove reachability.

The possibility of spurious solutions to Equation (4-1) — solutions which do not correspond to possible transition sequences — has resulted in only limited research on the matrix representation of Petri nets. The best research on this approach has been by Murata [1975; 1977a; 1977b].

4.3 Further Reading

Holt et al. [1968] and Holt and Commoner [1970] defined some of the early analysis problems for Petri nets — liveness and safeness — and these have continued as major problems for analysis. Liveness has been studied by Commoner [1972], Lautenbach [1975], and Lien [1976a]. Keller [1972] also considered liveness plus other problems. Lien [1976a] defined the conservation problem.

Karp and Miller [1968] first described the reachability tree construction and proved that it was finite. The coverability and reachability problems were defined by them, as were the equivalence and containment problems. These later problems were the subject of [Baker 1973b], while [Nash 1973] is a brief statement of the reachability problem. Hack [1974a] brings most of these problems together in one place and shows how the reachability tree can be used for some of them.

The matrix approach was considered by Peterson [1973] but was found to be of limited usefulness. Murata, with a better background in linear algebra, has done quite a bit more with this approach in [Murata and Church 1975; Murata 1975; Murata et al. 1975; Murata 1977a; Murata 1977b].

4.4 Topics for Further Study

1. Consider constructing a reachability graph rather than a tree. If a node x produces a successor z with $\mu[z] = \mu[y]$ for some nonfrontier node y, simply create an arc, labeled appropriately, from x to y. Note that a path from the root to a node need no longer be unique. Define the algorithm to construct the reachability graph, show that it terminates, and investigate its properties relative to the reachability tree.
2. The reachability tree cannot be used to solve the reachability problem because of the loss of information which results from the ω symbol. The ω symbol is introduced when we come to a marking μ' with a marking μ on the path from the root to μ', with $\mu' > \mu$. In this case, we can reach all markings of the form $\mu + n(\mu' - \mu)$. Investigate the possibility of using an expression $a + b \cdot n_i$ rather than ω to represent the value of the components. If you can define a reachability tree where all marking vectors are represented by expressions, then the reachability problem is simply the solution of a set of equations.
3. Extend the definition of conservation to allow negative weights. What would be a reasonable interpretation of a negative weight? Is it decidable if a Petri net is conservative if negative weights are allowed?
4. Use the matrix analysis approach to develop an algorithm for determining boundedness of a Petri net. (See [Crespi-Reghizzi and Mandrioli 1974].)
5. The major problem with matrix analysis is the lack of sequencing information and the existence of spurious solutions. However, the

reachability problem is the most important problem for Petri nets, and it is important to show that this problem is decidable, even if the solutions are computationally expensive. If the problem is decidable, then we can search for more efficient solution techniques, but we first need to show that a solution technique exists.

With this motivation, let Σ_f be the sum over all components of a firing vector f. That is, if $f = (f_1, f_2, \ldots, f_m)$, then $\Sigma_f = f_1 + f_2 + \cdots + f_m$. Now consider that if a sequence σ exists which will take a Petri net from a marking μ to a marking μ', then Equation (4-1) has a solution which is the firing vector $f(\sigma)$ for σ. The sequence σ can be determined from the firing vector $f(\sigma)$ by simply enumerating all possible sequences of length equal to $\Sigma_{f(\sigma)}$ and trying each to determine if it (1) is legal and (2) leads from μ to μ'. For a Petri net with m transitions, there are at most m^{Σ_f} possible sequences of length Σ_f. This can, in fact, be reduced further. Since we know how many times transition t_1 fires (f_1), how many times t_2 fires (f_2), and so on, we need examine no more than the $\Sigma_f!$ possible orderings of f_1 firings of t_1, f_2 firings of t_2, and so on.

This would seem to provide a decision procedure for determining if μ' is reachable from μ. First solve the matrix equation $\mu' = \mu + f \cdot D$. If there is no solution, μ' is not reachable from μ. If a solution f exists, then examine all $\Sigma_f!$ possible orderings of the transitions. If any of these transition sequences is legal, then μ' is reachable from μ, and we have the sequence of transitions which takes us from μ to μ'.

There is only one hitch. The solution f may not be unique but may be a (infinite) set of firing vectors represented by a set of expressions (as illustrated above for the analysis of Figure 4.27). Research is needed to determine if it can be shown that it is possible to determine reachability in this case. In the easiest case, it may be that either all solutions represented by an expression firing vector correspond to legal solutions, or none of the solutions do. In this case, we merely pick any solution and follow the above procedure of checking all possible orderings. More likely, however, some solutions may work while others fail. Since we cannot try all solutions (possibly infinite number), research must be done to determine which solutions should be tried.

5

Complexity and Decidability

In Chapter 4 we presented a number of problems which have been defined for Petri nets. These problems concern various properties of Petri net structure and behavior which, under appropriate circumstances, would be of interest to users of Petri nets.

Two solution techniques were also presented: the reachability tree and matrix equation approaches. These two techniques allow properties of safeness, boundedness, conservation, and coverability to be determined for Petri nets. Also, a necessary condition for reachability was established. However, these analysis techniques are not sufficient to solve several other problems, especially liveness, reachability, and equivalence. In this chapter we explore these problems, either to find solutions to them or at least to learn more about the properties of Petri nets.

5.1 Reducibility Between Analysis Problems

A fundamental concept which we use is *reducibility* [Karp 1972]. Solving a problem involves reducing it to another problem which we already know how to solve. For example, in the previous chapter, the problem of determining if a Petri net is conservative was reduced to solving a set of simultaneous linear equations. The problem of solving sets of simultaneous linear equations has in turn been reduced to a defined sequence of arithmetic operations (addition, subtraction, multiplication, division, and comparisons). Thus, since the simpler arithmetic operations can be computed, conservation can be determined.

Another example concerns the equality problem and subset problem for reachability sets.

DEFINITION 5-1 *Equality Problem* Given two marked Petri nets $C_1 = (P_1, T_1, I_1, O_1)$ with marking μ_1 and $C_2 = (P_2, T_2, I_2, O_2)$ with marking μ_2, is $R(C_1, \mu_1) = R(C_2, \mu_2)$?

DEFINITION 5-2 *Subset Problem* Given two marked Petri nets $C_1 = (P_1, T_1, I_1, O_1)$ with marking μ_1 and $C_2 = (P_2, T_2, I_2, O_2)$ with marking μ_2, is $R(C_1, \mu_1) \subseteq R(C_2, \mu_2)$?

These two problems can be very important if Petri nets are to be "optimized" or if the nets of two systems are to be compared. However, notice that if a solution to the subset problem can be found, the equality problem is also solved. If we wish to determine if $R(C_1, \mu_1) = R(C_2, \mu_2)$, we can first use the subset problem algorithm to determine if $R(C_1, \mu_1) \subseteq R(C_2, \mu_2)$, and then use the same algorithm to determine if $R(C_2, \mu_2) \subseteq R(C_1, \mu_1)$. $R(C_1, \mu_1) = R(C_1, \mu_2)$ if and only if $R(C_1, \mu_1) \subseteq R(C_2, \mu_2)$ and $R(C_2, \mu_2) \subseteq R(C_1, \mu_1)$. Thus, we can *reduce* the equality problem to the subset problem.

Two other considerations are of importance when considering analysis problems and reducibility. First, in trying to find a solution, we must consider the possibility that a problem has no solution technique; it is *undecidable*. Second, if a solution technique exists, we need to consider its cost: How much time and memory space are needed? For Petri nets to gain widespread general use, analysis problems must be solvable and by algorithms which are not excessively expensive in computer time or space.

Reducibility plays a role in both of these problems. Reducibility between problems is commonly used to show that a problem is decidable or undecidable. Our approach to *decidability theory* [Davis 1958; Minsky 1967] is based mainly on the work of Turing and on his model of computations, the Turing machine. The importance of the Turing machine is that it is a reasonable representation of a limited computing machine and that it can be shown that no algorithm exists which can solve certain Turing machine problems, especially the halting problem. From this basis, a collection of undecidable problems has been found. The importance of this theory is that it is not possible to produce a computer program which solves these problems. Thus, for practical analysis, these undecidable problems must be avoided, or the analysis questions will be unanswerable.

(An important distinction here is that undecidable problems produce questions which are not simply unanswered but *unanswerable*.

Questions can be unanswered but still answerable; this merely means that no one has yet found an answer but that the answer does exist. A famous example is Fermat's last theorem: Does the equation $x^n + y^n = z^n$ have solutions for $n > 2$ and nontrivial integer x, y, and z? This question has not been answered, but it is answerable. The answer is either yes or no. One way to answer the question is to produce numbers x,y,z, and n which satisfy the theorem. The other way would be to prove (logically deduce) that no such x, y, z, and n can exist. No one has yet done so.

However, assume that the problem were undecidable. Then it is not possible to decide whether x, y, z, and n exist which solve the equation. This means we could not logically deduce their nonexistence from the axioms of mathematics and that we cannot produce x, y, z, and n which solve the equation. But if we cannot produce x, y, z, and n, then they must not exist. If they did exist, we could set a computer to searching for them, and, eventually, it would find them. But if x, y, z, and n do not exist, then the answer to the question is no, and we have decided it. This contradicts our assumption that the question is undecidable, so the question is decidable.)

Now assume that a problem A is reducible to a problem B: An instance of problem A can be transformed into an instance of problem B. If problem B is decidable, then problem A is decidable, and the algorithm for problem B can be used to solve problem A. An instance of problem A can be solved by transforming it to an instance of problem B and applying the algorithm for problem B to determine the solution. Thus, if problem A is reducible to problem B and problem B is decidable, then problem A is decidable.

The contrapositive is also true: If problem A is reducible to problem B and problem A is undecidable, then problem B is undecidable; for if problem B were decidable, the above procedure is a decision technique for problem A, contradicting its undecidability. These two facts are central to most decidability techniques. To show that a problem is decidable, reduce it to a problem which is known to be decidable; to show that a problem is undecidable, reduce a problem which is known to be undecidable to it.

We shall make good use of this approach to reduce the amount of work we must do. For example, since the equality problem for reachability sets is reducible to the subset problem, we want to develop either (1) a solution procedure for the subset problem or (2) a proof that the equality problem is undecidable. If we can show (1), we have a solution technique for both problems; if we show (2), we know both problems are undecidable.

In some cases, we may be able to do even better. Two problems are *equivalent* if they are mutually reducible. That is, problem A is equivalent to problem B if problem A is reducible to problem B, and problem B is reducible to problem A. In this case, either both problems are decidable or both are undecidable, and we can work on either one. (Notice that this is not true in general. For example, if we were to show that the subset problem for reachability sets is undecidable, this would tell us nothing about the decidability or undecidability of the equality problem.)

The second consideration for investigating analysis problems is that if a solution technique exists it must be reasonably efficient. This requires that the amount of time and memory space needed by an algorithm to solve an instance of the problem not be excessive. The study of the cost of executing an algorithm is a part of *complexity theory.* Complexity theory deals with the amount of time and space needed to solve a problem. Obviously the amount of time and space will not be constant but will vary with the size of the problem to be solved. For Petri nets, time and space requirements would probably be a function of the number of places and transitions. Other factors which might influence things would be the number of tokens in the initial marking or the number of inputs and outputs for each transition and place (the number of arcs in the graph).

The time and space needed will vary with the particular instance of the problem to be solved. Therefore, complexity results may be in the form of a best case (lower bound) or worst case (upper bound) for an algorithm. Since it is not known in advance whether an instance will be a best case or worst case, the worst case is generally assumed, and the *complexity* of an algorithm is the worst case time or space requirements, as a function of the size of the input.

Complexity analysis is mainly concerned with the underlying problem complexity, and not concerned with a specific detailed implementation of any particular algorithm. Thus, complexity theory ignores constant factors. Complexity for a problem of size n is determined to be of order n^2 or e^n or $n \log n$ allowing for smaller terms and constant factors. In particular two general classes of algorithms are important: those with polynomial complexity (n, n^2, $n \log n$, n^8, and so on) and those with nonpolynomial complexity (especially exponential, 2^n, and factorial, $n!$).

Complexity analysis is generally applied to specific algorithms but can also be applied to general problems. In this case, a lower bound on the complexity of all algorithms to solve a problem is determined. This provides an algorithm-independent complexity result. It also can be useful in showing that a particular algorithm is optimal (within a constant) and when further work may produce a significantly better algorithm to solve a problem. For example, it is well-known that sorting n numbers is of complexity $n \log n$. Thus algorithms with $n \log n$ complexity cannot be significantly improved on (in the asymptotic worst case).

Reducibility can be useful in determining complexity. If a problem A can be reduced to a problem B and B has a complexity $f_B(n)$, then the complexity of A is at most the complexity of B plus the cost of the transformation from A to B (keeping in mind that the size of the problem may also change in the transformation). The complexity of the transformations is generally constant or linear and so is often ignored. Thus, reducing problem A to problem B gives either an upper bound for the complexity of A (if the complexity of B is known) or a lower bound for the complexity of B (if the complexity of A is known). Again by using as an example the equality and subset problems, the amount of work needed to solve the equality problem is no greater than twice the amount of work for the subset problem. Since this is a constant factor, the complexity of the subset problem should be the same as the complexity of the equality problem.

These two properties of Petri net analysis properties — decidability and complexity — are of major concern for the use of Petri nets. In this chapter we present some results which have been obtained. One of the techniques used is to reduce one Petri net problem to another.

5.2 Reachability Problems

The reachability problem is one of the most important problems for Petri net analysis. It is also open to a large amount of variation in definition. The following four reachability problems for a Petri net $C = (P,T,I,O)$ with initial marking μ have been posed.

DEFINITION 5-3 *The Reachability Problem* Given μ', is $\mu' \in R(C,\mu)$?

DEFINITION 5-4 *The Submarking Reachability Problem* For a subset $P' \subseteq P$ and a marking μ', does there exist $\mu'' \in R(C,\mu)$ such that $\mu''(p_i) = \mu'(p_i)$ for all $p_i \in P'$?

DEFINITION 5-5 *The Zero-Reachability Problem* Is $\mu' \in R(C,\mu)$ with $\mu'(p_i) = 0$ for all $p_i \in P$? [Is $0 \in R(C,\mu)$?]

DEFINITION 5-6 *The Single-Place Zero-Reachability Problem* For a given place $p_i \in P$, does there exist $\mu' \in R(C,\mu)$ with $\mu'(p_i) = 0$?

The submarking reachability problem restricts the reachability problem to considering only a subset of places, not caring about the markings of other places. The zero-reachability problem asks if the specific marking with zero tokens in all places is reachable. The single-place zero-reachability problem asks if it is possible to empty all the tokens out of a particular place.

Although these four problems are all different, they are all equivalent. Certain relationships are immediately obvious. The zero-reachability problem is reducible to the reachability problem; we simply set $\mu' = 0$ for the reachability problem. Similarly the reachability problem is reducible to the submarking reachability problem, by setting the subset $P' = P$. The single-place zero-reachability problem is reducible to the submarking reachability problem by setting $P' = \{p_i\}$ and $\mu' = 0$. More difficult to show is that the submarking reachability problem is reducible to the zero-reachability problem and that the zero-reachability problem is reducible to the single-place zero-reachability problem. This entire set of relationships is shown in Figure 5.1.

First, we show that the submarking reachability problem is reducible to the zero-reachability problem. Assume we are given a Petri net $C_1 = (P_1, T_1, I_1, O_1)$ with initial marking μ_1, a subset of places $P' \subseteq P_1$, and a marking μ'. We want to know if there exists $\mu'' \in R(C_1, \mu_1)$ with $\mu'(p_i) = \mu''(p_i)$ for all $p_i \in P'$. Our approach is to create a new Petri net $C_2 = (P_2, T_2, I_2, O_2)$ with initial marking μ_2 such that there exists $\mu'' \in R(C_1, \mu_1)$ with $\mu'(p_i) = \mu''(p_i)$ for all $p_i \in P'$ if and only if $0 \in R(C_2, \mu_2)$.

The construction of C_2 from C_1 is quite straightforward. We start

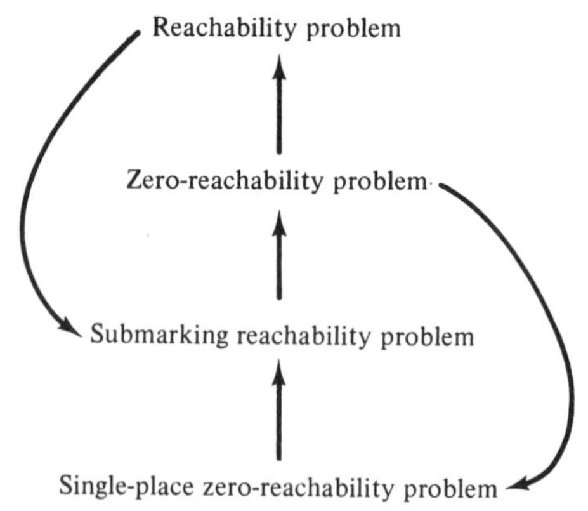

Figure 5.1 Reducibility among reachability problems. An arc from one problem to another indicates that the first is reducible to the second.

with C_2 the same as C_1. To allow any place p_i not in P' to become empty we add a transition t_i' with input $\{p_i\}$ and null output. This transition can fire whenever there is a token in p_i to drain off any tokens which may reside here. This allows us to ignore these places, being sure that they can always reach a zero marking.

For places p_i in P', we must assure that exactly $\mu'(p_i)$ tokens are in p_i. To assure this we create a new place p_i' for each $p_i \in P'$ with an initial marking of $\mu'(p_i)$ tokens and a transition t_i' with input $\{p_i, p_i'\}$ and null output. If there are exactly $\mu'(p_i)$ tokens in p_i, then this transition can fire exactly $\mu'(p_i)$ times, reducing the markings of p_i and p_i' to zero. If the number of tokens in p_i is not $\mu'(p_i)$, then the transition t_i' can only fire the minimum of the two markings, and so tokens will be left in either p_i or p_i', preventing the zero marking from being reached.

Figure 5.2 illustrates the two types of transitions introduced. Formally we define C_2 by

$$
\begin{aligned}
P_2 &= P_1 \cup \{p_i' \mid p_i \in P'\} \\
T_2 &= T_1 \cup \{t_i' \mid p_i \in P_1\} \\
I_2(t_j) &= I_1(t_j) \text{ for } t_j \in T_1 \\
I_2(t_i') &= \{p_i\} \text{ for } p_i \notin P' \\
&= \{p_i, p_i'\} \text{ for } p_i \in P' \\
O_2(t_j) &= O_1(t_j) \text{ for } t_j \in T_1 \\
O_2(t_i') &= \{\} \text{ for } p_i \in P_1
\end{aligned}
$$

with an initial marking

$$
\begin{aligned}
\mu_2(p_i) &= \mu_1(p_i), \ p_i \in P_1 \\
\mu_2(p_i') &= \mu'(p_i), \ p_i \in P'
\end{aligned}
$$

THEOREM 5-1 The submarking reachability problem is reducible to the zero-reachability problem.

Proof: We show that for the Petri net C_2 constructed above from C_1, $0 \in R(C_2, \mu_2)$ if and only if $\mu'' \in R(C_1, \mu_1)$ with $\mu''(p_i) = \mu'(p_i)$ for all $p_i \in P'$.

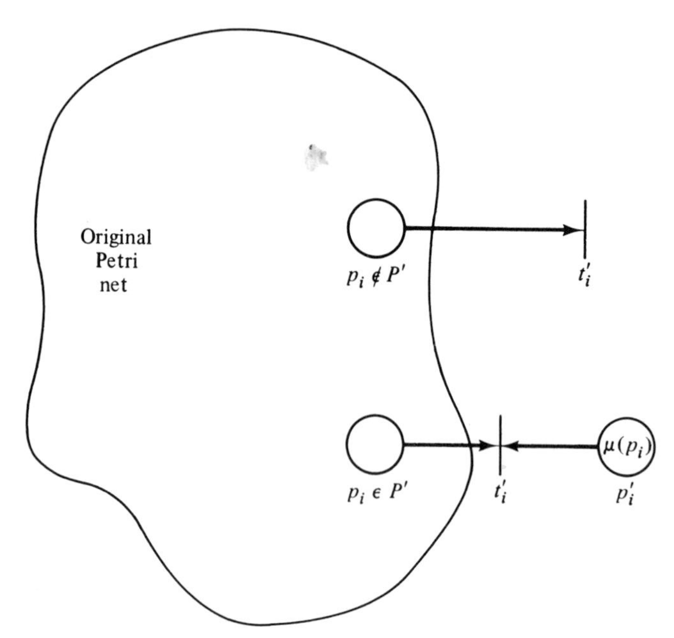

Figure 5.2 A Petri net showing that the submarking reachability problem can be reduced to the zero-reachability problem. The subset of places P' will have the marking μ in the original net if and only if the zero marking is reachable in the net as modified here.

To show that $0 \in R(C_2, \mu_2)$ if and only if there exists a $\mu'' \in R(C_1, \mu_1)$ with $\mu''(p_i) = \mu'(p_i)$ for $p_i \in P'$, assume first that μ'' exists in $R(C_1, \mu_1)$. Then in C_2 we can also reach the marking μ'' in the places $p_i \in P_1$ by firing only those transitions from T_1. Now for each $p_i \in P'$, we can fire t_i' exactly $\mu'(p_i)$ times, reducing both p_i and p_i' to zero. Then we can fire t_i' for each $p_i \notin P'$ as many times as necessary to put these to zero, so $0 \in R(C_2, \mu_2)$.

Now assume $0 \in R(C_2, \mu_2)$; then there exists a sequence of transition firings σ which leads from μ_2 to 0. This sequence will contain exactly $\mu'(p_i)$ firings of t_i' for $p_i \in P'$ (to remove the tokens from p_i') and some number of firings of t_i' for $p_i \notin P'$. Note that these transition firings only remove tokens from C_1, and since $\delta(\mu', t_j)$ is defined whenever $\delta(\mu, t_j)$ is defined for $\mu' \geqslant \mu$ (extra tokens never hurt), the sequence σ with all t_i' firings removed is also legal and will lead to a marking μ'' with exactly $\mu'(p_i)$ tokens in p_i for $p_i \in P'$. Thus if $0 \in R(C_2, \mu_2)$, then $\mu'' \in R(C_1, \mu_1)$ with $\mu''(p_i) = \mu'(p_i)$ for $p_i \in P'$.
□

Our next task is to show that the zero-reachability problem is reducible to the single-place zero-reachability problem. The proof of this statement again involves a construction. Given a Petri net $C_1 = (P_1, T_1, I_1, O_1)$ with initial marking μ_1, we wish to determine if $0 \in R(C_1, \mu_1)$. We construct, from C_1, a new Petri net C_2 with an additional place s ($P_2 = P_1 \cup \{s\}$) such that there exists a marking $\mu' \in R(C_2, \mu_2)$ with $\mu'(s) = 0$ if and only if $0 \in R(C_1, \mu_1)$.

The construction of C_2 defines s so that at all times the number of tokens in s is equal to the sum of the number of tokens in the places of C_1. Thus if $\mu'(s) = 0$, then there are zero tokens in all places of C_1 and vice versa. We define the initial marking μ_2 by

$$\mu_2(p_i) = \mu_1(p_i) \text{ for } p_i \in P_1$$

$$\mu_2(s) = \sum_{p_i \in P_1} \mu_1(p_i)$$

Now for each transition $t_j \in T_1$, the same transition is in C_2 but augmented by arcs to the place s. Define

$$d_j = \sum_{p_i \in P_1} \#(p_i, O(t_j)) - \#(p_i, I(t_j))$$

Then d_j is the change in the number of tokens which results from firing transition t_j. Now if $d_j > 0$, then d_j tokens must be added to place s, so we add d_j arcs from t_j to s; if $d_j < 0$, then we remove $-d_j$ tokens from s by $-d_j$ arcs from s to t_j.

If $d_j > 0$, then $\#(s, I(t_j)) = 0$; $\#(s, O(t_j)) = d_j$.
If $d_j < 0$, then $\#(s, I(t_j)) = -d_j$; $\#(s, O(t_j)) = 0$.
If $d_j = 0$, then $\#(s, I(t_j)) = 0$; $\#(s, O(t_j)) = 0$.

With this construction, any sequence of transition firings which leads C_1 to the marking 0 will lead C_2 to a marking μ' with $\mu'(s) = 0$ [$\mu'(p_i) = 0$ also] and vice versa.

THEOREM 5-2 The zero-reachability problem is reducible to the single-place zero-reachability problem.

Proof: The formal proof, based on the above construction, is left to the reader. □

With these two theorems, and the obvious observations, we can now conclude the following.

THEOREM 5-3 The following reachability problems are equivalent

1. The reachability problem
2. The zero-reachability problem
3. The submarking reachability problem
4. The single-place zero-reachability problem

These theorems and their proofs are mainly due to Hack [1975c].

5.3 Limited Petri Net Structures

The early work on Petri nets, and some current work, defined Petri nets in somewhat more restricted ways than the definition in Chapter 2. In particular, the following two restrictions are sometimes enforced.

RESTRICTION 5-1 The multiplicity of any place is limited to be less than or equal to 1. That is, $\#(p_i, I(t_j)) \leqslant 1$ and $\#(p_i, O(t_j)) \leqslant 1$ for all $p_i \in P$ and $t_j \in T$. This restricts the input and output bags to be sets.

RESTRICTION 5-2 No place may be both an input and an output of the same transition. $I(t_j) \cap O(t_j) = \varnothing$. This is often stated as $\#(p_i, I(t_j)) \cdot \#(p_i, O(t_j)) = 0$, for all p_i and t_j.

Petri nets which satisfy Restriction 1 are called *ordinary Petri nets.* Petri nets which satisfy Restriction 2 are called *self-loop-free Petri nets* or *nonreflexive Petri nets.* Petri nets satisfying both restrictions are called *restricted Petri nets.* These classes of Petri nets are related as shown in Figure 5.3.

These subclasses of the general Petri net model have been considered for several reasons. A major reason is that the propagation of Petri net concepts was informal in its earlier theory. The need for multiple arcs or self-loops did not occur in early modeling. Also, it was probably felt that the theory would be easier without these additional complications to the theory. As the theory has developed, however, it has become evident that the more general definitions have not been more difficult to work with. Current work using models with these restrictions is thus either the result of unnecessary timidity on the part of

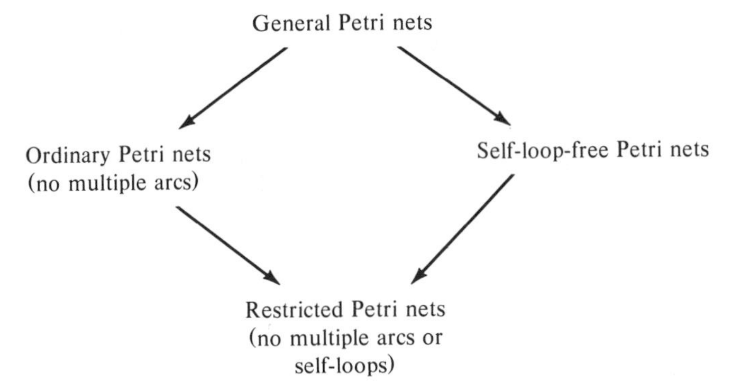

Figure 5.3 The relationships among the classes of Petri nets. An arc indicates containment; reducibility arcs would be directed in the opposite direction.

the researcher or the need for quicker exposition leading to simpler definitions.

However, these restrictions add nothing to our ability to analyze Petri nets. Consider the reachability problem for these classes of nets. To show the essential equivalence of these four classes of Petri nets, we prove the following.

THEOREM 5-4 The reachability problem for the following classes of Petri nets are equivalent.

1. General Petri nets
2. Ordinary Petri nets
3. Self-loop-free Petri nets
4. Restricted Petri nets

Proof: The following reducibilities are obvious from the definitions.

1. The reachability problem for ordinary Petri nets is reducible to the reachability problem for general Petri nets.
2. The reachability problem for self-loop-free Petri nets is reducible to the reachability problem for general Petri nets.
3. The reachability problem for restricted Petri nets is reducible to both the reachability problem for ordinary Petri nets and the reachability problem for self-loop-free Petri nets.

We show that general Petri nets can be transformed into restricted Petri nets in such a way as to reduce the reachability problem for general Petri nets to the reachability problem for restricted Petri nets. This then shows that these four reachability problems are equivalent.

To transform a general Petri net into a restricted Petri net, we use the following basic approach. Every place in the general Petri net is replaced by a *ring* of places in the restricted Petri net. Figure 5.4 shows the general form of a ring of places. Notice that a collection of tokens placed in the ring can freely move around the ring to any position at any time; they can all group into place $p_{i,1}$ or spread out uniformly to cover all k_i places in the ring. Thus a transition which needs three tokens from place p_i can pick up one from each of $p_{i,1}$, $p_{i,2}$, and $p_{i,3}$ rather than all three from p_i. Similarly a transition which uses p_i both as an input and as an output (a self-loop) may input from $p_{i,1}$ and output to $p_{i,2}$, eliminating the self-loop.

Formally, for a general Petri net $C_1 = (P_1, T_1, I_1, O_1)$ with marking μ_1, we define a restricted Petri net $C_2 = (P_2, T_2, I_2, O_2)$ with marking μ_2 as follows. First define, for each $p_i \in P_1$, an integer k_i by

$$k_i = \max_{t_j \in T_1} (\#(p_i, I(t_j)) + \#(p_i, O(t_j)))$$

The restricted Petri net C_2 is defined by

$$P_2 = \{p_{i,h} \mid p_i \in P_1, 1 \leqslant h \leqslant k_i\}$$

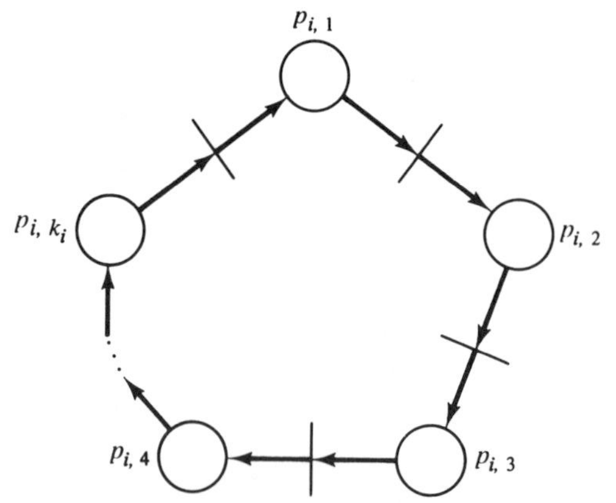

Figure 5.4 A ring of places to be used in a restricted Petri net to represent a place in a general Petri net. The number k_i of places representing a place p_i is determined by the sum of the maximum multiplicities of the place.

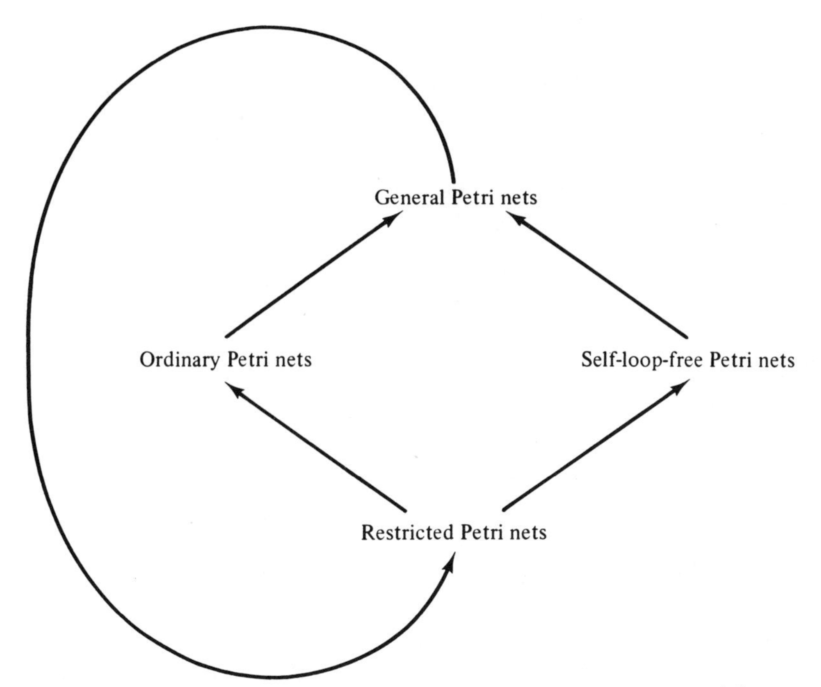

Figure 5.5 Reducibility of the reachability problem among classes of limited Petri nets.

$$T_2 = T_1 \cup \{t_{i,h} \mid p_{i,h} \in P_2\}$$

The input and output functions for "normal" transitions are defined such that

$$\begin{aligned} \#(p_{i,h}, I_2(t_j)) &= 1 \text{ if } 1 \leqslant h \leqslant \#(p_i, I_1(t_j)) \\ &= 0 \text{ otherwise} \end{aligned}$$

$$\begin{aligned} \#(p_{i,h}, O_2(t_j)) &= 1 \text{ if } \#(p_i, I_1(t_j)) < h \leqslant \#(p_i, I_1(t_j)) + \#(p_i, O_1(t_j)) \\ &= 0 \text{ otherwise} \end{aligned}$$

while for the "ring" transitions,

$$I_2(t_{i,h}) = \{p_{i,h}\}$$

$$O_2(t_{i,h}) = \{p_{i,n} \mid n = 1 + (h \bmod k_i)\}$$

The marking μ_2 is defined by

$$\mu_2(p_{i,1}) = \mu_1(p_i) \text{ for } p_i \in P_1$$
$$\mu_2(p_{i,h}) = 0 \text{ for } h > 1$$

By construction, for any marking μ which is reachable in C_1, there exists a marking μ' of C_2 such that

$$\sum_h \mu'(p_{i,h}) = \mu(p_i) \text{ for all } p_i \in P_1$$

In particular it is possible to move all tokens from $p_{i,h}$ to $p_{i,1}$ in C_2 at any time. Thus, we can define a marking μ' by

$$\mu'(p_{i,1}) = \mu(p_i) \text{ for } p_i \in P_1$$
$$\mu'(p_{i,h}) = 0 \text{ for } h > 1$$

and μ' is reachable in the restricted Petri net C_2 if and only if μ is reachable in C_1. □

Thus, from the point of view of analysis, general Petri nets and these three restricted classes of the general Petri net — ordinary Petri nets, self-loop-free Petri nets, and restricted Petri nets — are equivalent, each can be transformed into a similar net of another class, allowing a reachability problem in one to be reduced to a reachability problem in another. The constructions in this section are due to Hack [1974a].

5.4 Liveness and Reachability

Reachability is an important problem, but not the only remaining problem for Petri nets. Liveness is another problem which has received much attention in the Petri net literature. As pointed out in Section 4.1.4, liveness is related to deadlock. Two liveness problems for a Petri net $C = (P,T,I,O)$ with initial marking μ are of concern here. A Petri net is live if each transition is live. A transition t_j is live in a marking μ if for each $\mu' \in R(C,\mu)$ there exists a sequence σ such that t_j is enabled in $\delta(\mu',\sigma)$. A transition t_j is dead in a marking μ if there is no reachable marking in which it can fire.

DEFINITION 5-7 *Liveness Problem* For all transitions $t_j \in T$, is t_j live?

DEFINITION 5-8 *Single-Transition Liveness Problem* Given $t_j \in T$, is t_j live?

The liveness problem is obviously reducible to the single-transition liveness problem. To solve the liveness problem, we simply solve the single-transition liveness problem for each $t_j \in T$; if $|T| = m$, then we must solve m single-transition liveness problems.

The reachability problem can also be reduced to the liveness problem. Since the many variants of the reachability problem are equivalent, we use the single-place zero-reachability problem. If we have any of the other reachability problems, they can be reduced to the single-place zero-reachability problem as shown in Section 5.2. Now, if we wish to determine if place p_i can be zero in any reachable marking for a Petri net $C_1 = (P_1, T_1, I_1, O_1)$ with initial marking μ_1, we construct a Petri net $C_2 = (P_2, T_2, I_2, O_2)$ with initial marking μ_2, which is live if and only if the zero marking is *not* reachable from μ_1.

The Petri net C_2 is constructed from C_1 by the addition of two places, r_1 and r_2, and three transitions, s_1, s_2, and s_3. We first modify all transitions of T_1 to include r_1 as both an input and an output. The initial marking μ_2 will include a token in r_1. The place r_1 is a "run" place; as long as the token remains in r_1 the transitions of T_1 can fire normally. Thus any marking which is reachable in the places of P_1 in C_1 is also reachable in C_2. Transition s_1 is defined to have r_1 as its input and a null output. This allows the token in r_1 to be removed, disabling all transitions in T_1 and "freezing" the marking of P_1. (Note that all transitions of T_1 are in conflict and, by construction if not by definition, no more than one transition can fire at a time.)

The place r_1 and transition s_1 allow the net C_1 to reach any reachable marking and then for s_1 to fire and freeze the net at that marking. Now we need to see if place p_i is zero. We introduce a new place r_2 and a transition s_2 which has p_i as its input and r_2 as its output. If p_i can ever become zero, this transition is not live; in fact the entire net is dead if transition s_1 fires in that marking. Hence if p_i can be zero, the net is not live. If p_i cannot be zero, then s_2 can always fire, putting a token in r_2. In this case we must put a token back in r_1 and assure that all transitions in C_2 are live. We must be sure that C_2 is live even if C_1 is not live. This is accomplished by a transition s_3 which "floods" the net C_2 with tokens, assuring that every transition is live if a token is ever put in r_2. Transition s_3 has r_2 as its input and every place of C_2 (all p_i in C_1 and r_1 and r_2) as output. This construction is illustrated in Figure 5.6.

Now, if a marking μ is reachable in $R(C_1, \mu_1)$ with $\mu(p_i) = 0$, then the net C_2 can also reach this marking on the place of P_1 by exe-

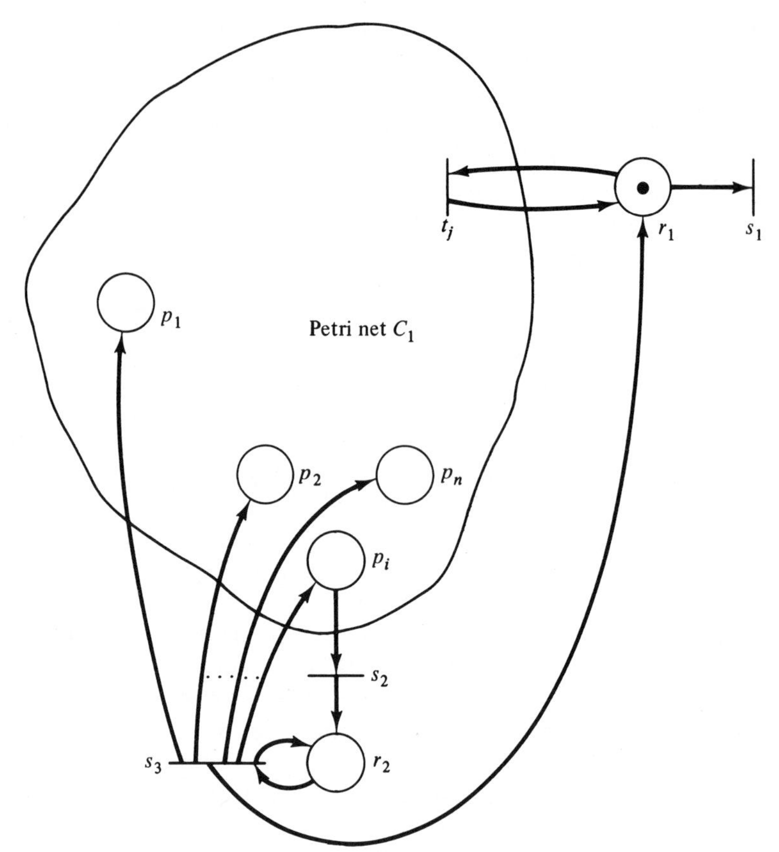

Figure 5.6 A construction converting the single-place zero-reachability problem [is a marking reachable with $\mu(p_i)=0$?] to the liveness problem [is this net live?].

cuting the same sequence of transition firings. Then s_1 can fire, freezing the C_1 subset. Since $\mu(p_i) = 0$, transition s_2 cannot fire and C_2 is dead. Thus if p_i can become zero, then C_2 is not live.

Conversely, if C_2 is not live, then a marking μ must be reachable in which $\mu(r_2) = 0$ and there is no reachable state in which r_2 has a token. [If r_2 has a token, s_3 is enabled, and s_3 can be fired repeatedly enough times to enable any (or all) transitions, and so the net is live.] If r_2 has no token and cannot get any, then the marking of p_i must also be zero. Thus if C_2 is not live, then a marking is reachable in which the marking of p_i is zero.

On the basis of this construction, we have the following.

THEOREM 5-5 The reachability problem is reducible to the liveness problem.

Now we need to show the following.

THEOREM 5-6 The single-transition liveness problem is reducible to the reachability problem.

The proof that the single-transition liveness problem is reducible to the reachability problem rests on testing for the reachability of any of a finite set of *maximal* t_j*-dead submarkings.* A Petri net is not live for a transition t_j if and only if some marking is reachable in which the transition t_j is not fireable and cannot become fireable. A marking of this sort is called t_j*-dead.* For any marking μ we can test if it is t_j-dead by constructing the reachability tree with μ as the root and testing if transition t_j can fire anywhere in the tree. If it cannot then μ is t_j-dead. Checking for liveness of t_j then requires checking if any t_j-dead marking is reachable.

In general, however, there may be an infinite number of t_j-dead markings and an infinite set of markings in which to find the t_j-dead markings. The set of markings which must be checked for reachability is reduced to a finite number by noting two properties. First, if a marking μ is t_j-dead, then any marking $\mu' \leqslant \mu$ is also t_j-dead. (Any firing sequence possible from μ' is also possible from μ, so if μ' could lead to the firing of t_j, so could μ.) Second, the markings of some places will not affect the t_j-deadness of a marking, and so the markings of these places are "don't-cares"; they can be arbitrary. Borrowing from the reachability tree construction, we replace these "don't-care" components by ω to indicate that an arbitrarily large number of tokens can be in this place without affecting the t_j-deadness of the marking. Now since any $\mu' \leqslant \mu$ is t_j-dead if μ is t_j-dead, we need not consider those places p_i with $\mu(p_i) = \omega$. This means we use the submarking reachability problem with $P' = \{p_i \mid \mu(p_i) \neq \omega\}$.

As an example, consider the Petri net of Figure 5.7. The markings (2, 0), (1, 0), (0, 0), (0, 1), (0, 2), (0, 3), ... are t_2-dead, but they can be finitely represented by the set $(2,0), (1,0), (0,\omega)\}$.

Hack [1974c; 1975c] has shown that there exists for a Petri net C a finite set D_t of markings (extended to include ω) such that C is live if and only if no marking in D_t is reachable. If a marking of D_t contains ω, submarking reachability is implied.

Further, D_t can be effectively computed. Since D_t is finite, the non-ω-components of the markings have an upper bound b. This bound b is characterized as the smallest number such that for any marking μ

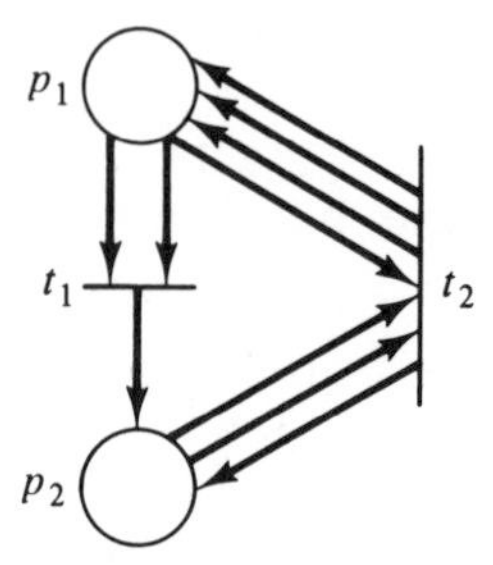

Figure 5.7 A Petri net to illustrate t_j-dead markings.

with $\mu(p_i) \leqslant b + 1$ for all p_i, if μ is t_j-dead, then the submarking μ', with $\mu'(p_i) = \mu(p_i)$ if $\mu(p_i) \leqslant b$ and $\mu'(p_i) = \omega$ if $\mu(p_i) = b + 1$, is t_j-dead. With this characterization of b, we can construct D_t as follows.

1. Compute b. Start with $b = 0$, and increase b until the first b is found which satisfies the characterization of the bound defined above. Testing for b requires checking all $(b + 2)^n$ markings with components less than or equal to $b + 1$.
2. Compute D_t by testing all markings and submarkings with components less than or equal to b or equal to ω. D_t is the set of t_j-dead markings from this set of $(b + 2)^n$ markings.

Once we have constructed D_t, we then apply the submarking reachability problem for each element of D_t. If any element of D_t is reachable from the initial marking, the Petri net is not live; if no element of D_t is reachable, the Petri net is live.

From these two theorems, we have the following.

THEOREM 5-7 The following problems are equivalent:

1. The reachability problem
2. The liveness problem
3. The single-transition liveness problem

More formal proofs of the reducibility of liveness to reachability can be found in [Hack 1974c; Hack 1975c].

5.5 Undecidable Results

In Section 5.4 we have shown that a number of problems in reachability and liveness are equivalent, but no result has been obtained yet on the decidability of these problems. To show decidability, it is necessary to reduce a Petri net problem to a problem with a known solution, or to show undecidability, to reduce a problem which is known to be undecidable to a Petri net problem. The first important result of this kind was by Rabin [Baker 1973b]. Rabin showed that for two Petri nets C_1 with marking μ_1 and C_2 with marking μ_2 it is undecidable if $R(C_1,\mu_1) \subseteq R(C_2,\mu_2)$. Hack [1975a] later strengthened this to show that it is undecidable if $R(C_1,\mu_1) = R(C_2,\mu_2)$. The proof of these statements is based on *Hilbert's tenth problem.* (In 1900, D. Hilbert presented 23 problems to a conference of mathematicians; this was the tenth in his list.)

DEFINITION 5-9 Given a polynomial P over n variables with integer coefficients, does there exist a vector of integers, $(x_1,x_2, \ldots,x_n)$ such that

$$P(x_1,x_2, \ldots,x_n) = 0 ?$$

The equation $P(x_1,x_2, \ldots,x_n) = 0$ is a *Diophantine* equation. In general it will be a sum of terms

$$P(x_1,x_2, \ldots,x_n) = \sum_i R_i(x_1,x_2, \ldots,x_n)$$

$$R_i(x_1,x_2, \ldots,x_n) = a_i \cdot x_{s_1} \cdot x_{s_2} \cdot \cdots \cdot x_{s_h}$$

Diophantine equations include $x_1 = 0$, $3x_1 \cdot x_2 + 6x_3 = 0$, and so on.

In 1970, Matijasevic proved that Hilbert's tenth problem was undecidable [Davis 1973; Davis and Hersh 1973]: There is no general algorithm to determine if an arbitrary Diophantine equation has a root (a set of values for which the polynomial is zero). This forms the basis of the proof that the equality problem for Petri net reachability sets is undecidable. The strategy is to construct for a Diophantine polynomial a Petri net which (in some sense) computes all values of the polynomial.

5.5.1 The Polynomial Graph Inclusion Problem

The proof of the undecidability of the equality problem is in three parts (Figure 5.8). First, Hilbert's tenth problem is reduced to the *poly-*

nomial graph inclusion problem. Then the polynomial graph inclusion problem is reduced to the *subset problem for Petri net reachability sets.* Finally, the subset problem for Petri net reachability sets is reduced to the *equality problem for Petri net reachability sets.* This shows that Hilbert's tenth problem, known to be undecidable, is reducible to the equality problem, which must therefore also be undecidable.

DEFINITION 5-10 The *graph* $G(P)$ of a Diophantine polynomial $P(x_1, \ldots, x_n)$ with nonnegative coefficients is the set

$$G(P) = \{(x_1, \ldots, x_n, y) \mid y \leqslant P(x_1, \ldots, x_n) \text{ with } 0 \leqslant x_1, \ldots, x_n, y\}$$

DEFINITION 5-11 The *polynomial graph inclusion problem* is to determine for two Diophantine polynomials A and B if $G(A) \subseteq G(B)$.

We first show that Hilbert's tenth problem is reducible to the polynomial graph inclusion problem. This proves the following.

THEOREM 5-8 The polynomial graph inclusion problem is undecidable.

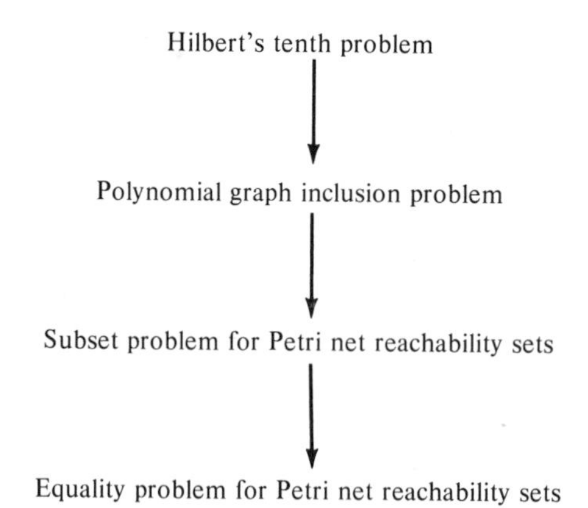

Figure 5.8 The reducibilities showing that the equality (and subset) problem for Petri net reachability sets is undecidable.

***Proof*:**

1. We limit our proof to problems with nonnegative solutions. If $(x_1, \ldots, x_n)$ is a solution to $P(x_1, \ldots, x_n) = 0$, with $x_i < 0$, then $(x_1, \ldots, -x_i, \ldots, x_n)$ is a solution to $P(x_1, \ldots, -x_i, \ldots, x_n) = 0$. Thus, for an arbitrary polynomial, we need only test each of the 2^n polynomials which result from changing the sign of some subset of variables for nonnegative solutions to determine the solution for the arbitrary polynomial.
2. Similarly, since $P^2(x_1, \ldots, x_n) = 0$ if and only if $P(x_1, x_2, \ldots, x_n) = 0$, we need only consider polynomials whose value is nonnegative.
3. Now we can separate any polynomial $P(x_1, x_2, \ldots, x_n)$ into two polynomials $Q_1(x_1, \ldots, x_n)$ and $Q_2(x_1, \ldots, x_n)$ such that $P(x_1, \ldots, x_n) = Q_1(x_1, x_2, \ldots, x_n) - Q_2(x_1, \ldots, x_n)$ by putting all terms with positive coefficients in Q_1 and all terms with negative coefficients in Q_2. Now since $P(x_1, \ldots, x_n) \geqslant 0$ (by 2 above), we have $Q_1(x_1, \ldots, x_n) \geqslant Q_2(x_1, \ldots, x_n)$ and $P(x_1, \ldots, x_n) = 0$ if and only if $Q_1(x_1, x_2, \ldots, x_n) = Q_2(x_1, \ldots, x_n)$.
4. Consider the two polynomial graphs

$$G(Q_1) = \{(x_1, \ldots, x_n, y) \mid y \leqslant Q_1(x_1, \ldots, x_n)\}$$

$$G(Q_2 + 1) = \{(x_1, \ldots, x_n, y) \mid y \leqslant 1 + Q_2(x_1, \ldots, x_n)\}$$

Now, $G(Q_2 + 1) \subseteq G(Q_1)$ if and only if for all nonnegative $x_1, \ldots, x_n$ and y, $y \leqslant 1 + Q_2(x_1, \ldots, x_n)$ implies that $y \leqslant Q_1(x_1, \ldots, x_n)$. This is true if and only if there does not exist $x_1, \ldots, x_n$ and y such that

$$Q_1(x_1, \ldots, x_n) < y \leqslant 1 + Q_2(x_1, \ldots, x_n)$$

But from 3 above, $Q_1 \geqslant Q_2$ so that

$$Q_1(x_1, \ldots, x_n) < y \leqslant 1 + Q_2(x_1, \ldots, x_n) \leqslant 1 + Q_1(x_1, \ldots, x_n)$$

and, since all quantities are integers,

$$y = 1 + Q_2(x_1, \ldots, x_n) = 1 + Q_1(x_1, \ldots, x_n)$$

which is true if and only if $Q_1 = Q_2$. Thus, we see that $G(Q_2 + 1) \subseteq G(Q_1)$ if and only if there does not exist $x_1, \ldots, x_n$

such that $Q_1(x_1, \ldots, x_n) = Q_2(x_1, \ldots, x_n)$, which is to say there does not exist $x_1, \ldots, x_n$ such that $P(x_1, \ldots, x_n) = 0$.

5. Therefore to determine that the equation $P(x_1, x_2, \ldots, x_n) = 0$ has a solution, we need only to show that it is not the case that $G(Q_2 + 1) \subseteq G(Q_1)$. □

5.5.2 Weak Computation

Now we need to show that Petri nets can (in some sense) compute the value of a polynomial $Q(x_1, x_2, \ldots, x_n)$. We have carefully limited the polynomial Q to having a nonnegative value, nonnegative coefficients, and nonnegative variables. This allows us to encode the values of the variables and the value of the polynomial as the number of tokens in places in a Petri net. Figure 5.9 shows the general scheme. The input values $x_1, \ldots, x_n$ are encoded by x_i tokens in p_i for $i = 1, \ldots, n$. Initially a token also resides in the "run" place. The execution of the net will terminate by placing a token in the "quit" place. At this time the "output" place will have y tokens, where $y \leqslant Q(x_1, \ldots, x_n)$.

This Petri net will *weakly compute* the value $Q(x_1, \ldots, x_n)$. Weak computation means that the value computed will not exceed

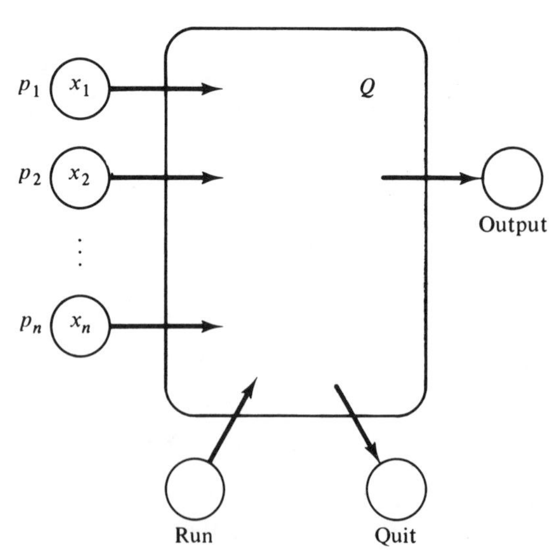

Figure 5.9 Basic structure of a Petri net to weakly compute the value of a polynomial, $Q(x_1, x_2, \ldots, x_n)$.

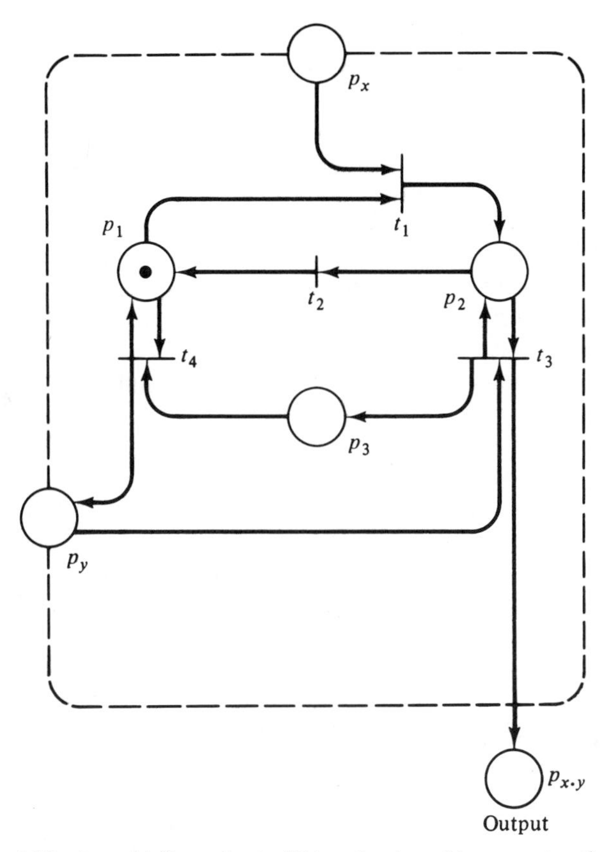

Figure 5.10 A multiplier subnet. This subnet weakly computes the product of x and y.

$Q(x_1, \ldots, x_n)$ but may be any (nonnegative) value less than $Q(x_1, \ldots, x_n)$. Weak computation is necessary for Petri nets because of the *permissive* nature of transition firings; a Petri net cannot be forced to finish. The definition of a polynomial graph $G(Q)$ was made specifically with this in mind.

What we show now is that subnets can be constructed which weakly compute the function of (binary) multiplication. From this, we can construct a composite net which weakly computes the value of each term of a polynomial by successive multiplication subnets. The output of the subnet for each term will be deposited in the output place for the polynomial. Thus the number of tokens in the output place will be the sum of the outputs for each term.

The multiplication subnet is shown in Figure 5.10. This net will weakly compute the product of the numbers, x and y, of tokens in its two inputs and place this many tokens in its output. The operation of the net is quite simple. To compute the product of x and y, transition t_1 first fires, moving one token from p_x to p_2. This token enables transition t_3, which can now copy y tokens from place p_y, putting them in p_3 and putting y tokens in $p_{x \cdot y}$, the output place. Now t_2 can fire, putting the token in p_2 back into p_1. This enables t_4, which can copy the y tokens from p_3 back into p_y. This entire process can be repeated exactly x times, each time putting y tokens in $p_{x \cdot y}$. Then the marking of place p_x has been reduced to zero, and the net must stop. The total number of tokens in place $p_{x \cdot y}$ is then the product of x and y.

The above case is the best case, in the sense that the number of output tokens is exactly $x \cdot y$. However, the token in p_2 enables both transitions t_3 and t_2, and it is possible for t_2 to fire before all y tokens have been copied from p_y to p_3 and been added to $p_{x \cdot y}$. In this case, the number of tokens deposited in $p_{x \cdot y}$ will be less than $x \cdot y$. Since t_3 can fire no more than y times for each firing of t_1 and t_1 can fire no more than x times, we can guarantee that the number of tokens in $p_{x \cdot y}$ never exceeds $x \cdot y$, but because of the permissive nature of transition firings, we cannot guarantee that the number of tokens in $p_{x \cdot y}$ will actually equal $x \cdot y$; it could be less. Thus, this Petri net weakly computes the product of x and y. Now to weakly compute a term R_i which is the product $a_i x_{s_1} x_{s_2} \cdots x_{s_h}$ we construct a Petri net of the form shown in Figure 5.11. Since each subnet weakly computes the product of two terms, the entire subnet weakly computes the value of the term R_i.

Figure 5.12 then shows how a polynomial $P = R_1 + R_2 + \cdots + R_k$ can be weakly computed. Each subnet is of the form of Figure 5.11 and weakly computes the value of one term. The outputs of the k subnets for each term have been merged together, giving a total value which is the sum of each term.

Now some control transitions and places are added to create the specific reachability sets needed. First we need to be able to produce an arbitrary value for each of the variables $(x_1, \ldots, x_n)$ and record that value in the places $p_1, \ldots, p_n$. A transition t_i is created for each p_i with null input and outputs to p_i and every place which is an input corresponding to x_i in a term R_j which uses x_i. Thus, in the polynomial $x_1 + x_1 x_2$ we would have a transition t_1 with outputs to p_1 and to the x_1 inputs of the two terms, x_1 and $x_1 x_2$, which use x_1; t_2 would output to p_2 and to the x_2 input of the term $x_1 x_2$.

These transitions can fire an arbitrary number of times, creating any value in $p_1, \ldots, p_n$. Thus, for every $y \leqslant P(x_1, \ldots, x_n)$ a marking μ is reachable with $\mu(p_1) = x_1, \ldots, \mu(p_n) = x_n$ and $\mu(output) = y$. The value $y = P(x_1, \ldots, x_n)$ can be achieved by first firing t_1 x_1 times, put-

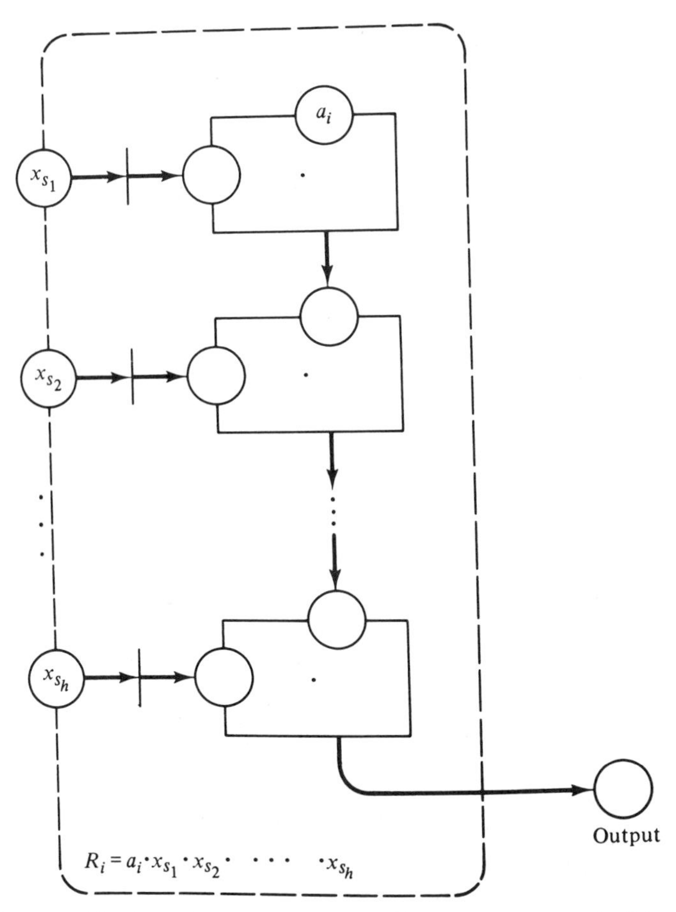

Figure 5.11 A Petri net structure to weakly compute a term of a Diophantine polynomial. Each box is a net of the form of Figure 5.10.

ting x_1 tokens in p_1, then firing t_2 x_2 times, and so on until t_n has fired x_n times. The subnet for each term R_i of the polynomial can then execute, with the resulting polynomial value put in the output place.

To reduce the polynomial graph inclusion problem to the subset problem for Petri net reachability sets, we perform the following steps. For polynomials A and B, we wish to determine if $G(A) \subseteq G(B)$.

1. We construct the Petri net C_A which weakly computes $A(x_1, \ldots, x_n)$ and the Petri net C_B which weakly computes $B(x_1, \ldots, x_n)$.
2. If the number of places in the two nets is not equal, we add places to the smaller to equalize the number of places. These places are ini-

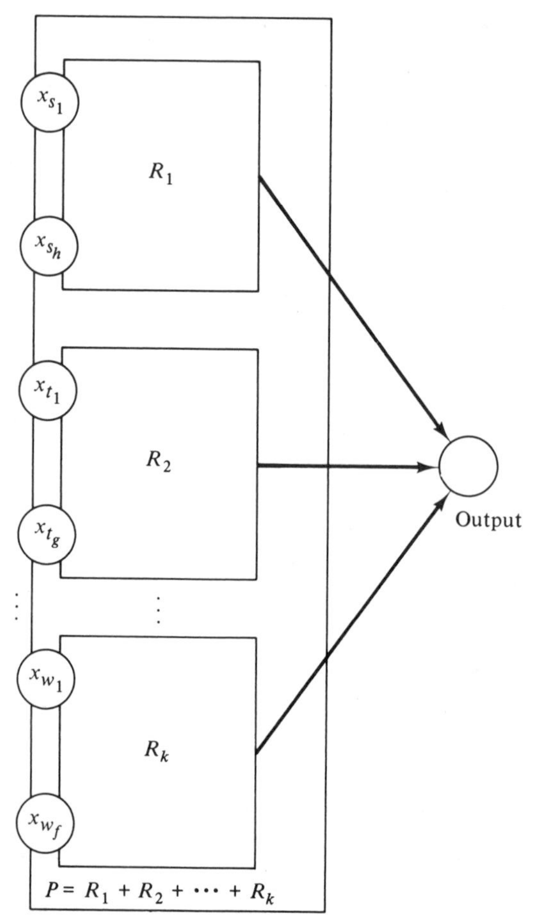

Figure 5.12 A Petri net to weakly compute $P(x_1, x_2, \ldots, x_n)$ by using a collection of subnets of the form in Figure 5.11.

tially unmarked and are not used by any of the transitions in the net.

3. Now we must eliminate the effects of all internal places on the reachability sets. A set of $n+1$ places are distinguished in both C_A and C_B, the places corresponding to the values of $x_1, \ldots, x_n$ and the output of each net. All other places are internal places, whose markings are unimportant. However, we may find that for an internal place p_i in C_A and corresponding p_i' in C_B that there exists a marking μ in $R(C_A, \mu_A)$ with no equal marking μ' in $R(C_B, \mu_B)$, because $\mu(p_i) \neq \mu''(p_i)$ for all μ'' in $R(C_B, \mu_B)$.

To prevent this problem we add two new places q and r to C_A (giving C_A') and q' and r' to C_B (giving C_B'). In C_A', q and r are not used for any transitions, and initially r is empty and q is marked with one token. In C_B', r' is a "run" place. It is initially marked, and every transition in C_B' is modified to include r' as both an input and an output. Thus, as long as the token remains in r', the net C_B' can function as before. A new transition transfers the enabling token from r' to q', disabling all transitions in C_B' and "freezing" the marking. Now we add two new transitions for each internal place in C_B'.

For each internal place p_i whose marking is unimportant, one transition has places q' and p_i as inputs and only q' as an output (allowing the marking in p_i to be decreased by one), and another transition has q' as input and both q' and p_i as outputs (allowing the marking in p_i to be increased by one). These transitions allow the marking of each internal place to be made arbitrary by an appropriate sequence of increasing or decreasing firings.

4. The new construction is illustrated in Figure 5.13. For these two Petri nets, C_A' and C_B' with initial marking μ_A' and μ_B', respectively, $G(A) \subseteq G(B)$ if and only if $R(C_A', \mu_A') \subseteq R(C_B', \mu_B')$.

The reachability sets of C_A' and C_B' are as follows. For C_A',

$\mathbf{p}_1 \ldots \mathbf{p}_n$	***Output***	r	q	***Internal Places***
$x_1 \ldots x_n$	$y \leqslant A(x_1, \ldots, x_n)$	0	1	Some arbitrary marking

For C_B',

$\mathbf{p}_1 \ldots \mathbf{p}_n$	***Output***	r	q	***Internal Places***
$x_1 \ldots x_n$	$y \leqslant B(x_1, \ldots, x_n)$	1	0	Some arbitrary marking
$x_1 \ldots x_n$	$y \leqslant B(x_1, \ldots, x_n)$	0	1	All arbitrary markings

Thus, if $G(A) \subseteq G(B)$, then $R(C_A', \mu_A') \subseteq R(C_B', \mu_B')$, and conversely, if $R(C_A', \mu_A') \subseteq R(C_B', \mu_B')$, then $G(A) \subseteq G(B)$.

This concludes our demonstration of the following.

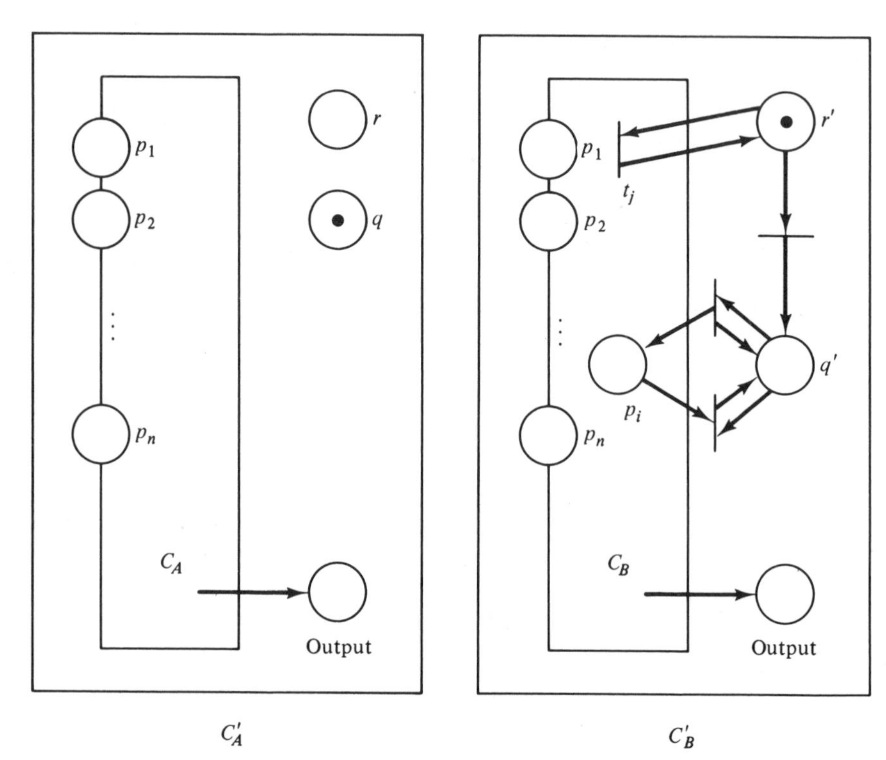

Figure 5.13 The constructed Petri nets to test for polynomial graph inclusion.

THEOREM 5-9 The polynomial graph inclusion problem is reducible to the subset problem for Petri net reachability sets.

This proof is from [Hack 1975a; Hack 1975c].

5.5.3 The Equality Problem

We now have only to show that the subset problem for Petri net reachability sets is reducible to the equality problem.

Assume that we have two Petri nets A and B and wish to determine if $R(A,\mu_A) \subseteq R(B,\mu_B)$ (the subset problem). We now show that two Petri nets D and E can be defined such that $R(A,\mu_A) \subseteq R(B,\mu_B)$ if and only if $R(D,\mu_D) = R(E,\mu_E)$. The basis for this construction is the fact that

$$R(A,\mu_A) \subseteq R(B,\mu_B) \text{ if and only if } R(B,\mu_B) = R(A,\mu_A) \cup R(B,\mu_B)$$

Both D and E are constructed from a common subnet, C. The net

C encodes the reachability sets of both A and B in such a way as to produce their union. Figure 5.14 illustrates the basic construction. The n places $p_1, \ldots, p_n$ act as either the n places of net A or the n places of net B. Originally they are unmarked. Two new places r_A and r_B are added as "run" places for net A and net B, respectively. All transitions of net A are modified to include r_A as both an input and an output; all transitions of net B are modified to include r_B as both an input and an output.

Now, one more place, s, is added and two new transitions, t_A and t_B. The initial marking for this entire net (including A and B as subnets with shared places; places r_A, r_B, and s; and transitions t_A and t_B) is one token in s and zero tokens elsewhere. Transition t_A has place s as

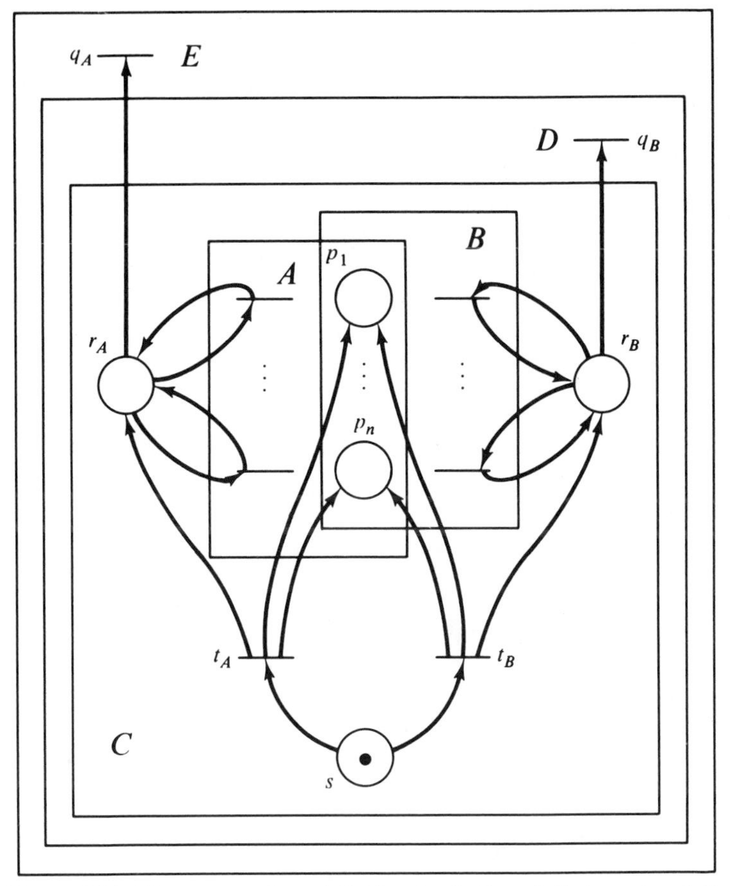

Figure 5.14 The construction of Petri nets C, D, and E from A and B. This construction is used to show that the subset problem is reducible to the equality problem for reachability sets.

its input and as output produces the initial marking for net A plus a token in r_A; transition t_B has place s as its input and produces the initial marking for net B plus a token in r_B. Thus, if t_A fires, then the subnet A has its initial marking, and all of its transitions can fire as normal since there is a token in r_A. However, subnet B is completely disabled, since there is no token in r_B. If t_B fires first, then the subnet B can operate, and A is disabled. The set of firing sequences for C is then any sequence of the form

$$t_A, < \textit{any sequence of firings from } A >$$

or any sequence of the form

$$t_B, < \textit{any sequence of firings from } B >$$

The net D is obtained from C by adding one new transition, q_B. Transition q_B has place r_B as its input and no output. Notice that q_B can fire *only* if transition t_B was the first to fire; if transition t_A fires first, then r_B will be empty, and t_B cannot fire.

The net E is constructed from D by adding a new transition, q_A. Transition q_A has place r_A as its input and no output. Transition q_A can fire only if t_A was the first to fire: Notice that net E is constructed from D, *not* (directly) from C. So E has both transition q_A and transition q_B.

Now let us examine the reachability sets of the Petri nets C, D, and E. The reachability set of C is all markings of the form

s	r_A	r_B	$p_1, \ldots, p_n$
1	0	0	$0, \ldots, 0$
0	1	0	Any $\mu \in R(A, \mu_A)$ (if t_A fires)
0	0	1	Any $\mu \in R(B, \mu_B)$ (if t_B fires)

Petri net D adds one other class of markings to this set:

s	r_A	r_B	$p_1, \ldots, p_n$
1	0	0	$0, \ldots, 0$
0	1	0	Any $\mu \in R(A, \mu_A)$ (if t_A fires)
0	0	1	Any $\mu \in R(B, \mu_B)$ (if t_B fires)
0	0	0	Any $\mu \in R(B, \mu_B)$ (if q_B fires)

And Petri net E adds one more class to this:

s	r_A	r_B	$p_1, \ldots, p_n$
1	0	0	$0, \ldots, 0$
0	1	0	Any $\mu \in R(A, \mu_A)$ (if t_A fires)
0	0	1	Any $\mu \in R(B, \mu_B)$ (if t_B fires)
0	0	0	Any $\mu \in R(B, \mu_B)$ (if q_B fires)
0	0	0	Any $\mu \in R(A, \mu_A)$ (if q_A fires)

Now, if $R(A, \mu_A) \subseteq R(B, \mu_B)$, the last class in $R(E, \mu_E)$ [markings of the form $(0,0,0,\mu)$ with $\mu \in R(A, \mu_A)$] is included in the last class of $R(D, \mu_D)$ [markings of the form $(0,0,0,\mu)$ with $\mu \in R(B, \mu_B)$]. Since all other markings are the same,

$$R(D, \mu_D) = R(E, \mu_E) \text{ if } R(A, \mu_A) \subseteq R(B, \mu_B)$$

Similarly, if $R(D, \mu_D) = R(E, \mu_E)$, then we must have $R(A, \mu_A) \subseteq R(B, \mu_B)$, since for each $(0,0,0,\mu)$ with $\mu \in R(A, \mu_A)$ in $R(E, \mu_E)$ there must exist an equal marking in $R(D, \mu_D)$. But all markings with $\mu(s, r_A, r_B) = (0,0,0)$ are of the form $(0,0,0,\mu)$ with $\mu \in R(B, \mu_B)$, so $R(A, \mu_A) \subseteq R(B, \mu_B)$.

Thus, this construction shows the following.

THEOREM 5-10 The subset problem for Petri net reachability sets is reducible to the equality problem for Petri net reachability sets.

These three theorems then lead to the following.

THEOREM 5-11 The following problems are undecidable.

1. The polynomial graph inclusion problem
2. The subset problem for Petri net reachability sets
3. The equality problem for Petri net reachability sets

These theorems and their proofs are due to Hack [1975a; 1975c].

5.6 Complexity of the Reachability Problem

The undecidability of the subset and equality problems for Petri net reachability sets creates the possibility that the reachability problem itself is also undecidable. However, at the moment, the decidability (or undecidability) of the reachability problem is *open.* There is currently neither an algorithm to solve the reachability problem nor a proof that such an algorithm cannot exist.

In 1977, a "proof" of the decidability of the reachability problem was presented at the ACM Symposium on Theory of Computing [Sacerdote and Tenney 1977]. However, this "proof" has several serious flaws, and attempts to correct them, to produce a correct proof, have been unsuccessful. Still the prevailing feeling is that the reachability problem is decidable — it is believed that an algorithm does exists and will be discovered in time.

Assuming that an algorithm to solve the reachability problem does exist, it is likely to be very complex. The obvious question is: If an algorithm to solve the reachability problem exists, how complex must it be? Some bounds on this complexity can be established without reference to any specific algorithm.

Lipton [1976] has shown that any algorithm to solve the reachability problem will require at least an exponential (2^{cn}) amount of space for storage and an exponential amount of time. The exponent (n) is a measure of the size of the problem and in Lipton's case reflects the number of places and their interconnections to transitions.

Lipton proved that exponential space is necessary by showing that a Petri net can be constructed in which a place acts as a counter of the numbers $0, 1, \ldots, 2^{2^n}$. Representing this in the reachability problem algorithm would require at least $\log_2(2^{2^n}) = 2^n$ bits. Just as important is that his construction uses at most $h \cdot n$ places (for some constant h).

Lipton's proof hinges on the ability to create a net to count to 2^{2^n} in only $h \cdot n$ places. Part of the constraints is a need to test this place for zero. Petri nets, of course, have been designed so that there is no direct way to test for zero. However, a common technique used with Petri nets to allow zero testing is to use two places p and p' such that $\mu(p) + \mu(p')$ is a constant. If we know that $\mu(p) + \mu(p') = k$, then we can test for $\mu(p)$ being zero by testing if $\mu(p')$ has k tokens; if $\mu(p')$ has k tokens, then $\mu(p)$ has zero tokens and vice versa. A place can be tested for nonzero by using it in a self-loop. Note that to maintain this ability we must maintain the constant nature of $\mu(p) + \mu(p')$; that is, the net must be conservative, at least with respect to these two places.

For small numbers k one can test if the marking of a place is k by having the place be an input to a transition k times (Figure 5.15). However, these arcs contribute to the size of the problem, and so we cannot do this in general. Lipton showed that if the constant sum of two places (p_k, p'_k) is k and k is a product of two smaller integer factors $k = k_1 \cdot k_2$ which are the constant sums of two other pairs of places $(p_{k_1}, p'_{k_1}$ and $p_{k_2}, p'_{k_2})$ and we can test $\mu(p_{k_1}) = 0$ and $\mu(p_{k_2}) = 0$, then we can test if $\mu(p_k) = 0$. This allowed Lipton to build subnets such as Figure 5.16. These nets are then used to control multiplication nets, similar to the nets used to weakly compute the polynomial graph (see Figure 5.10). The test-for-zero subnet allows the Petri net to compute the exact product (not a weak product which is merely bounded by the real product).

These simple nets allow Lipton to build a net, for a given n, which can generate exactly 2^{2^n} tokens in a place (p) with zero tokens in p' and the ability to test $\mu(p)$ for zero. The number of places used is only a constant factor times n. The existence of a Petri net like this shows that the reachability problem requires at least exponential time and space and hence will be very expensive to solve.

The construction of a Petri net which can count up to 2^{2^n} has a

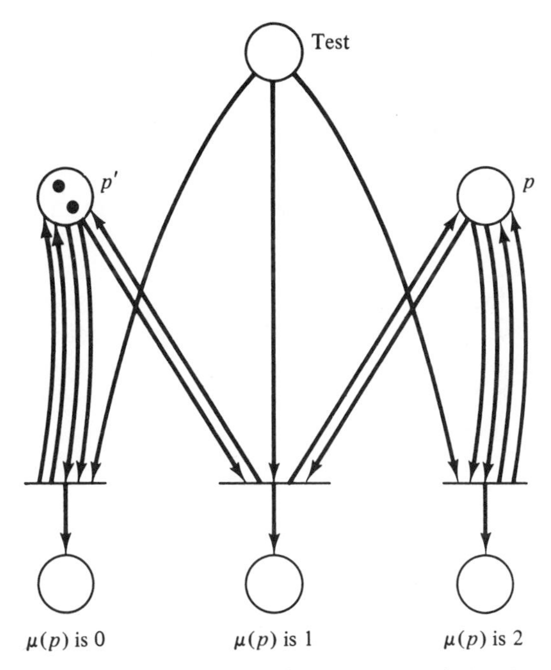

Figure 5.15 Testing a bounded place for a marking of 0, 1 or 2. All transitions must maintain the sum of the markings.

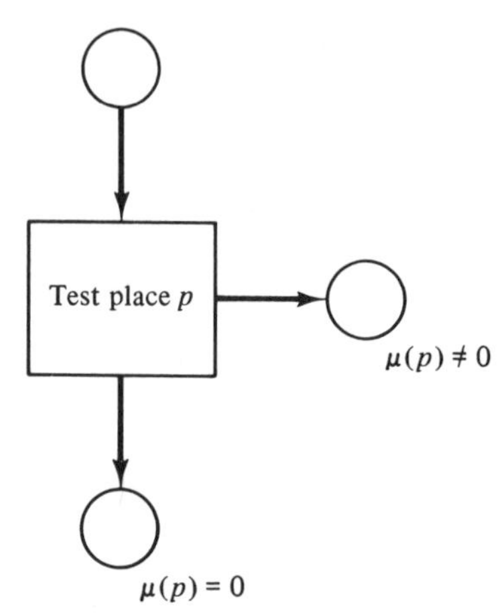

Figure 5.16 The form of the Petri nets which Lipton uses to construct a larger net which can test a larger counter for zero..

very important corollary, too. The Petri net which is constructed is bounded, since the number of tokens in any given place cannot exceed 2^{2^n}. This means that any algorithm to determine boundedness of a Petri net must also require exponential time and space. Thus, even simple problems for Petri nets, while decidable, may require large amounts of time and space for solution.

It should be remembered that these are lower bounds on the worst-case behavior of an algorithm. It may be the case that many interesting problems can be decided for most Petri nets relatively efficiently. These complexity results show that even if an algorithm works very well most of the time, there *exists* a Petri net which will take lots of time and space to analyze.

Although these are worst-case complexity results (which means the average case may be much better), they are also lower-bound results. We know that the reachability problem requires exponential space, *at least.* It may be that reachability is even worse than exponential. Rackoff [1976] has developed an algorithm for determining boundedness in exponential time, so the boundedness problem is known to be of exponential complexity. However, the reachability problem is simply known to be at least exponentially complex (and may not even be decidable).

A recent result by Mayr [1977] showed that the subset and equality problems for *bounded* Petri net reachability sets are of nonprimitive recursive complexity. These results indicate that some problems for Petri nets, while decidable, are computationally intractable.

Exercises

1. Show that the reachability problem for ordinary Petri nets is equivalent to the single-place zero-reachability problem for self-loop-free Petri nets.
2. For a Petri net $C_1 = (P_1, T_1, I_1, O_1)$, define a new Petri net $C_2 = (P_2, T_2, I_2, O_2)$ with

$$P_2 = P_1 \cup \{p_j' \mid t_j \in T_1\}$$
$$T_2 = T_1$$
$$I_2 = I_1$$
$$O_2(t_j) = O_1(t_j) \cup \{p_j'\}$$

This introduces one extra place as an output of each transition.

(a) What is the meaning of the number of tokens in each of these places? For a live Petri net, what is the bound on the marking of these places?

(b) Suppose we add one extra transition with each p_j' as an input and no output. Show that the net is live if and only if this new transition is live.

5.7 Further Reading

Computability theory is an early part of the theory of computation and developed from the work of Turing, Kleene, Godel, and Church. Davis [1958] and Minsky [1967] offer good explanations of this work. Karp [1972] shows how reducibility can be used for decidability and complexity results.

The reachability problem first arose in [Karp and Miller 1968]; it was reported as a research question in [Nash 1973]. Preliminary results were reported in [Van Leeuwen 1974; Hopcroft and Pansiot 1976], but these do not generalize.

Most of the results in this chapter are due to the work of Hack [1974a; 1974c; 1975a; 1975c]. Hack has been one of the major researchers on decision problems for Petri nets. Other work on decision properties includes [Araki and Kasami 1976; Araki and Kasami 1977; Mayr 1977]. Complexity results have been produced by Lipton [1976], Rackoff [1976], and Jones et al. [1976] among others. Some related work not directly tied to Petri nets is [Cardoza 1975; Cardoza et al. 1976].

5.8 Topics for Further Study

1. A Petri net is *reversible* if for every transition $t_j \in T$ there exists $t_k \in T$ such that

$$\# (p_i, I(t_j)) = \# (p_i, O(t_k))$$
$$\# (p_i, O(t_j)) = \# (p_i, I(t_k))$$

   That is, for every transition there is another transition with inputs and outputs reversed. This allows any sequence of transitions to be "undone" by firing their complementary transitions in the opposite order. It has been stated [Hopcroft and Pansiot 1976] that the reachability and equivalence problems are decidable for reversible Petri nets. This theorem is based on work with commutative semigroups [Cardoza 1975]. Follow this statement up, showing the relationship between reversible Petri nets and commutative semigroups, and establish the decidability of reachability and equivalence for reversible Petri nets. Also consider the liveness problem, complexity issues, and the languages of reversible Petri nets to develop a theory of reversible Petri nets.
2. There would seem to be a very useful connection between Petri nets and *Presburger arithmetic.* Presburger arithmetic is a theory of arithmetic using addition and subtraction with integers. It has been shown that it is possible to determine the truth or falseness of all statements formed from first-order quantifiers, equality, the operations of addition and subtraction, and integers. The original proof was presented in [Presburger 1929] and has been used as the basis of theorem-proving programs [Davis 1957; Cooper 1971]. The connection of Presburger arithmetic to semilinear sets was mentioned in [Ginsburg 1966; Ginsburg and Spanier 1966], and the relationship of semilinear sets to Petri net reachability has been mentioned in [Van Leeuwen 1974; Crespi-Reghizzi and Mandrioli 1974; Landweber and Robertson 1975; Hopcroft and Pansiot 1976; Jaffe 1977]. I suspect that Presburger arithmetic can be used to solve analysis problems for Petri nets. Investigate the usefulness of Presburger arithmetic in the analysis of Petri nets.

6

Petri Net Languages

The discussions of Chapters 4 and 5 have concentrated on problems related to the reachability problem, dealing mainly with problems of reachable markings. A related but quite different approach is to focus not on *what* markings are reachable but rather on *how* we reach them. The primary object of interest is then the transition and particularly the sequences of transitions which lead us from one marking to another in a Petri net.

A sequence of transitions is a string, and a set of strings is a language. Thus, in this chapter, we concentrate on the languages defined by Petri nets and their properties.

6.1 Motivation

Petri nets are being developed for two major reasons: (1) the description of proposed or existing systems and (2) the analysis of systems which can be modeled by a Petri net description. The analysis of a Petri net attempts to determine the properties of the net and the system which it models. One of the most important properties of a system is the set of actions which can occur. The set of all possible sequences of actions *characterizes* the system.

In a Petri net, actions are modeled by transitions, and the occurrence of an action is modeled by the firing of a transition. Sequences of actions are modeled by sequences of transitions. Thus the set of allowable transition sequences characterizes a Petri net and (to the degree that the Petri net correctly models a system) the modeled system.

These sequences of transitions can be of extreme importance in using Petri nets. Assume that a new system has been designed to replace an existing one. The behavior of the new system is to be identical to the old system (but the new system is cheaper, faster, easier to repair, or something). If both systems are modeled as Petri nets, then the behaviors of these two nets should be identical, and so their languages should be equal. Two Petri nets are *equivalent* if their languages are equal. This provides a formal basis for stating the equivalence of two systems.

A particular instance in which equivalence is important is *optimization.* Optimizing a Petri net involves creating a new Petri net which is equivalent (languages are equal) but for which the new net is "better" than the old one (for some definition of better). For example, if a Petri net is to be directly implemented in hardware, then a Petri net with fewer places, transitions, and arcs would be less expensive to build, since it has fewer components. Thus one optimization problem might be to reduce $|P| + |T|$ without changing the behavior of the net.

For purposes of optimization, a set of *language-preserving transformations* might be useful. If a transformation applied to a Petri net produces a new Petri net with the same language, then it is language preserving. An optimal Petri net can be produced by applying language-preserving transformations to a nonoptimal Petri net. Practical use of a Petri-net-based system for modeling and analysis would require a collection of language-preserving transformations.

Petri net languages can also be useful for analysis of Petri nets. Techniques have been developed in Chapter 4 for determining specific properties of Petri nets, such as safeness, boundedness, conservation, liveness, reachability, and coverability. Although these are important (and difficult) properties to establish, they are not the only properties for which a Petri net might be analyzed. It may be necessary to establish the correctness of a modeled system by showing that system-specific properties are satisfied. Thus, either new techniques must be developed for each new property, or a general analysis technique for Petri nets is needed.

A large class of questions can be posed in terms of the sequences of actions which are possible in the system. If we define the set of possible sequences of actions as the language of the system, then we can analyze the system by analyzing the language of the system. Now problems may be answered by considering the emptiness question (Will any sequence get me from here to there?) or the membership question (Is a sequence of this form possible?). This may provide us with a general technique for analyzing arbitrary systems for system-specific properties.

Riddle [1972] has investigated analysis based on the language of the modeled system.

Another use of Petri net languages would be in the specification and automatic *synthesis* of Petri nets. If the behavior which is desired can be specified as a language, then it may be possible to automatically synthesize a Petri net whose language is the specified language. This Petri net can be used as a controller, guaranteeing that all and only the sequences specified are possible. *Path expressions* [Lauer and Campbell 1974] have been developed to define allowable sequences of actions. Techniques have been developed for automatically creating a Petri net from a path expression.

Another motivation for the study of Petri net languages comes from a desire to determine decidability results for Petri nets. The decidability of many properties of Petri nets is unknown. The decidability of a few basic questions for Petri nets, such as reachability, is the object of much current research. One area in which decidability questions have been considered is *formal language theory.* By consideration of the languages of Petri nets, the concepts and techniques of formal language theory may be brought to bear on the problems of Petri nets. This may produce new results concerning Petri nets and their decidability questions. It is also possible to use Petri net methods to obtain new useful results about formal languages.

6.2 Related Formal Language Theory Concepts

The Petri net language theory which has developed so far is similar to development of other parts of formal language theory. Several books present the classical theory of formal languages [Hopcroft and Ullman 1969; Salomaa 1973; Ginsburg 1966]. Many of the basic concepts of Petri net language theory have been borrowed from the classical theory of formal languages.

An *alphabet* is a finite set of symbols. A *string* is any sequence of finite length of symbols from the alphabet. The *null* or *empty string* λ is the string of no symbols and zero length. If Σ is an alphabet, then Σ^* is the set of all strings of symbols from Σ, including the empty string. Σ^+ is the set of all nonempty strings over an alphabet Σ. $\Sigma^* = \Sigma^+ \cup \{\lambda\}$.

A language is a set of strings over an alphabet. Languages may in general be infinite; thus representing the language is a problem. Two approaches have been developed for representing languages. One approach is to define a machine which when executed generates a string from the language, and all strings of the language are generated by some execution. Alternatively, a *grammar* may be defined which specifies how to generate a string of a language by successively applying the productions of the grammar.

Restrictions on the forms of the machines or grammars which generate the languages define *classes* of languages. The traditional classes of languages are regular, context-free, context-sensitive, and type-0 languages, corresponding to finite state machines, pushdown automata, linear bounded automata, and Turing machines. Each of these classes of languages is generated by the appropriate class of automata. This provides an excellent means of linking Petri net theory to formal language theory: We define the class of Petri net languages as the class of languages generated by Petri nets. The details of the definition should be similar to the details for any of the other classes of languages.

As an example, let us consider finite state machines and regular expressions. A finite state machine C is a five-tuple $(Q, \delta, \Sigma, s, F)$, where Q is a finite set of states, Σ is an alphabet of symbols, $\delta: Q \times \Sigma \rightarrow Q$ is a next state function, $s \in Q$ is the *start state*, and $F \subseteq Q$ is a finite set of *final states.* The next state function δ is extended to a function from $Q \times \Sigma^*$ to Q in the natural way. The language $L(C)$ generated by the finite state machine is the set of strings over Σ defined by

$$L(C) = \{\alpha \in \Sigma^* | \delta(s, \alpha) \in F\}$$

With each finite state machine is associated a language, and the class of all languages which can be generated by finite state machines is called the class of regular languages. The language of a finite state machine is a characterizing feature of the machine. If two finite state machines have the same language, they are *equivalent.*

6.3 Definitions of Petri Net Languages

The same basic concepts used to produce a regular language from a finite state machine can be applied to Petri nets to produce a theory of Petri net languages. In addition to the Petri net structure defined by the sets of places and transitions (which correspond roughly to the state set and next-state function of the finite state machine), it is necessary to define an initial state, an alphabet, and a set of final states. The specification of these features for a Petri net can result in different classes of Petri nets. We consider each of these in turn.

6.3.1 Initial State

The initial state of a Petri net can be defined in several ways. The most common definition is to allow an arbitrary marking μ to be specified as an initial state. However, this definition can be modified in several ways. One convenient limitation is to restrict the initial state to a marking with one token in a *start place* and zero tokens elsewhere [Peterson 1976]. Another more general definition allows a set of initial markings rather than simply one marking.

These three definitions are essentially the same. Certainly the start place definition is a special case of the initial marking definition, which is a special case of the set of initial markings definition. However, if a set of initial markings $M = \{\mu_1, \mu_2, \ldots, \mu_k\}$ is needed with a start place definition, we can simply augment the Petri net with an extra place p_0 and a set of transitions $\{t_1', \ldots, t_k'\}$. Transition t_j' would input a token from p_0 and would output marking μ_j. Thus the behavior of the augmented net would be identical to a Petri net with a set of initial markings, except that each transition sequence would be preceded with t_j' if marking μ_j were used to start the execution.

We see then that these three definitions of start states for a Petri net are essentially equivalent. Out of a sense of tradition, we define a Petri net language as starting from a single arbitrary marking μ.

6.3.2 Labeling of Petri Nets

Labeling Petri nets is another situation where several definitions can be used. We must define both what the alphabet Σ of a Petri net should be and how it is to be associated with the Petri net. We have indicated that the symbols are associated with the transitions, so that a sequence of transition firings generates a string of symbols for the language. The association of symbols to transitions is made by a *labeling* function, $\sigma: T \rightarrow \Sigma$. Variations in language definition may result from various restrictions placed on the labeling function.

A *free-labeled* Petri net is a labeled Petri net where all transitions are labeled distinctly [i.e., if $\sigma(t_i) = \sigma(t_j)$, then $t_i = t_j$]. The class of free Petri net languages is a subset of the class of Petri net languages with a more general labeling function which does not require distinct labels. An even more general labeling function has been considered which allows null labeled transitions, $\sigma(t_j) = \lambda$. These λ-labeled transitions do not appear in a sentence of a Petri net language, and their occurrence in an execution of a Petri net thus goes unrecorded. These three classes of labeling functions (free, λ-free, and with λ-transitions) define three variations of Petri net languages.

Without further investigation, it is not obvious which of these three labeling definitions is most reasonable. Perhaps each of the three is the most useful labeling function for some application. Thus, we need to consider the languages resulting from each possible definition of the labeling function.

6.3.3 Final States of a Petri Net

The definition of final states for a Petri net has a major effect upon the language of a Petri net. Four major different definitions of the final state set of a Petri net have been suggested. Each of these may produce different Petri net languages.

One definition is derived from the analogous concept for finite state machines. It defines the set of final states F as a finite set of final markings. For this definition we define the class of *L-type* Petri net languages.

DEFINITION 6-1 A language L is an *L-type Petri net language* if there exists a Petri net structure (P,T,I,O), a labeling of the transitions $\sigma: T \rightarrow \Sigma$, an initial marking μ, and a finite set of final markings F such that $L = \{\sigma(\beta) \in \Sigma^* | \beta \in T^* \text{ and } \delta(\mu,\beta) \in F\}$.

The class of L-type Petri net languages is rich and powerful, but it has been suggested that the requirement for a sentence to result in *exactly* a final state in order to be generated is contrary to the basic philosophy of a Petri net. This comment is based on the observation that if $\delta(\mu,t_j)$ is defined for a marking μ and a transition t_j, then $\delta(\mu',t_j)$ is defined for any $\mu' \geqslant \mu$. From this comment we define a new class of languages, the class of *G-type* Petri net languages, by the following.

DEFINITION 6-2 A language L is a *G-type Petri net language* if there exists a Petri net structure (P,T,I,O), a labeling $\sigma: T \rightarrow \Sigma$, an initial marking μ, and a finite set of final markings F, such that $L = \{\sigma(\beta) \in \Sigma^* | \beta \in T^*$ and there exists $\mu_f \in F$ such that $\delta(\mu,\beta) \geqslant \mu_f\}$.

A third class of Petri net languages is the class of *T-type* Petri net languages. These are defined by identifying the set of final states used in the definition of L-type languages with the (not necessarily finite) set of *terminal states*. A state μ_t is terminal if $\sigma(\mu_t,t_j)$ is undefined for all $t_j \in T$. Thus the class of T-type Petri net languages is as follows.

DEFINITION 6-3 A language L is a *T-type Petri net language* if there exists a Petri net structure (P,T,I,O), a labeling $\sigma: T \rightarrow \Sigma$, and an initial marking μ such that $L = \{\sigma(\beta) \in \Sigma^* | \beta \in T^*$ and $\delta(\mu, \beta)$ is defined but for all $t_j \in T$, $\delta(\delta(\mu, \beta), t_j)$ is undefined $\}$.

Still a fourth class of languages is the class of *P-type* Petri net languages whose final state set includes all reachable states. These languages are *prefix* languages since if $\alpha \in \Sigma^*$ is an element of a P-type language, then for all prefixes β of α ($\alpha = \beta x$ for some $x \in \Sigma^*$) β is an element of that same language.

DEFINITION 6-4 A language L is a *P-type Petri net language* if there exists a Petri net structure (P,T,I,O), a labeling $\sigma: T \rightarrow \Sigma$, and an initial marking μ such that $L = \{\sigma(\beta) \in \Sigma^* | \beta \in T^*$ and $\delta(\mu, \beta)$ is defined$\}$.

6.3.4 Classes of Petri Net Languages

In addition to the four classes of languages based on different specifications of the final state set, we have the previously mentioned variations due to the labeling function. Figure 6.1 lists the 12 classes of languages which result from the cross product of the four types of final state specification and the three types of labeling functions. Each cell of Figure 6.1 lists the notation which is used to denote each class of Petri net language.

	Free	Non-λ	λ-Transitions
L-type	L^f	L	L^λ
G-type	G^f	G	G^λ
T-type	T^f	T	T^λ
P-type	P^f	P	P^λ

Figure 6.1 The 12 classes of Petri net languages.

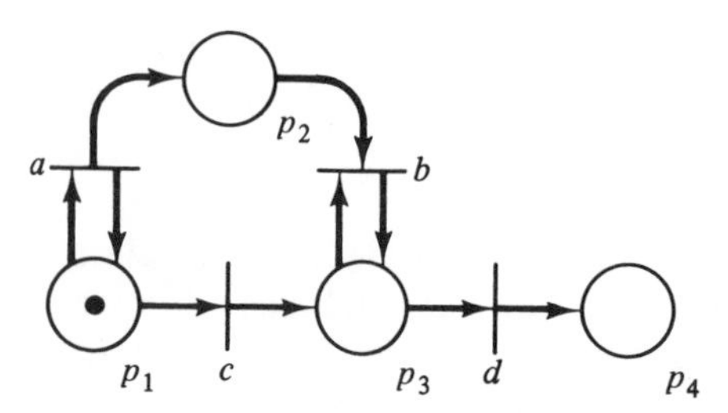

Figure 6.2 An example Petri net to illustrate the different classes of languages. Each transition is labeled with its label.

To specify a particular Petri net language, four quantities must be defined: the Petri net structure, $C = (P,T,I,O)$; the labeling function, $\sigma: T \longrightarrow \Sigma$; the initial marking, $\mu: P \longrightarrow N$; and the set of final markings, F (for L- and G-type languages). We define $\gamma = (C, \sigma, \mu, F)$ to be a *labeled Petri net* with Petri net structure C, labeling σ, initial marking μ, and final state set F. For a given labeled Petri net γ, 12 languages can be defined: $L(\gamma)$, $G(\gamma)$, $T(\gamma)$, $P(\gamma)$, $L^f(\gamma)$, $G^f(\gamma)$, $T^f(\gamma)$, $P^f(\gamma)$, $L^\lambda(\gamma)$, $G^\lambda(\gamma)$, $T^\lambda(\gamma)$, $P^\lambda(\gamma)$.

The different definitions of Petri net languages can associate different languages with a given Petri net. Consider, for example, the Petri net of Figure 6.2. The initial marking of (1, 0, 0, 0) is given on the net, and each transition t_j is labeled by $\sigma(t_j)$. If we define $F = \{(0, 0, 1, 0)\}$ (one token in place p_3), the L-type language is $\{a^n cb^n | n \geqslant 0\}$, the G-type language is $\{a^m cb^n | m \geqslant n \geqslant 0\}$, the T-type language is $\{a^m cb^n d | m \geqslant n \geqslant 0\}$, and the P-type language is $\{a^m | m \geqslant 0\} \cup \{a^m cb^n | m \geqslant n \geqslant 0\} \cup \{a^m cb^n d | m \geqslant n \geqslant 0\}$. For this example, all four types of languages are different. The labeling function given is a free labeling, but by using different labeling functions, other languages can also be produced.

Despite the differences in definitions, the classes of Petri net languages are closely related. For instance, the set of free labelings is a subset of the set of non-λ labelings, which is a subset of the set of λ-labelings. Thus,

$$L^f \subseteq L \subseteq L^\lambda$$

$$G^f \subseteq G \subseteq G^\lambda$$

$$T^f \subseteq T \subseteq T^\lambda$$

$$P^f \subseteq P \subseteq P^\lambda$$

Also, every P-type language is a G-type language where $F = \{(0, 0, \ldots, 0)\}$. Thus,

$$P^f \subseteq G^f$$

$$P \subseteq G$$

$$P^\lambda \subseteq G^\lambda$$

We can also show that each language of type G or G^λ is also a language of type L or L^λ, respectively. Let G be a G-type language for a Petri net structure (P,T,I,O), initial marking μ, and final set F. Construct a new labeled Petri net with the same places but with additional transitions as defined by the following:

For each $t_j \in T$, let B_j be the set of all proper subbags of $O(t_j)$. Each subbag in B_j is used to define a new transition with the same label and inputs as t_j but with the subbag as output. These new transitions are added to the previous set of transitions.

For example, if we consider the transition labeled a in the Petri net of Figure 6.2, its input bag is $\{p_1\}$, and its output bag is $\{p_1,p_2\}$. The subbags of $\{p_1,p_2\}$ are $\{p_1\}, \{p_2\}$, and $\{\}(=\emptyset)$. This transition would result in three new transitions being added to the net. All of these new transitions would be labeled a and have input bag $\{p_1\}$, but the output bags would be the three subbags listed above (one transition for each subbag). New transitions would also be added for the transitions labeled b, c, and d, which would have the same inputs but null outputs (since the current outputs are singleton bags and hence the only subbag is $\emptyset$). This new net is shown in Figure 6.3.

This new net has been modified in such a way that the "extra" tokens which exceed a final state in F need not be produced, if one selects the appropriate new transition which has fewer outputs. Thus the L-type language of the new net is the same as the G-type language of the old net.

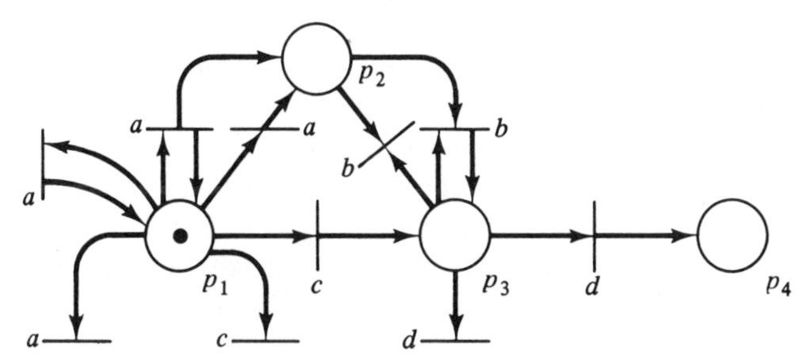

Figure 6.3 A Petri net whose L-type language is the same as the G-type language of the Petri net of Figure 6.2.

This construction requires creating new transitions with the same label as other transitions, so no conclusion about the relationship of G^f and L^f can be drawn from this construction.

$$G \subseteq L$$

$$G^\lambda \subseteq L^\lambda$$

The above construction can also be modified slightly to show that a generalization of the definition of L-type languages (to permit incompletely specified final markings) does not change the classes L and L^λ. Let a final marking for a Petri net with n places be an n-vector over $N \bigcup \{\omega\}$. If a component of a final marking is ω, this means we do not care what the value of that component is in a final state. A state s is a final state if there exists a final marking f such that for all $i, 1 \leqslant i \leqslant n$, if $f_i \neq \omega$, then $s_i = f_i$. This is obviously a more general definition than the definition of L-type languages given earlier.

Now consider a language which is defined by a Petri net and an incompletely specified final marking, f. Let τ be the set of all places for which $f_i = \omega$. For each transition $t_j \in T$ for which $O(t_j) \bigcap \tau \neq \varnothing$, let $\rho_j = O(t_j) \bigcap \tau$ and $\psi_j = O(t_j) - \rho_j$. The bag ρ_j is all places whose markings are don't-cares, while the bag ψ_j is all places whose markings must be exactly as specified in the final marking f. We add new transitions to the net whose label and input bag are the same as the label and input bag for t_j but whose output bag is $\psi_j + \xi$, where ξ ranges over all subbags of ρ_j. This construction does not in any way modify the behavior of those places which are not in τ but allows those places which are don't-cares to have any number of tokens less than or equal to the number of tokens which would have appeared in the original net. Thus, for each sentence in the generalized language of the original net, it is possible to pick appropriate transitions to fire so that when the net reaches a state s such that $s_i = f_i$ for $f_i \neq \omega$, $s_i = 0$ for $f_i = \omega$. Thus the L-type language for the constructed Petri net with a final marking f' (where all ω in the incompletely specified marking f have been replaced by 0) is the same as the generalized language of the original net (as defined by the incompletely specified final marking f).

For a language defined by a set of incompletely specified final markings (as opposed to only one such marking as just discussed), we use the fact that L (and L^λ) languages are closed under union (see Section 6.5.2) to show that the language is still an L (or L^λ) language.

With the introduction of incompletely specified final markings, we can show that T-type languages are also L-type languages (except possibly for free T-type languages).

A final state μ for a T-type language is such that no transition t_j

can fire [i.e., $\mu \ngeq I(t_j)$ for all $t_j \in T$]. The condition which specifies a final state for a T-type language is thus exactly the opposite of the condition specifying a final state for a G-type language. (We might call the T-type languages *inverse* G-type languages.) It is not difficult to see that such a set of markings can be described by a finite set of incompletely specified markings (as was done in Section 5.4). For example, the condition [$\mu \ngeq (2,0)$ and $\mu \ngeq (1,1)$] is equivalent to [$\mu = (0,\omega)$ or $\mu = (1,0)$]. A T-type language (or, more generally, an inverse G-type language) can thus be rewritten as a generalized (i.e., incompletely specified) L-type language and then as an L-type language. Thus $T \subseteq L$, and $T^\lambda \subseteq L^\lambda$.

It is known that every L^λ-type language can be generated by a Petri net in which every transition has an input place, and where the unique final marking is the zero marking, at which no transition is fireable [Hack 1975b]. If we add to every place a λ-transition with a single input and a single output of that same place (i.e., a self-loop), then the language is not changed, and the zero marking becomes the only terminal marking. Hence, $L^\lambda \subseteq T^\lambda$, and from our previous demonstration that $T^\lambda \subseteq L^\lambda$, these two classes of languages are identical, $T^\lambda = L^\lambda$.

Figure 6.4 represents graphically the relationships among the classes of Petri net languages which we have just derived. An arc between classes indicates the containment of one class of Petri net languages within another.

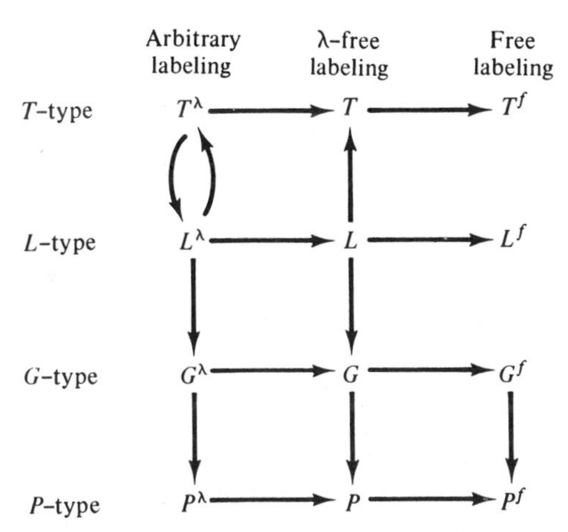

Figure 6.4 The known relations among the classes of Petri net languages. An arc from a class A to a class B means that class A contains class B.

6.4 Properties of Petri Net Languages

The study of Petri net languages is just beginning, and at present, knowledge of the properties of these newly defined classes of languages is limited. The power of Petri nets is reflected in the variety of potentially different classes of Petri net languages which can be defined. The newness of this research is reflected by our inability to either show the complete relationships between these languages, or to argue that only a few of these classes are of importance. This results in a vast field of study, with the need to develop the properties of 12 different classes of languages.

It is obviously not possible to develop all 12 classes of languages in this volume, and so we limit ourselves here to considering only one class of Petri net languages, the L-type languages. The major reasons for this limitation are space and the fact that this language has been investigated in the literature [Peterson 1973; Hack 1975b; Peterson 1976]. Some results have also been obtained by Hack [1975b] for the prefix (P-type) languages and will be presented in our summary. The G-type and T-type languages have been defined, but no work has been done on their development. Remember also that the class of L-type languages includes the classes of T-type, G-type, and P-type languages. Thus, L-type languages are in some sense the largest class of Petri net languages and so are appropriately investigated first.

Our investigation of the properties of L-type Petri net languages will focus on two aspects. First we present *closure* properties of Petri nets under certain common operations (concatenation, union, concurrency, intersection, reversal, complement, indefinite concatenation, and substitution). Then we consider the relationship between Petri net languages and the classical formal languages: regular, context-free, context-sensitive, and type-0. This presentation provides an understanding of the power and limitations of (L-type) Petri net languages. It also indicates how the other classes of Petri net languages can be investigated.

Although we are interested in the entire class of L-type Petri net languages, we limit our discussion to only a limited set of *standard form* Petri nets. This limitation is made in order to ease the proofs and constructions; it does not limit the class of Petri net languages. For every Petri net language, there may be many Petri nets which generate that language; we choose to work only with those nets with certain properties. To show that this does not reduce the set of languages, we show that for every L-type Petri net language there exists a standard form Petri net which generates it.

First we define a standard form Petri net.

DEFINITION 6-5 A labeled Petri net $\gamma = (C, \sigma, \mu, F)$ with language $L(\gamma)$ in *standard form* satisfies the following properties:

1. The initial marking μ consists of exactly one token in a start place p_s and zero tokens elsewhere. $p_s \notin O(t_j)$ for all $t_j \in T$.
2. There exists a place $p_f \in P$ such that
   (a) $F = \{p_f\}$ [if $\lambda \notin L(\gamma)$] or $F = \{p_f, p_s\}$ [if $\lambda \in L(\gamma)$].
   (b) $p_f \notin I(t_j)$ for all $t_j \in T$.
   (c) $\delta(\mu', t_j)$ is undefined for all $t_j \in T$, and $\mu' \in R(C, \mu)$ which have a token in p_f [i.e., $\mu'(p_f) > 0$].

The execution of a Petri net in standard form begins with one token in the start place. The first transition which fires removes this token from the start place, and after this firing the start place is always empty. Eventually the Petri net may stop by placing one token in the final place. This token cannot be removed from the final place both because no place has an input from the final place and because all transitions are disabled. Thus the execution of a Petri net in standard form is quite simple and limited. This is of great use when compositions of Petri nets are constructed. To show that Petri nets in standard form are not less powerful than general Petri nets, we prove the following theorem.

THEOREM 6-1 Every Petri net is equivalent to a Petri net in standard form.

***Proof*:** The proof is by construction. Let $\gamma = (C, \sigma, \mu, F)$ be a labeled Petri net with $C = (P, T, I, O)$. We show how to construct an equivalent $\gamma' = (C', \sigma', \mu', F')$ with $C' = (P', T', I', O')$ which is in standard form (Figure 6.5).

To start, we define three new places p_r, p_f, and p_s which are not in p. Place p_s is the start place, place p_f is the final place, and p_r is a "run" place; a token must be in p_r to allow any transition in T to fire. The initial marking of C' will have one token in p_s; the final marking will consist of one token in p_s [if $\lambda \in L(\gamma)$] or p_f [for $\lambda \notin L(\gamma)$].

Now we wish to be sure that every transition sequence in C leading from the initial marking to a final marking is "mimicked" in C'. To this end we consider three types of strings in $L(\gamma)$. First, the empty string λ is properly handled by the definition of F'. We can determine if $\lambda \in L(\gamma)$ by checking if the initial marking μ is a final marking $\mu \in F$.

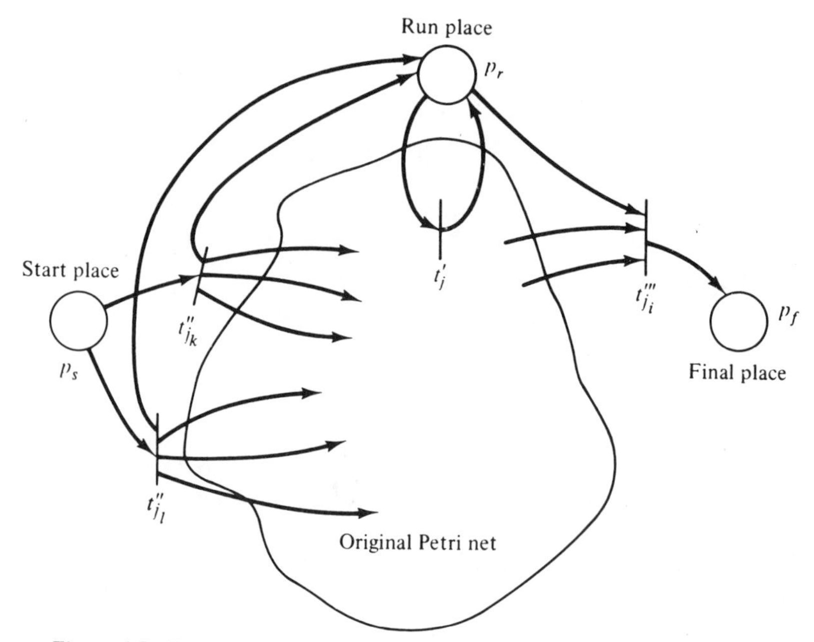

Figure 6.5 The constructed form of a Petri net in standard form. The execution of the net is the same as the original.

Second, for all strings of length 1 in $L(\gamma)$, we include a special transition from p_s to p_f in C' as follows: For $\alpha \in \Sigma$ with $\alpha \in L(\gamma)$, define $t_\alpha \in T'$, with $I(t_\alpha) = \{p_s\}$ and $O(t_\alpha) = \{p_f\}$. The label for t_α is α. We can determine if $\alpha \in L(\gamma)$ by checking all transitions $t_j \in T$, with $\sigma(t_j) = \alpha$, to see if $\delta(\mu, t_j) \in F$.

Finally, consider all strings of length greater than 1. These strings result from a transition sequence

$$t_{j_1} t_{j_2} \cdots t_{j_i}$$

of transitions in T. We would like to define a sequence

$$a t_{j_1}' \cdots t_{j_i}' b$$

with new transitions a and b. The new transition a would input a token from p_s and output the initial marking μ of C plus a token in p_r. Every transition t_j' in T' is the same as t_j in T except that p_r is an input and an output. This allows us to disable all transitions in T' — by removing the token in p_r. Finally, the transition b would remove the token from p_r and a final marking of C and output a token to p_f. With this construction, the token in the start place would move to the final place in C' only as a result of a sequence

$$a t_{j_1}' \cdots t_{j_i}' b$$

which corresponds to a sequence

$$t_{j_1} \cdots t_{j_i}$$

leading from μ to a final marking in C.

Unfortunately this would produce a sequence which is too long since the extra symbols corresponding to transitions a and b would exist for C' but not for C. One solution would be a null labeling for a and b, but the L-type languages do not allow null labelings. So to solve this problem we are forced to collapse transitions a and t_{j_1}' into one transition t_{j_1}'' and collapse transition b and t_{j_i}' into t_{j_i}'''. Thus for all $t_j \in T$ we define the following transitions in T':

1. Define $t_j' \in T'$, with $I(t_j') = I(t_j) \cup \{p_r\}$ and $O(t_j') = O(t_j) \cup \{p_r\}$.
2. If $I(t_j) \subseteq \mu$ (i.e., the inputs to t_j are a subset of the initial marking, so that t_j could be the first transition to fire), define a transition t_j'' with $I(t_j'') = \{p_s\}$ and $O(t_j'') = \mu - I(t_j) + O(t_j) \cup \{p_r\}$.
3. If $O(t_j) \subseteq \mu'$ for some $\mu' \in F$ (so that t_j could be the final transition which fires, leading to a final marking), define a transition t_j''' with $I(t_j''') = \mu' - O(t_j) + I(t_j) \cup \{p_r\}$ and $O(t_j''') = \{p_f\}$.

Now we define the labeling σ' by

$$\sigma'(t_j') = \sigma'(t_j'') = \sigma'(t_j''') = \sigma(t_j)$$

Any string $\alpha \in L(\gamma)$ is by definition generated by a sequence $t_{j_1}t_{j_2} \cdots t_{j_i}$ with $\alpha = \sigma(t_{j_1}t_{j_2} \cdots t_{j_i})$. By construction

$$\alpha = \sigma'(t_{j_1}''t_{j_2}' \cdots t_{j_{i-1}}'t_{j_i}''')$$

and so $\alpha \in L(\gamma)$. Thus, since $L(\gamma) = L(\gamma')$, the two nets γ and γ' are equivalent. By construction γ' is in standard form. □

Figure 6.6 gives a simple Petri net which is not in standard form. Applying the construction of the proof to this Petri net produces the standard form Petri net of Figure 6.7.

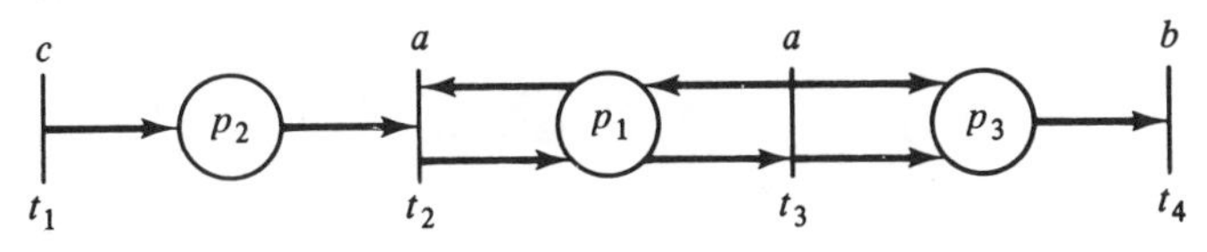

Figure 6.6 A Petri net which is not in standard form. Place p_1 is both the start and final place.

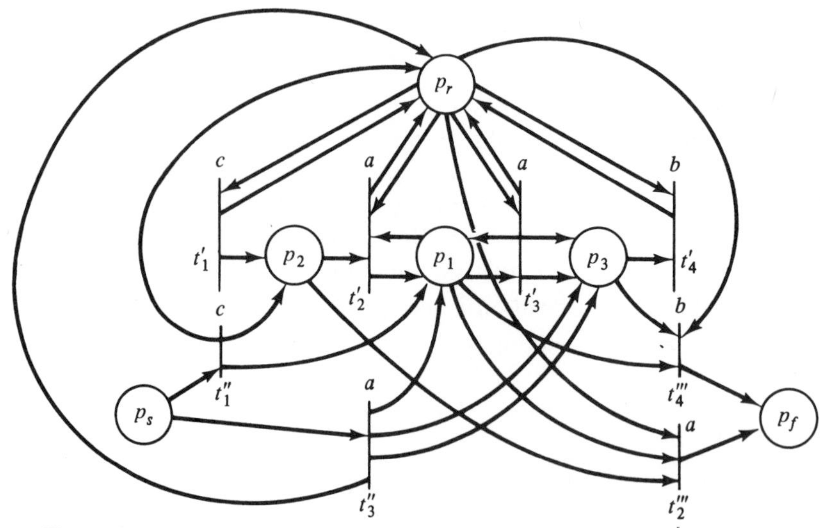

Figure 6.7 A standard form Petri net equivalent to the Petri net of Figure 6.6.

6.5 Closure Properties

We now examine the closure properties of Petri net languages under several forms of composition (union, intersection, concatenation, concurrency, and substitution) and under some operations (reversal, complement, and indefinite concatenation). The motivation for this examination is twofold. First, it increases our understanding of the properties and limits of Petri net languages as languages. Second, many of these compositions reflect how large systems are designed and constructed by the composition of smaller systems into larger systems. Thus, this investigation may help in the development of synthesis techniques.

Most of the closure properties deal with compositions of Petri net languages. For this we assume two Petri net languages L_1 and L_2. We know that each of these languages is generated by some Petri net in standard form. Thus, we consider the two standard form labeled Petri nets $\gamma_1 = (C_1, \sigma_1, \mu_1, F_1)$ and $\gamma_2 = (C_2, \sigma_2, \mu_2, F_2)$ with $L_1 = L(\gamma_1)$ and $L_2 = L(\gamma_2)$. Since they are in standard form, γ_1 has a start place $p_{s_1} \in P_1$ and γ_2 has $p_{s_2} \in P_2$. Also $F_1 = \{p_{s_1}, p_{f_1}\}$ or $\{p_{f_1}\}$ and $F_2 = \{p_{s_2}, p_{f_2}\}$ or $\{p_{f_2}\}$.

From these two labeled Petri nets we show how to construct a new labeled Petri net $\gamma' = (C', \sigma', \mu', F')$ with language $L(\gamma')$ which is the desired composition of L_1 and L_2. We illustrate these constructions with example Petri nets.

We begin by considering the composition of two Petri net languages by concatenation, union, intersection, and concurrency. Then we consider the reversal and complement of a Petri net language and, finally, substitution.

6.5.1 Concatenation

Many systems are composed of two sequential subsystems. Each of the subsystems may separately be expressed as a Petri net with its own Petri net language. When the two systems are combined sequentially, the resulting execution is the *concatenation* of an execution from the first Petri net language with an execution from the second. The concatenation of two languages can be formally expressed as

$$L_1L_2 = \{x_1x_2 | x_1 \in L_1,\ x_2 \in L_2\}$$

THEOREM 6-2 If L_1 and L_2 are Petri net languages, then L_1L_2 is a Petri net language.

Proof: Define γ' such that the final place of γ_1 overlaps with the start place of γ_2. Then the transition which places a token in p_{f_1}, signaling an end to the execution of γ_1, also signals the start of an execution of γ_2. Any string in the concatenation of L_1 and L_2 thus has a path from p_{s_1} to $p_{f_1} = p_{s_2}$ and then to p_{f_2} in the composite Petri net and is an element of $L(\gamma')$. Similarly, if a string is generated by C', it must be composed of a string generated by γ_1, followed by a string in γ_2.

The formal definition of γ' must take into consideration the empty string and so is more complex but can be defined as the union of γ_1 and γ_2 with the following extra transitions. For each transition $t_j \in T_2$ with $I_2(t_j) = \{p_{s_2}\}$, we introduce a new transition t_j' with $I'(t_j') = \{p_{f_1}\}$ and $O'(t_j') = O_2(t_j)$. If $p_{s_1} \in F_1$, then we also add t_j'' with $I'(t_j'') = \{p_{s_1}\}$ and $O'(t_j'') = O_2(t_j)$. We define $\sigma'(t_j') = \sigma_2(t_j)$ and $\sigma'(t_j'') = \sigma_2(t_j)$. The new final set F' is $F_1 \cup F_2$ if $p_{s_2} \in F_2$; otherwise $F' = F_2$. □

This shows that Petri net languages are closed under concatenation. Figure 6.8 illustrates this construction for $L_1 = (a + b)^+$ and $L_2 = a^nb^n$.

6.5.2 Union

Another common method of combining systems is *union*. In this composition either one, but only one, of the subsystems will be exe-

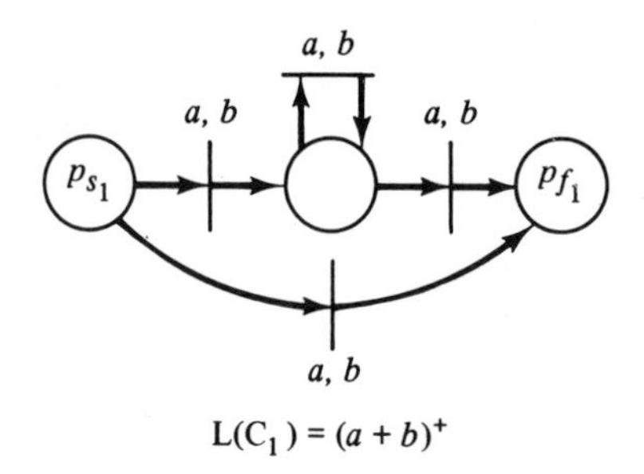

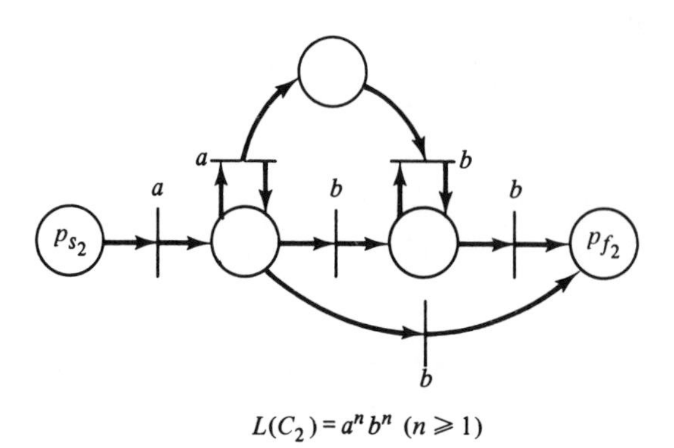

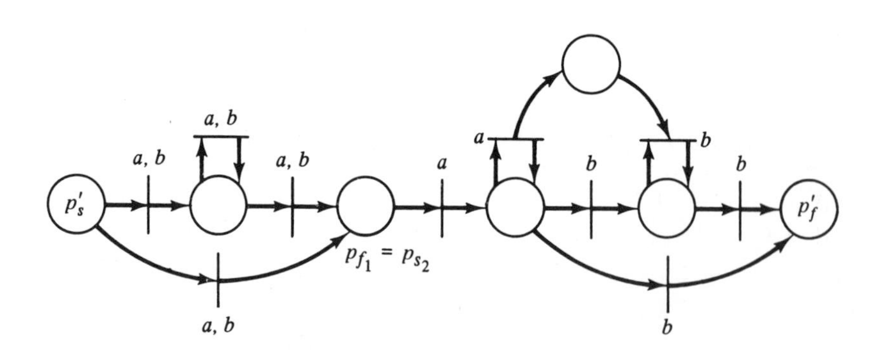

Figure 6.8 Illustrating the concatenation of two Petri net languages.

cuted. This is similar to set union and is a common composition of languages. It can be defined by

$$L_1 \cup L_2 = \{x \mid x \in L_1 \text{ or } x \in L_2\}$$

THEOREM 6-3 If L_1 and L_2 are Petri net languages, then $L_1 \cup L_2$ is a Petri net language.

***Proof*:** The proof of this theorem is similar to the proof of the previous theorem. We define a new Petri net γ' such that $L(\gamma') = L_1 \bigcup L_2$. The new Petri net combines the two start places into one new start place. Then the first transition which fires removes the token from the start place, and either the Petri net γ_1 (if the transition was in T_1) or the Petri net γ_2 (if the transition was in T_2) continues to operate exactly as it had before.

Formally we define a new start place p_s' and new transitions t_{j_1}' for each $t_{j_1} \in T_1$ with $I(t_{j_1}) = \{p_{s_1}\}$ and t_{j_2}' for each $t_{j_2} \in T_2$ with $I(t_{j_2}) = \{p_{s_2}\}$. We define $I'(t_{j_1}') = \{p_s'\}$ and $I'(t_{j_2}') = \{p_s'\}$ with $O'(t_{j_1}') = O_1(t_{j_1})$ and $O'(t_{j_2}') = O_2(t_{j_2})$. The labeling function $\sigma'(t_{j_1}') = \sigma_1(t_{j_1})$ and $\sigma'(t_{j_2}) = \sigma_2(t_{j_2})$. The new initial marking has one token in p_s'; the new final marking set $F' = F_1 \bigcup F_2$. If $p_{s_1} \in F_1$ or $p_{s_1} \in F_2$, then $p_s' \in F'$ also. □

Figure 6.9 illustrates the construction with $L_1 = a(a + b)b$ and $L_2 = a^m b^n$ $(m > n > 1)$.

6.5.3 Concurrency

Still another way of combining two Petri nets is by allowing the executions of two systems to occur *concurrently.* This allows all possible interleavings of an execution of one Petri net with an execution of the other Petri net. Riddle [1972] has introduced the $\|$ operator to represent this concurrency. It can be defined for $a,b \in \Sigma$ and $x_1,x_2 \in \Sigma^*$ by

$$ax_1 \| bx_2 = a(x_1 \| bx_2) + b(ax_1 \| x_2)$$

$$a \| \lambda = \lambda \| a = a$$

The concurrent composition of two languages is then

$$L_1 \| L_2 = \{x_1 \mid x_2 \| x_1 \in L_1,\ x_2 \in L_2\}$$

As a simple example, if $L_1 = \{ab\}$ and $L_2 = \{c\}$, then $L_1 \| L_2$ is equal to $\{abc, acb, cab\}$.

It is easily shown that regular, context-sensitive, and type-0 languages are closed under concurrency by demonstrating that the cross product of two finite state machines, linear bounded automata, or Turing machines is still a finite state machine, linear bounded automaton, or Turing machine, respectively. Since the cross product of two pushdown stack automata cannot be transformed into another pushdown stack automaton, context-free languages are not closed under concurrency. For Petri net languages, we have the following theorem.

THEOREM 6-4 If L_1 and L_2 are Petri net languages, then $L_1 \| L_2$ is a Petri net language.

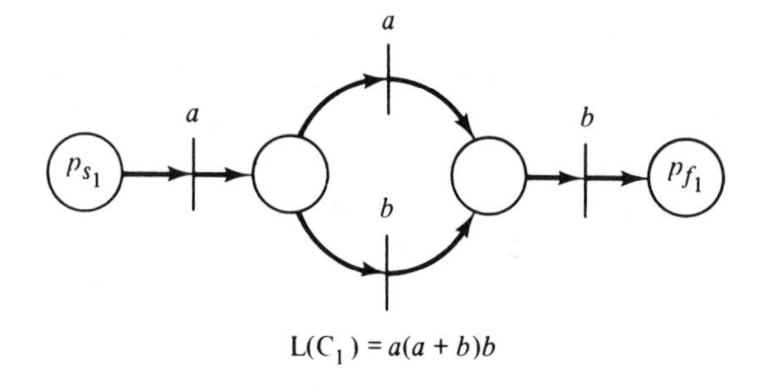

$L(C_1) = a(a + b)b$

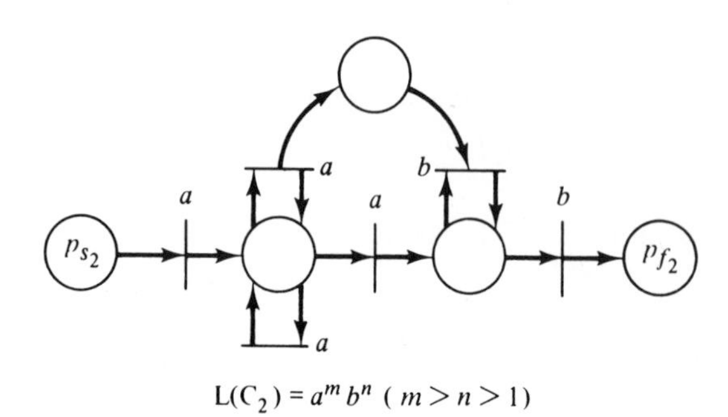

$L(C_2) = a^m b^n \ (m > n > 1)$

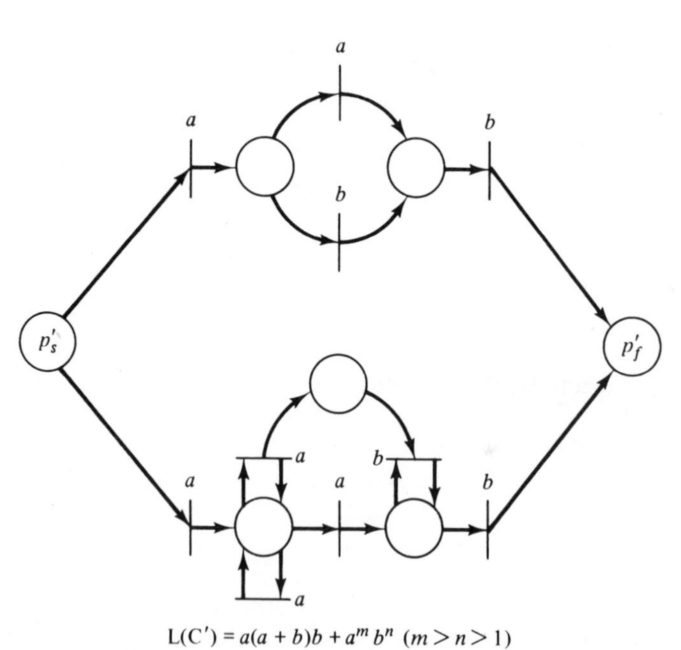

$L(C') = a(a + b)b + a^m b^n \ (m > n > 1)$

Figure 6.9 Illustrating the union of two Petri net languages.

Proof: The construction of a Petri net to generate the concurrent composition of L_1 and L_2, given Petri nets generating these Petri net languages, is basically a Petri net which places tokens in the start places of both γ_1 and γ_2, as an initial marking, and defines the new final marking set to be any marking which is in F_1 (over the places of P_1) and is in F_2 (over the places of P_2). □

This construction is illustrated in Figure 6.10 for $L_1 = a(a + b)^+$ and $L_2 = ca^{3n}cb^{2n}c$.

6.5.4 Intersection

As with union, the intersection composition is similar to the set theory definition of intersection and is given for a Petri net language by

$$L_1 \cap L_2 = \{x \mid x \in L_1 \text{ and } x \in L_2\}$$

THEOREM 6-5 If L_1 and L_2 are Petri net languages, then $L_1 \cap L_2$ is a Petri net language.

The construction of a Petri net to generate the intersection of two Petri net languages is rather complex. At a given point in the generated string, if a transition fires in one Petri net, there must be a transition in the other Petri net with the same label which fires also. When more than one transition exists in each Petri net with the same label, we must consider all possible pairs of transitions from the two Petri nets. For each of these pairs, we create a new transition which fires if and only if both transitions fire in the old Petri nets. This is done by making the input (output) bag of the new transition the bag sum (see the appendix) of the input (output) bags of the pair of transitions from the old Petri nets. Thus if $t_j \in T_1$ and $t_k \in T_2$ are such that $\sigma(t_j) = \sigma(t_k)$, then we have a transition $t_{j,k} \in T'$ with $I'(t_{j,k}) = I_1(t_j) + I_2(t_k)$ and $O'(t_{j,k}) = O_1(t_j) + O_2(t_k)$. Some of these transitions have inputs which include the start place. If for a transition $t_{j,k}$ in T', as defined above, $I'(t_{j,k}) = \{p_{s_1}, p_{s_2}\}$, then we replace this transition with a new transition $t_{j,k}'$, where $I'(t_{j,k}') = \{p_s'\}$. This construction essentially places the two original Petri nets in a lockstep identical execution mode. This composite Petri net generates the intersection of L_1 and L_2. The construction is demonstrated in Figure 6.11.

6.5.5 Reversal

Unlike the previous composition operations which we have examined, the operation of reversal seems to have only academic interest.

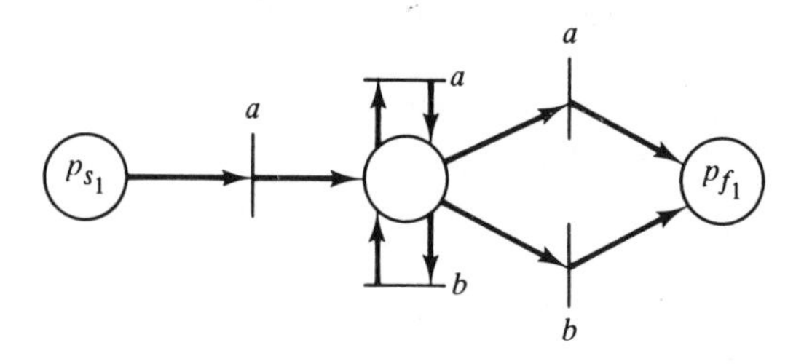

$L(C_1) = a(a + b)^+$

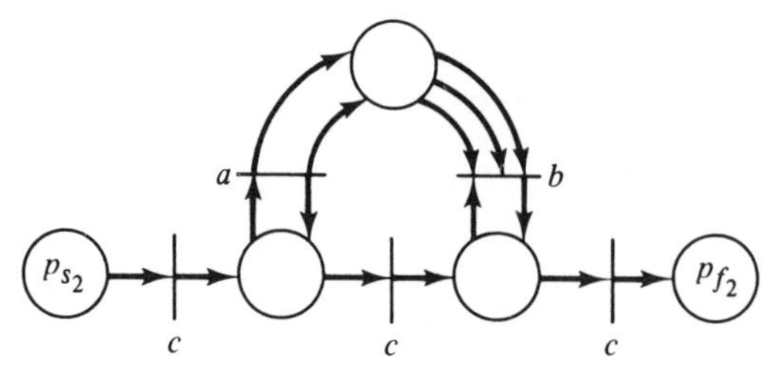

$L(C_2) = ca^{3n}cb^{2n}c$

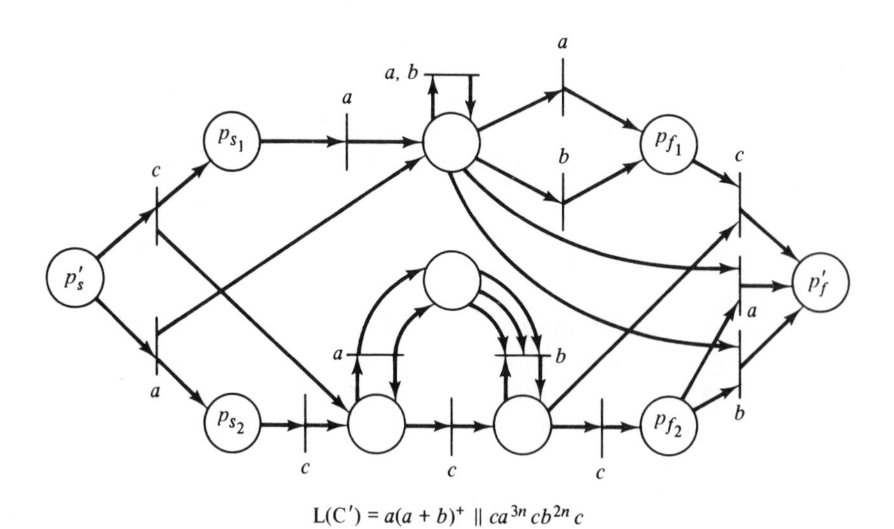

$L(C') = a(a + b)^+ \parallel ca^{3n}cb^{2n}c$

Figure 6.10 Illustrating the concurrent composition of two Petri net languages.

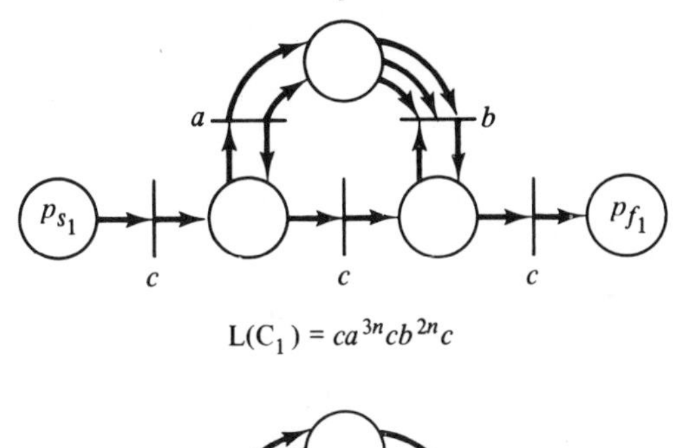

$L(C_1) = ca^{3n}cb^{2n}c$

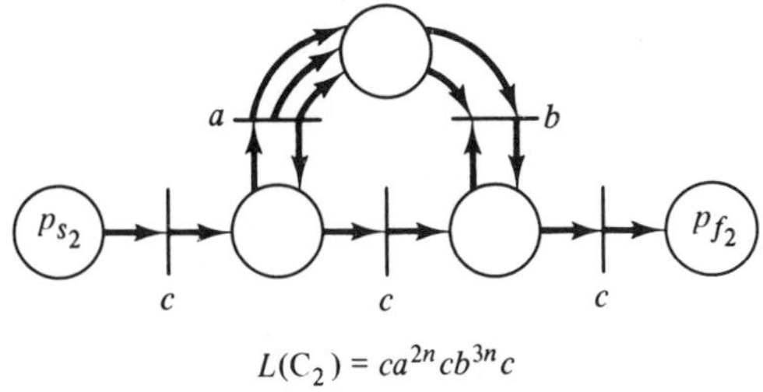

$L(C_2) = ca^{2n}cb^{3n}c$

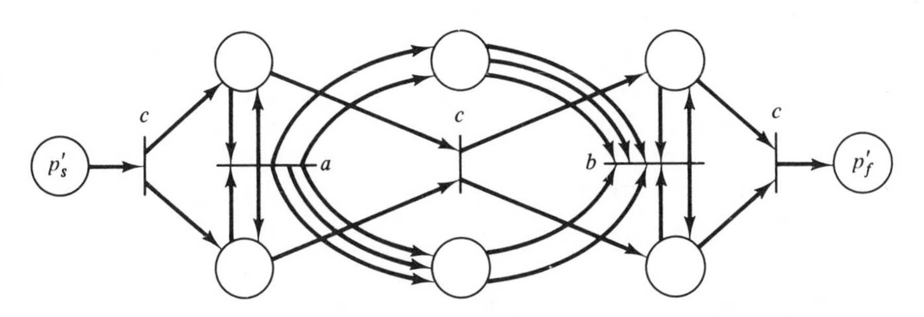

$L(C') = ca^{3n}cb^{2n}c \cap ca^{2n}cb^{3n}c$

Figure 6.11 Illustrating the intersection of two Petri net languages.

The *reversal* of a sentence x^R is the sentence x with its symbols in the reverse order. We define this recursively by

$$a^R = a$$

$$(ax)^R = x^R a$$

for $a \in \Sigma$, $x \in \Sigma^+$. Then for a language we have

$$L^R = \{x^R | x \in L\}$$

THEOREM 6-6 If L is a Petri net language, then L^R is a Petri net language.

The construction is trivial. The start and final markings are interchanged, and the input and output bags for each transition are also interchanged. Then $L(\gamma') = L(\gamma)^R$. This merely runs the Petri net backwards and reverses all generated strings.

6.5.6 Complement

The complement $\bar{L}$ of a language L over an alphabet Σ is the set of all strings in Σ^* which are not in L. This can be expressed as

$$\bar{L} = \Sigma^* - L$$

or

$$\bar{L} = \{x \in \Sigma^* | x \notin L\}$$

This operation on a Petri net language may be useful in the analysis of Petri nets since in checking for forbidden states or forbidden sequences it may be easier to check for existence in the complement than check for nonexistence in the Petri net language.

Closure under complement for the L-type Petri net languages is an open problem. However, Crespi-Reghizzi and Mandrioli [1977] have shown that some Petri net languages are not closed under complement; that is, there are Petri net languages whose complement is *not* a Petri net language.

6.5.7 Repeated Composition

So far we have considered the operations of union, intersection, concatenation, concurrency, reversal, and complement. Except for the last operation, we were able to give constructions that show that Petri net languages are closed under these operations. From these results we can immediately derive the following corollary.

COROLLARY 6-1 Petri net languages are closed under any finite number of applications, in any order, of the operations of union, intersection, reversal, concurrency, and concatenation.

The proof follows from the theorems above.

A new operation can be defined by removing the constraint that only a finite number of compositions be allowed. The *indefinite concatenation* (Kleene star) of a language is the set of all concatenations (of any length) of elements from that language. This is represented by L^* and is defined by

$$L^* = L \cup LL \cup LLL \cup \cdots$$

This operation may be of practical use. Indefinite concatenation is similar to the modeling of a loop. Also it is known that regular, context-free, context-sensitive, and type-0 languages are closed under indefinite concatenation [Hopcroft and Ullman 1969]. Thus it is unexpected and perhaps unfortunate that Petri net languages are not closed under indefinite concatenation.

This appears to be due to the combination of the finite nature of Petri nets (finite number of places and transitions) and the permissive nature of the state changes [transitions are allowed (permitted) to fire but cannot be required to do so]. To construct a Petri net to generate the indefinite concatenation of a Petri net language would, in general, require the reuse of some portion of the Petri net. This allows tokens to be generated and left in some of the places which are reused. At a later repetition of the Petri net, these tokens may be used to allow transitions to fire when they should not.

The proof that Petri nets are not closed under indefinite concatenation appears to be very difficult. Perhaps a flavor of the approach can be given by considering an example. We have already seen that $a^n b^n$ $(n > 1)$ is a Petri net language. We claim that $(a^n b^n)^*$ is not a Petri net language. All generators of $(a^n b^n)^*$ must have some place or set of places that encode the number n for each portion of the string. These tokens control the generation of the b symbols. For a Petri net to generate $(a^n b^n)^*$ it is necessary to use these places more than once. But since the net is permissive, there is no way to guarantee that these places are empty before they are reused. Thus for any Petri net which attempts to generate $(a^n b^n)^*$ there exists some i, j, k such that the Petri net also generates some string of the following form:

$$a^{n_1} b^{n_1} \cdots a^{n_i} b^{n_j} \cdots a^{n_j} b^{n_i} \cdots a^{n_k} b^{n_k}$$

with $n_i > n_j$. Kosaraju [1973] has given the basis for a formal proof that Petri net languages are not closed under indefinite concatenation.

For readers familiar with the formal language theory of Abstract Families of Languages (AFL) [Ginsburg 1975], it is easy to prove that Petri net languages are not closed under indefinite concatenation. It is well known that the smallest full AFL closed under intersection and containing $\{a^n b^n\}$ contains every recursively enumerable set. Thus, since Petri net languages are closed under intersection and $\{a^n b^n\}$ is a

Petri net language, if they were closed under indefinite concatenation, they would be such an AFL. However, we know that $\{ww^R\}$ is not a Petri net language (see Section 6.6.2), and so Petri net languages are not closed under indefinite concatenation. This argument is due to Mandrioli.

A subclass of Petri net languages exists which is closed under indefinite concatenation, however. This is the class of Petri net languages for *properly terminating* Petri nets. Proper termination was defined by Gostelow [1971] for complex bilogic graph models of computation. We borrow his definition and rephrase it in terms of Petri nets as follows.

DEFINITION 6-6 A Petri net is *properly terminating* if whenever it terminates it is guaranteed that (1) only one token remains in the Petri net and it is in the final place and (2) the number of tokens used in the Petri net is finite.

We notice first that the second part of the definition is not actually a restriction since if the Petri net terminates, then it terminates in a finite amount of time and hence generates only a finite number of tokens. But the first part of the definition is a restriction. We may view the Petri net as a machine which generates strings of symbols. We place a token in the input to the machine, and a string of symbols is printed on a piece of tape for us. Eventually a light may come on saying that the machine has halted (i.e., there are no enabled transitions). In normal use, before we can use the printed output, we must look inside the machine and check that a final marking has been reached. If a final marking has not been reached, then we must reject the output and try again. If the Petri net is properly terminating, then we need not look inside the machine; we are guaranteed that a final marking has been reached.

We can see then how a properly terminating Petri net can be used to construct a Petri net generating the indefinite concatenation of its language. We simply connect the final place to the start place. Since the Petri net is known to be empty whenever a token appears at the final place, no tokens will be left in the Petri net to cause spurious transitions when the Petri net is reused.

Unfortunately, this subclass of Petri nets is not very interesting since we can show that all properly terminating Petri nets are finite state machines and vice versa. Hence the Petri net languages of properly terminating Petri nets are regular languages, and it is already known that this class of languages is closed under indefinite concatenation. Thus we see that the properties of systems modeled by Petri net languages are limited to finite repetitions of smaller subsystems or indefinite repetitions of smaller finite subsystems.

6.5.8 Substitution

We have mentioned that systems can be designed and modeled hierarchically by Petri nets. This involves first specifying a rough outline of the system and then successively refining this outline by substituting for operations the definitions of these operations in terms of other operations. With Petri nets, this refinement may take the form of substituting a complete Petri net for a transition or a place. The latter is very straightforward; we therefore restrict our attention to considering the problem of substituting a subnet for a transition (or operation) of a Petri net.

When it is desired to substitute a Petri net for an operation in another Petri net, this can be considered as a composition of the Petri net languages of the two Petri nets. Since the operation is represented by a symbol from Σ, *substitution* of a Petri net language L_2 for a symbol σ in another Petri net language L_1 is defined as the replacement of all occurrences of σ in L_1 by the set of strings L_2. Petri net languages are closed under substitution if the result of a substitution involving a Petri net language is also a Petri net language. Variations on substitution include *finite substitution*, where L_2 must be a finite set of strings, and *homomorphism*, where L_2 is a single string.

Again, unfortunately, we have a negative result. Petri net languages are not closed under general substitution. This is shown immediately by letting $L_1 = c^*$ and substituting $L_2 = a^n b^n$ for c in L_1. The problem is again caused by the possible reuse of a Petri net. For finite substitution and homomorphism, however, we see that L_2 is a regular language, and hence a properly terminating Petri net can be constructed to generate it. This yields the following theorem and corollary.

THEOREM 6-7 If L_2 is a regular language and L_1 is a Petri net language, then the result of substituting L_2 for a symbol in L_1 is a Petri net language.

COROLLARY 6-2 Petri net languages are closed under finite substitution and homomorphism.

6.6 Petri Net Languages and Other Classes of Languages

Having considered the properties of Petri net languages as a class of languages, we turn to investigating the relationship between Petri net languages and other classes of languages. In particular, we consider the classes of regular, context-free, context-sensitive, and type-0 languages.

6.6.1 Regular Languages

One of the simplest and most studied classes of formal languages is the class of regular languages. These languages are generated by regular grammars and finite state machines. They are characterized by regular expressions. Problems of equivalence or inclusion between two regular languages are decidable, and algorithms exist for their solution [Hopcroft and Ullman 1969]. With such a desirable set of properties, it is encouraging that we have the following theorem.

THEOREM 6-8 Every regular language is a Petri net language.

The proof of this theorem is based on the fact that every regular language is generated by some finite state machine and we have shown (Section 3.3.1) that every finite state machine can be converted to an equivalent Petri net.

The converse to this theorem is not true. Figure 6.12 displays a Petri net which generates the context-free language $\{a^n b^n | n > 1\}$. Since this language is not regular, we know that not all Petri net languages are regular languages.

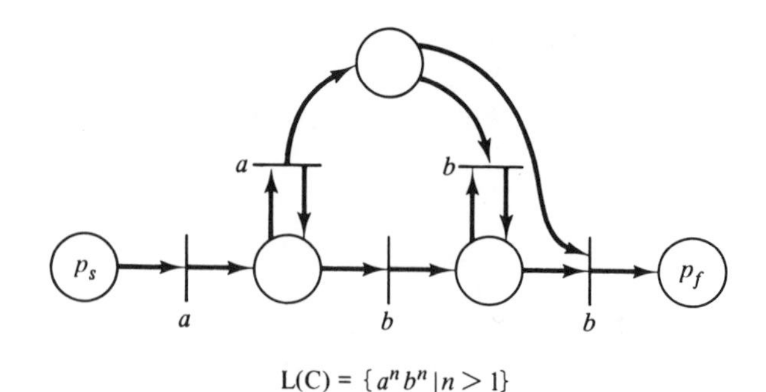

Figure 6.12 A context-free Petri net language which is not a regular language.

6.6.2 Context-Free Languages

Figure 6.12 demonstrated that not all Petri net languages are regular languages by exhibiting a Petri net language which was context-free but not regular. Figure 6.13 shows that not all Petri net languages are context-free by exhibiting a Petri net language which is context-sensitive but not context-free. Unlike the situation with regular languages, however, there also exist context-free languages that are not Petri net languages. An example of such a language is the context-free language $\{ww^R | w \in \Sigma^*\}$. This results in the following theorem.

THEOREM 6-9 There exist context-free languages which are not Petri net languages.

Proof: Assume there exists an n-place, m-transition Petri net which generates the language ww^R. Let k be the number of symbols in Σ, $k > 1$. For an input string xx^R, let $r = |x|$, the length of x. Since there are k^r possible input strings x, the Petri net must have k^r distinct reachable states after r transitions in order to remember the complete string x. If we do not have this many states, then for some strings x_1 and x_2 we have $\delta(\mu, x_1) = \delta(\mu, x_2)$ for $x_1 \neq x_2$. Then we see that

$$\begin{aligned} \delta(\mu, x_1 x_2^R) &= \delta(\delta(\mu, x_1), x_2^R) \\ &= \delta(\delta(\mu, x_2), x_2^R) \\ &= \delta(\mu, x_2 x_2^R) \in F \end{aligned}$$

and the Petri net will incorrectly generate $x_1 x_2^R$.

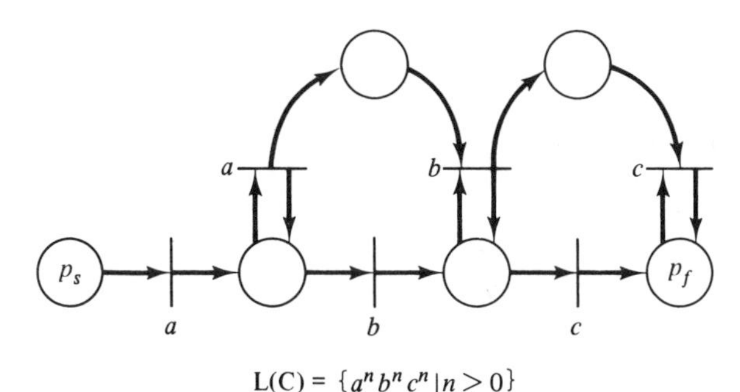

L(C) = $\{a^n b^n c^n | n > 0\}$

Figure 6.13 A context-sensitive but not context-free Petri net language.

In Section 4.2.2, we showed that for each transition t_j there exists a vector $v_j = e[j] \cdot D$ such that if $\delta(q,t_j)$ is defined, then the value of $\delta(q,t_j)$ is $q + v_j$. Thus after r inputs, we have a state q_r where

$$q_r = \mu + \sum_j v_{i_j}$$

for a sequence of transitions $t_{i_1}t_{i_2}\cdots t_{i_r}$. Another way of expressing this sum is as

$$q_r = \mu + \sum_j f_j v_j$$

where f_j is the number of times transition t_j occurs in the sequence $t_{i_1}t_{i_2}\cdots t_{i_r}$ [$f = (f_1,f_2, \ldots,f_m)$ is the firing vector]. We also have the constraint that

$$\sum_j f_j = r$$

At best the vectors $v_1,v_2, \ldots,v_m$ will be linearly independent, and each firing vector $(f_1,f_2, \ldots,f_m)$ will represent a unique state q_r. Since the sum of the coefficients is r, the vector $(f_1,f_2, \ldots,f_m)$ is a *partition* of the integer r into m parts. Knuth [1973] has shown that the number of partitions of an integer r into m parts is

$$\binom{r + m - 1}{m - 1}$$

Now since

$$\binom{r + m - 1}{m - 1} = \frac{(r + m - 1)\cdots(r + 1)}{(m - 1)!} < (r + m)^m$$

there are strictly less than $(r + m)^m$ reachable states in the reachability set after r inputs. For large enough r, we have then that

$$\binom{r + m - 1}{m - 1} < (r + m)^m < k^r$$

and it is impossible for there to be k^r distinct states for each of the k^r possible input strings. Thus, it is impossible for a Petri net to generate the language ww^R. □

The proof that ww^R is not a Petri net language sheds some light on the limitations of Petri nets as automata and hence on the nature of

Petri net languages. Petri nets are not capable of remembering arbitrarily long sequences of arbitrary symbols. From the proof, we see that Petri nets can remember sequences of bounded length (but so can finite state machines). Another feature of Petri nets is the ability to remember the number of occurrences of a symbol, such as in $a^n b^n c^n$, to an extent that regular and context-free generators cannot. However, a Petri net cannot simulate the power of a pushdown stack, which is necessary to generate context-free languages. The rate of growth of the reachable state space for a Petri net is combinatorial with the length of the input and not exponential as needed for a pushdown stack.

The reason that Petri nets are able to generate languages which a pushdown stack cannot generate despite the smaller state space is because of the more flexible interconnections between states in the Petri net compared to the restrictive paths between states allowed in a pushdown stack automaton. This results from restricting the pushdown stack automaton to accessing only the top of the stack, while the Petri net can access any of its counters at any time.

Having shown that not all context-free languages are Petri net languages and not all Petri net languages are context-free, the question arises as to what is the class of languages which are both context-free and Petri net languages? At present we cannot fully answer this question, but we can give an indication of some of the members of this intersection. One subset of both context-free and Petri net languages is, of course, the class of regular languages. Another subset is the set of *bounded context-free languages* investigated by Ginsburg [1966].

6.6.3 Bounded Context-free Languages

A context-free language L is *bounded context-free* if there exist words (or strings) $w_1, w_2, \ldots, w_m$ from Σ^* such that

$$L \subseteq w_1^* w_2^* \cdots w_m^*$$

The adjective "bounded" refers to the finite number of words $w_1, w_2, \ldots, w_m$. Ginsburg has developed a fairly detailed examination of the properties of bounded context-free languages. He mentions that, at the time of his research, there were no known unsolvable questions of interest concerning bounded context-free languages. There were some open questions.

Bounded context-free languages are characterized by the following theorem of Ginsburg [1966, Theorem 5.4].

THEOREM 6-10 The family of bounded context-free languages is the smallest family of sets defined by

1. If W is a finite subset of Σ^*, then W is bounded context-free.
2. If W_1 and W_2 are bounded context-free, then $W_1 \cup W_2$ and $W_1 W_2$ are bounded context-free.
3. If W is bounded context-free and x, $y \in \Sigma^*$, then $\{x^i W y^i | i \geq 0\}$ is bounded context-free.

We have already shown (Section 3.3.1) that every finite state machine and hence every regular language and every finite subset of Σ^* is a Petri net language. In Sections 6.5.1 and 6.5.2 we showed that Petri net languages are closed under concatenation and union. Thus we have only to show that 3 above is satisfied for Petri net languages. To show this, we would construct a Petri net $\gamma' = (C', \Sigma, \mu', F')$ which generates the Petri net language $\{x^i W y^i | i \geq 0\}$ given standard form Petri nets $\gamma_x = (C_x, \Sigma, \mu_x, F_x)$, $\gamma_y = (C_y, \Sigma, \mu_y, F_y)$, and $\gamma_W = (C_W, \Sigma, \mu_W, F_W)$ that generate x, y, and W, respectively. γ' combines the Petri nets γ_x, γ_y, and γ_W and a new place p so that each time γ_x is executed, a token is placed in p. The place p counts the number of times γ_x is executed. After γ_x has executed as many times as it wishes, γ_W is executed, and finally γ_y is executed repeatedly, removing one token from p for each repetition. Since the input string is correctly generated only if the Petri net is empty (except for F' which is defined to be F_y), we are assured that the number of executions of γ_x and γ_y are equal.

This construction, which is demonstrated in Figure 6.14, for $x = ab$, $y = b(a + b)$, and $W = b^+a$, shows that $\{x^i W y^i | i \geq 0\}$ is a Petri net language. Thus any bounded context-free language is a Petri net language.

Are there context-free languages which are also Petri net languages but not bounded? Unfortunately, the answer is yes. Ginsburg shows that the regular expression $(a + b)^+$ is not a bounded context-free language. Since this language is a context-free Petri net language, we see that bounded context-free languages are only a proper subset of the family of context-free Petri net languages. We also exhibit the language $\{(a + b)^+ a^n b^n | n > 1\}$ which is a context-free Petri net language but is neither bounded nor regular. Further research is needed to completely characterize the set of context-free Petri net languages.

The presence of both regular sets and bounded context-free languages as subsets of the class of Petri net languages is encouraging since both classes of languages have several desirable properties and some nice analysis characteristics.

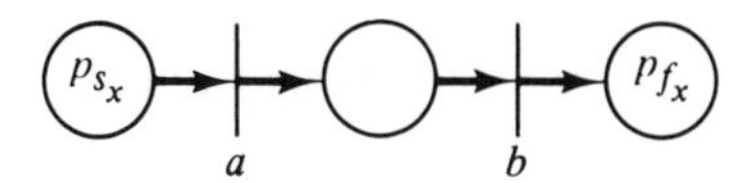

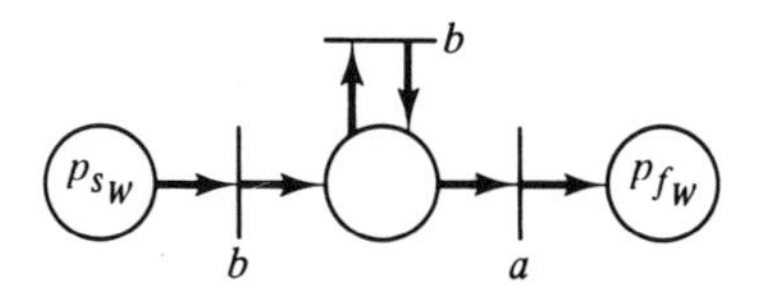

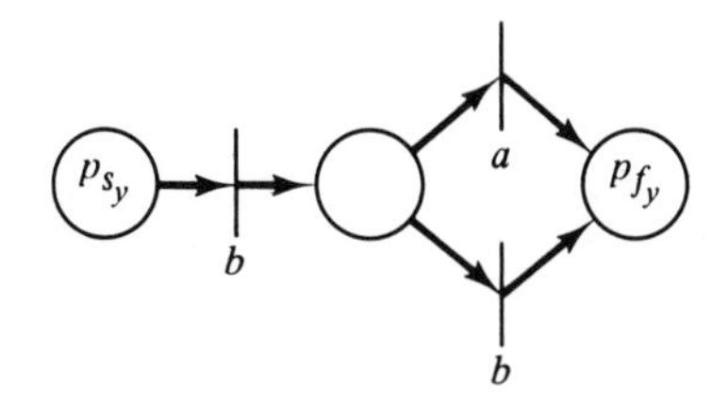

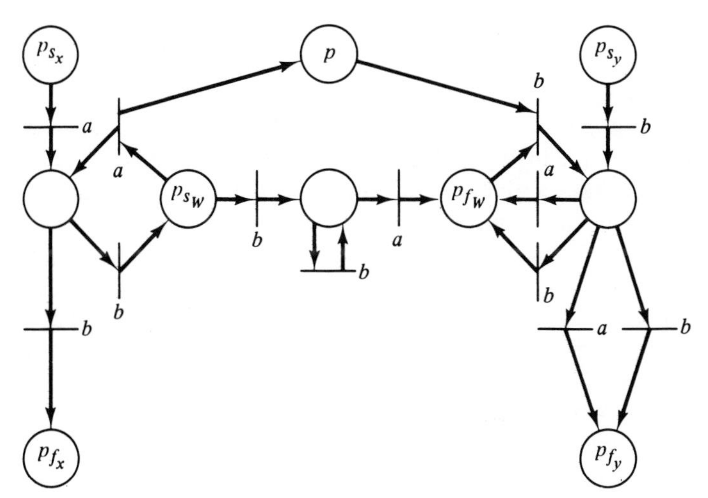

Figure 6.14 A Petri net to generate the language $\{x^iWy^i | i \geq 0\}$. This construction shows that all bounded context-free languages are also Petri net languages.

6.6.4 Context-sensitive Languages

We have investigated the relationship between Petri net languages and regular languages and between Petri net languages and context-free languages. Now we turn to context-sensitive languages. Figure 6.13 has shown that some Petri net languages are context-sensitive; below we prove that all Petri net languages are context-sensitive. Since we know that all context-free languages are also context-sensitive and that there exist context-free languages which are not Petri net languages, we know that there exist context-sensitive languages which are not Petri net languages. Thus the inclusion is proper.

THEOREM 6-11 All Petri net languages are context-sensitive languages.

The proof that all Petri net languages are context-sensitive is rather complex. There are two ways to show that a language is context-sensitive: Construct a context-sensitive grammar which generates it, or specify a nondeterministic linear bounded automaton which generates it. Peterson [1973] has shown how to define a context-sensitive grammar to generate a Petri net language; here we indicate why a linear bounded automaton can generate a Petri net language.

A linear bounded automaton is similar to a Turing machine. It has a finite state control, a read/write head, and a (two-way infinite) tape. The limiting feature which distinguishes it from a Turing machine is that the amount of tape which can be used by a linear bounded automaton to generate a given input string is bounded by a linear function of the length of the input string. In this sense it is similar to the pushdown stack automaton used to generate context-free languages (since the maximum length of the stack is bounded by a linear function of the input string) except that the linear bounded automaton has random access to its memory, while the pushdown automaton has access to only one end of its memory.

To generate a Petri net language, one can simulate the Petri net by remembering, after each input, the number of tokens in each place. How fast can the number of tokens grow as a function of the input? Consider the number of tokens after r transition firings. This number, denoted by c, is, for a transition sequence $t_{i_1} t_{i_2} \cdots t_{i_r}$,

$$c = 1 + \sum_j |O(t_{i_j})| - |I(t_{i_j})|$$

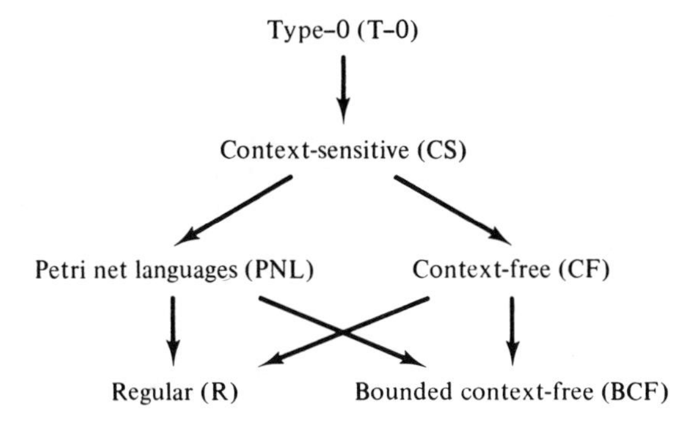

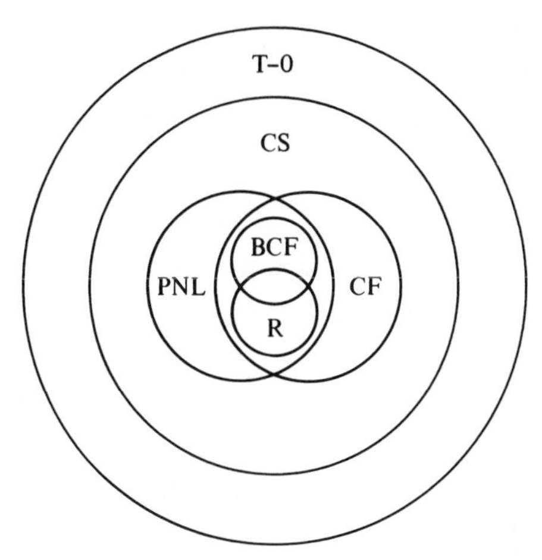

Figure 6.15 Summary illustration of the relationship of Petri net languages to the traditional classes of phrase structure languages.

Since the numbers $\#(p_k, O(t_j))$ and $\#(p_k, I(t_j))$ and hence $|O(t_j)|$ and $|I(t_j)|$ (the cardinality of the output and input bags) are fixed by the structure of the Petri net, there is a maximum value l for

$$l = \max_{t_j \in T} |O(t_j)| - |I(t_j)|$$

and thus

$$\begin{aligned} c &= 1 + \sum_j |O(t_{i_j})| - |I(t_{i_j})| \\ &\leqslant 1 + r \cdot l \end{aligned}$$

Closure under	L^f	L	L^λ	P^f	P	P^λ
Union	No	Yes	Yes	?	Yes	Yes
Intersection	Yes	Yes	Yes	Yes	Yes	Yes
Concatentation	?	Yes	Yes	?	Yes	Yes
Concurrency	?	Yes	Yes	?	Yes	Yes
Regular substitution	No	λ-free	Yes	?	Prefix	Prefix
Inverse homomorphism	Yes	Yes	Yes	Yes	Yes	Yes
Kleene star (iteration)	?	No	No	?	?	?
Complement	No	?	No	No	?	No
Contains regular languages	No	Yes	Yes	No	Prefix	Prefix
Symmetric difference with context-free languages (non-empty)	Yes	Yes	Yes	Yes	Yes	Yes
Is context-sensitive	Yes	Yes	?	Yes	Yes	?

Figure 6.16 Summary of properties of some Petri net languages (from Peterson [1973] and [Hack 1975b]).

The number of tokens, and hence the amount of memory needed to remember them, is bounded by a linear function of the input length. Hence, Petri net languages can be generated by a linear bounded automaton.

With this proof, we have shown that all Petri net languages are context-sensitive languages. We summarize the results of our investigation into the relationship between the class of Petri net languages and other classes of languages in the graph and Venn diagram of Figure 6.15.

Figure 6.16 summarizes the properties of Petri net languages. It is drawn from [Peterson 1976] and [Hack 1975b].

6.7 Additional Results

Most of the results presented here have been developed in both [Peterson 1976] and [Hack 1975b]. In addition, Hack has investigated a number of decidability problems for Petri net languages. The membership problem [Is a string α an element of a language $L(\gamma)$?] is decidable, while the emptiness problem [Is the language $L(\gamma)$ empty?] can be easily seen to be equivalent to the reachability problem. It is undecidable if two Petri net languages are equal or if one is contained within the other (the equivalence and inclusion problems).

Figure 6.17 summarizes these results.

Decision Problems	L	L	L^λ	P^f	P	P^λ
Membership	D	D	?	D	D	D
Emptiness	?	?	?	D	D	D
Finiteness	?	?	?	D	D	D
Equivalence and inclusion	?	U	U	?	U	U

Figure 6.17 Summary table of the decision properties of the L-type and P-type Petri net languages (D is decidable; U is undecidable).

A different approach to the study of Petri nets by the use of formal language theory has been considered by Crespi-Reghizzi and Mandrioli [1974]. They noticed the similarity between the firing of a transition and the application of a production in a derivation, thinking of places as nonterminals and tokens as separate instances of the nonterminals. The major difference is, of course, the lack of ordering information in the Petri net which is contained in the sentential form of a derivation. This resulted in the definition of the *commutative grammars*, which are isomorphic to Petri nets. In addition Crespi-Reghizzi and Mandrioli considered the relationship of Petri nets to matrix [Abraham 1970], scattered context, nonterminal bounded, derivation bounded, equal-matrix, and Szilard languages. For example, it is not difficult to see that the class L^f is the set of Szilard languages of matrix context-free languages [Crespi-Reghizzi and Mandrioli 1977; Hopner and Opp 1977].

6.8 Further Reading

There are three good studies on Petri net languages. [Peterson 1976] or [Peterson 1977] is probably the easiest to get, although [Hack 1975b] is a more rigorous and complete investigation. [Crespi-Reghizzi and Mandrioli 1974] is rather hard to find, but nonetheless it is an excellent piece of work, quite inventive and well explained.

6.9 Topics for Further Study

Although a good start on research into the properties of Petri nets has been made, much still remains to be done. Of the classes of languages which have been defined, only two, the classes of P-type and L-type languages, have been studied and these only for generalized Petri nets. Several subsets of the set of general Petri nets have been defined, including marked graphs, conflict-free nets, restricted nets, and free-choice nets, as we see in Chapter 7. Each of these classes of Petri nets presumably has its own class of languages with its own distinctive properties. Some investigation of these classes has been done. It is known [Hack 1975b] that the classes L, L^λ, G, G^λ, P, and P^λ for restricted Petri nets [no self-loops, no multiple arcs; i.e., all bags are sets, and for every t_j, $I(t_j) \cap O(t_j) = \emptyset$] are identical to the corresponding classes for generalized Petri nets. It is also not difficult to see that the classes L^λ, G^λ, and P^λ are not changed by restricting the nets to free-choice nets (see Section 7.4.3). This still leaves many interesting cases to be studied. In particular, the languages generated by marked graphs, or by conflict-free nets in general, seem to have a structure reminiscent of deterministic context-free languages, and their study should be very promising.

Another important open problem concerns the distinction between λ-free languages (L, P, ...) and unrestricted languages (L^λ, P^λ, ...). It is not known whether $L = L^\lambda$ or not, for example.

1. Pick a class of Petri net languages other than the L-type languages and develop its theory.
2. Develop a complete set of interrelations between the 12 classes of Petri net languages. Specifically, for each pair of Petri net language classes, either show that one class is contained in another, or exhibit a language which is in one class and not in the other.
3. Consider the possibility of associating labels with the *places* of a Petri net, not the transitions. This was the approach used in [Gostelow 1971]. Determine the feasibility of this approach, and if feasible, develop a new theory of Petri net place languages.
4. Develop the theory of Petri net languages where Petri nets are considered to be recognizers, not generators, of the language.

7

Extended and Restricted Petri Net Models

In Chapter 3 we demonstrated that Petri nets can be used to model a wide variety of systems: computer hardware and software, chemical, social, and so on. However, this large set of examples only shows that Petri nets can model some systems; there may be systems which cannot be properly modeled as Petri nets. That is, there may be limits on the modeling power of Petri nets.

In addition, in Chapter 5 we have shown that not all analysis questions are decidable for Petri nets. The inclusion and equivalence problems for Petri net reachability sets and for Petri net languages are undecidable, but these problems may be very important for producing optimal Petri nets. Even those analysis questions which are decidable are very difficult, meaning that they require large amounts of computation.

In this chapter, we investigate the suggestions which have been made to overcome these two limitations on Petri nets: limitations on Petri net *modeling power* and limitations on Petri net *decision power*. We look first at some of the extensions to the Petri net model which have been suggested. Extending the Petri net model should increase the modeling power of Petri nets but may also decrease the decision power. The effects of any extension on the decision power of the extended model must be carefully considered.

Having then seen that extending the Petri net model may tend to reduce decision power, we also consider how decision power can be increased by restricting the Petri net model. Various subclasses of the Petri net model have been suggested. These models are produced by restrictions on the structure of the Petri net. We must examine the effect of these restrictions on both modeling power and decision power.

This investigation shows how modeling and decision power can be traded, one for the other, and also indicates the limits of both for the Petri net model.

7.1 Limitations of Modeling with Petri Nets

Several researchers who have used Petri nets to model systems have found them too simple and limited to easily model real systems. Thus, there has been a marked tendency to extend the model to make it easier to use. These extensions have been of several types.

Patil [1970a] suggested extension of Petri nets to include *constraints.* A constraint is a set of places. The firing rule is modified to allow a transition to fire if and only if the resulting marking does not have all of the places which are in a constraint simultaneously marked. For example, if $\{p_1, p_4\}$ is a constraint set, then either p_1 or p_4 must be empty at all times; if p_1 is marked, then a token cannot be put in p_4 until all tokens in p_1 are removed and vice versa.

Noe, in his model of the CDC 6400 operating system [Noe 1971], introduced a different extension: the *exclusive-OR transition* (Figure 7.1). Normally, a Petri net transition fires when all of its inputs have tokens; this is called *AND logic*, since we must have tokens in the first input *and* tokens in the second input *and* tokens in the third input and so on. The exclusive-OR transition can fire if and only if exactly one of its inputs has tokens and all the others have zero tokens. Thus, the enabling rule is that the first input have a token or the second input have a token (but not both). When the transition fires, it removes a token only from the input with tokens.

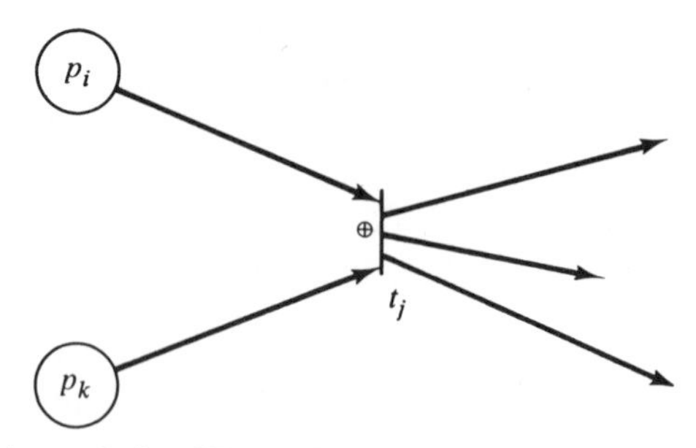

Figure 7.1 An exclusive-OR transition. Transition t_j can fire only if either p_i or p_k has a token and the other does not.

A similar extension was used by Baer in his model of a compiler [Baer 1973b]. Baer introduced *switches* (Figure 7.2). A switch is a special transition with a special input called the switch input and exactly two outputs (one labeled *e* for empty and the other labeled *f* for full). A switch transition fires when it is enabled (ignoring the state of the special switch input). When it fires a token is put in the output labeled *e* if the switch input is empty or a token is put in the output labeled *f* if the switch input is full. Thus, firing a switch transition will result in either one of two markings, depending on the state of the switch. A token is removed from the switch input if it had one, so that the switch input is empty after the switch transition is fired.

These extensions to Petri nets were created for solving specific problems which the researchers encountered in their attempts to model real systems. However, the emphasis of their work was modeling, not the theoretical power of Petri nets, so no attempt was made to show that these extensions were either necessary or sufficient to handle general modeling problems. In fact, in all of these cases, the nets involved were safe, and hence the reachability set is finite, meaning that these

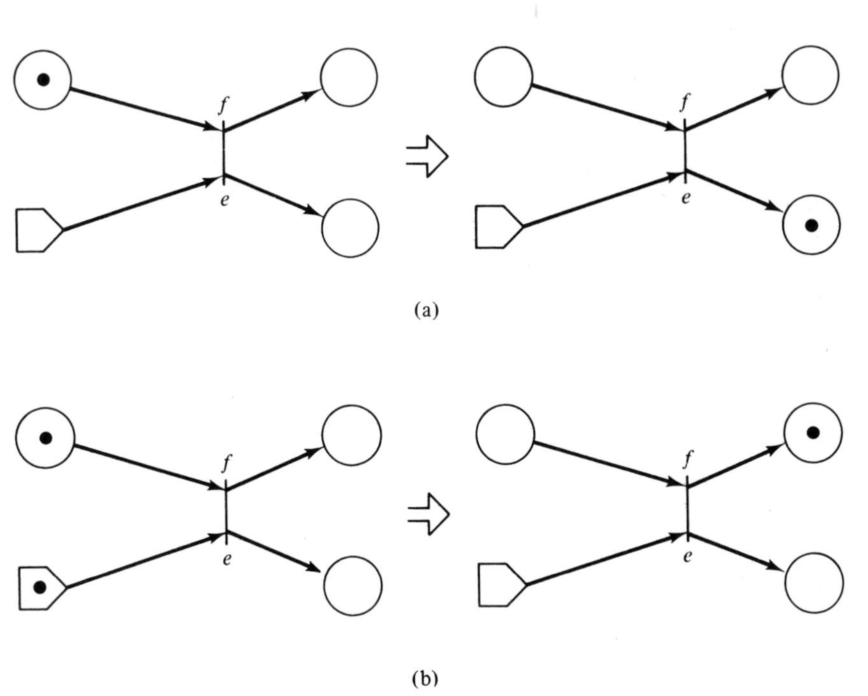

Figure 7.2 Firing a switch transition. The box-shaped place is the switch input. (a) Empty switch. (b) Full switch.

nets could be represented by finite state machines, which we have seen (Section 3.3.1) can be easily represented as ordinary Petri nets. Thus, in these cases, these "extensions" were not necessary, merely convenient. We have also seen, in Section 5.3, that many "extensions" to the restricted Petri net model, such as multiple arcs and self-loops, are not in fact extensions but merely convenient.

So the question remains as to what, if any, are the limitations of Petri nets. The answer to this question was found as the result of asking similar questions about Dijkstra's P and V operations on semaphores.

Dijkstra defined his P and V operations on semaphores to facilitate synchronization and communication in systems of cooperating processes [Dijkstra 1968]. A semaphore can be thought of as an integer variable which only takes on nonnegative values. A V operation on a semaphore S increases the value of the semaphore by one, $S = S + 1$. A P operation, on the other hand, decrements S by one as long as the result is nonnegative; the process must wait until S can be decremented before it continues. The relationship between semaphores and Petri nets was examined in Section 3.4.8.

Since P and V operations were proposed as the mechanism for solving all synchronization problems between programs, questions naturally arose as to their *completeness*, that is, their ability to solve all possible coordination problems. Patil [1971] set out to prove that P and V operations are not powerful enough to solve all coordination problems. His approach was quite simple: He created a synchronization problem which cannot be solved with P and V operations. The problem posed was the *cigarette smokers' problem.*

The cigarette smokers' problem consists of (at least) four processes: an agent and three smokers. Each smoker continuously makes a cigarette and smokes it. But to make a cigarette, three ingredients are needed: tobacco, paper, and matches. One of the processes has paper, another tobacco and the third has matches. The agent has an infinite supply of all three. The agent places two of the ingredients on the table. The smoker who has the remaining ingredient can then make and smoke a cigarette, signaling the agent upon completion. The agent then puts out another two of the three ingredients and the cycle repeats.

In terms of semaphores, the problem is posed by associating a semaphore with each ingredient. These semaphores are initially zero. The agent will increment two of the three semaphores by V operations, and then wait on a *done* semaphore. The appropriate smoker process must decrement the two semaphores (by P operations), make and smoke the cigarette, and increment the *done* semaphore. The problem is to define the code for the smoker processes to determine which of the three processes should proceed; the agent is fixed and cannot be changed.

Figure 7.3 illustrates the obvious "solution." The problem with this "solution" is quite simple: Suppose the agent puts out tobacco and paper $[V(t); V(p))]$. Then the smoker with paper may grab the tobacco $[P(t)]$ and the smoker with tobacco may grab the paper $[P(p)]$, resulting in a deadlock.

Patil proved that *no* sequence of P and V operations can correctly solve this problem. This was shown by proving that all P and V "solutions" can be modeled as Petri nets of a certain kind (each transition has at most two inputs) but that the solution is a Petri net of another kind, and there is no way to convert a net of one kind into a net of the other kind without introducing the danger of deadlock.

There were some problems with Patil's solution — specifically dealing with arrays of semaphores (see [Parnas 1972]) — but the concept is correct. Kosaraju [1973] and Agerwala and Flynn [1973] followed up on Patil's work to produce a problem which cannot be solved by P and V operations or by Petri nets. This same limitation had been earlier discovered by Keller [1972].

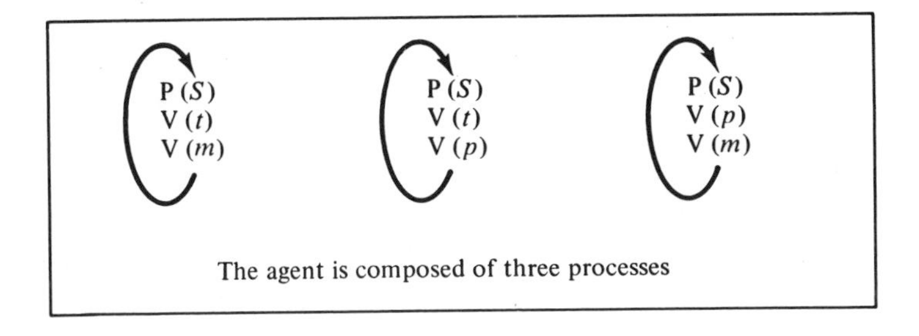

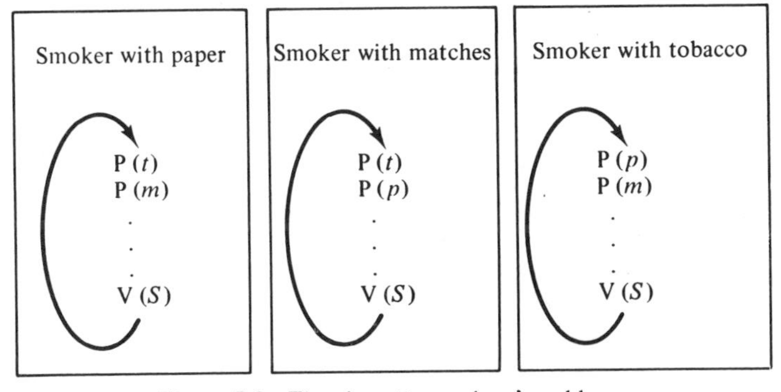

Figure 7.3 The cigarette smokers' problem.

The problem posed by Kosaraju and Agerwala and Flynn is quite real. Assume that we have two producer processes and two consumer processes. One producer process P_1 creates items for the first consumer process C_1, and the other producer P_2 produces for the other consumer C_2. Items which are produced but not yet consumed are placed in a buffer, buffer B_1 for (P_1,C_1) and B_2 for (P_2,C_2). The transmission from the buffers to the consumers is over a shared channel. The channel can only transmit one item at a time but can transmit from either buffer to either consumer. The producers merely put items into the buffers; the consumers must coordinate the use of the channel between themselves. The controlling consumer tells the channel to fetch an item from the appropriate buffer. This is sketched in Figure 7.4.

The major problem with this system is the allocation of the channel. The producer/consumer pair (P_1,C_1) are required to have *priority* over (P_2,C_2) for use of the channel. Specifically, the channel is never to transmit items from buffer B_2 to consumer C_2 as long as buffer B_1 is nonempty. This priority rule prevents this system from being modeled as a Petri net.

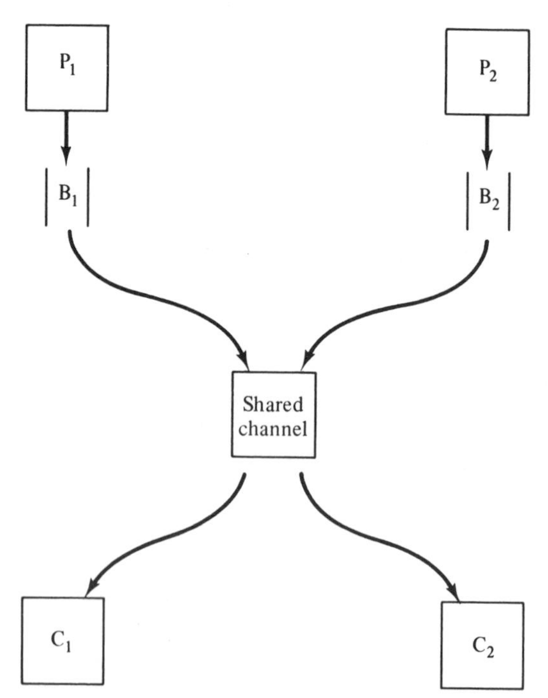

Figure 7.4 The buffered producer/consumer with a shared channel.

The proof is relatively straightforward. Assume that we are in a state μ with items in both buffer B_1 and buffer B_2. Now, if producer P_1 takes a break, then all items from buffer B_1 will eventually be transmitted to consumer C_1, and buffer B_1 will be empty. This will allow an item from buffer B_2 to be moved to consumer C_2. Thus there exists a path from μ to a state μ' in which consumer C_2 can use the channel. Now, if instead the producer P_1 produces an extra k items, then we will be in state $\mu + k$ rather than μ. But because of the permissive nature of Petri net firings, the sequence of firings which took us from μ to μ' will still be enabled and will take us from $\mu + k$ to $\mu' + k$. And since consumer C_2 could use the channel in μ' and Petri nets are permissive, consumer C_2 can still use the channel, despite the presence of k items in buffer B_1. Thus, the permissive nature of Petri net firings does not allow this priority system to be modeled correctly.

More specifically, the limitation on Petri net modeling is precisely the inability to test for exactly a specific marking in an unbounded place and take action on the outcome of the test. This is commonly stated as an inability to test for a zero marking in a place, and so this property is known as *zero testing* [Keller 1972]. Petri nets cannot test an unbounded place for zero. [If the place is bounded, zero can be tested. For a bounded place p_i with bound k, we can create a complement place p_i' such that the sum $\mu(p_i) + \mu(p_i')$ is constant at k for all reachable markings. This allows us to test if $\mu(p_i)$ is zero by testing if $\mu(p_i')$ is k (see Section 5.6).]

Exercises

1. Draw a Petri net model of the "solution" to the cigarette smokers' problem of Figure 7.3. Show that this net is not live. Create a Petri net which can solve the cigarette smokers' problem. How does this second net differ from the first?

7.2 Extensions

How does this limitation of Petri net modeling power relate to the extensions of Petri nets which have been suggested? All of the suggested extensions are directed at creating an ability in Petri nets for zero testing.

The simplest extension to Petri nets which allows zero testing is *inhibitor* arcs. An inhibitor arc is illustrated in Figure 7.5. An inhibitor arc from a place p_i to a transition t_j has a small circle rather than an

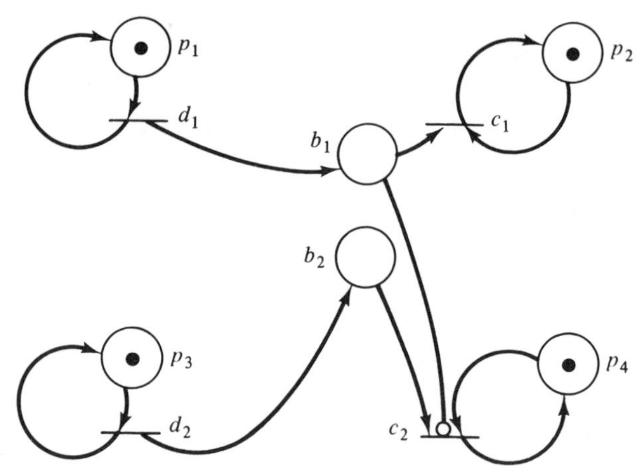

Figure 7.5 An extended Petri net with an inhibitor arc.

arrowhead at the transition. This notation is borrowed from switching theory where the small circle means "not." The firing rule is changed as follows: A transition is enabled when tokens are in all of its (normal) inputs and zero tokens are in all of its inhibitor inputs. The transition fires by removing tokens from all of its (normal) inputs.

Thus in the extended Petri net of Figure 7.5, transition C_2 can fire only if there is a token in b_2 and p_4 and zero tokens in b_1. This net is a solution to the priority shared channel problem which Kosaraju defined to show the limitations of Petri nets.

Petri nets with inhibitor arcs are intuitively the most direct approach to increasing the modeling power of Petri nets. It is also the case that all other extensions to Petri nets which have been suggested either are not true extensions (that is, they are in fact equivalent to ordinary Petri nets) or are equivalent to Petri nets with inhibitor arcs. We discuss several suggested extensions below to illustrate this point.

7.2.1 Constraints

Constraints were proposed by Patil [1970a] to improve the modeling power of Petri nets. In Patil's context, constraints were only meant to make the modeling easier but not to increase modeling power, because all the places in Patil's work were bounded. However, the definition of constraints is not limited to bounded Petri nets, and for the more general class of Petri nets, they are equivalent to inhibitor arcs.

To show the equivalence of constraints and inhibitor arcs, assume we have a Petri net $C = (P,T,I,O)$ with a constraint $Q \subseteq P$. We must guarantee that all places in Q are not marked in any reachable marking. The only way this could happen would be for a transition t_j to fire putting tokens in those places of the constraint which were not marked before the transition fired. Thus, for each transition t_j with output places which are in the constraint, we must guarantee that at least one of the members of the constraint will not be marked after the transition fires. To assure this, we create a new transition $t_{j,k}$ for every place p_k in the constraint Q but not in $O(t_j)$. The transition $t_{j,k}$ is identical to t_j except that it also has an inhibitor arc from p_k to $t_{j,k}$. The effect of firing $t_{j,k}$ is the same as firing t_j, and if t_j can be fired without violating a constraint, then at least one of the $t_{j,k}$ can fire.

As an example of this construction, consider the Petri net of Figure 7.6. If we impose the constraint $\{p_3,p_7\}$ (that is, no marking should have p_3 and p_7 simultaneously marked), then the equivalent Petri net with inhibitor arcs is shown in Figure 7.7.

The transformation from inhibitor arcs to constraints is somewhat more difficult. We cannot simply insist that no output of a transition can be marked at the same time as an inhibitor input, since tokens can

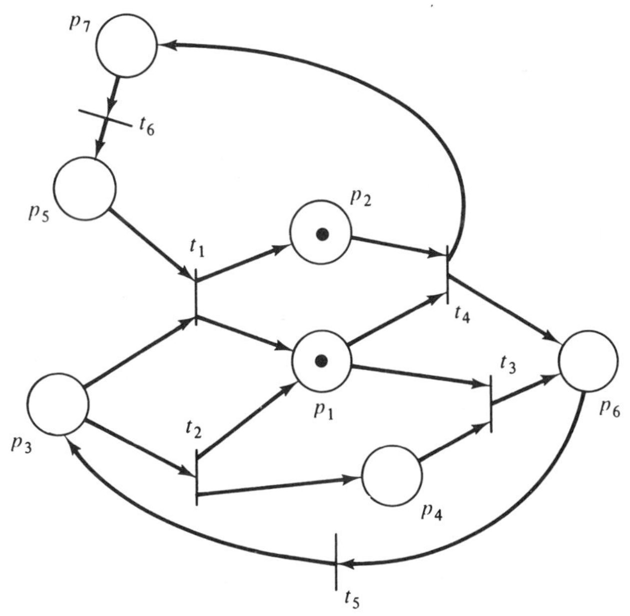

Figure 7.6 A Petri net with a constraint $\{p_3,p_7\}$. The constraint means that tokens are not allowed in both p_3 and p_7 in any reachable marking.

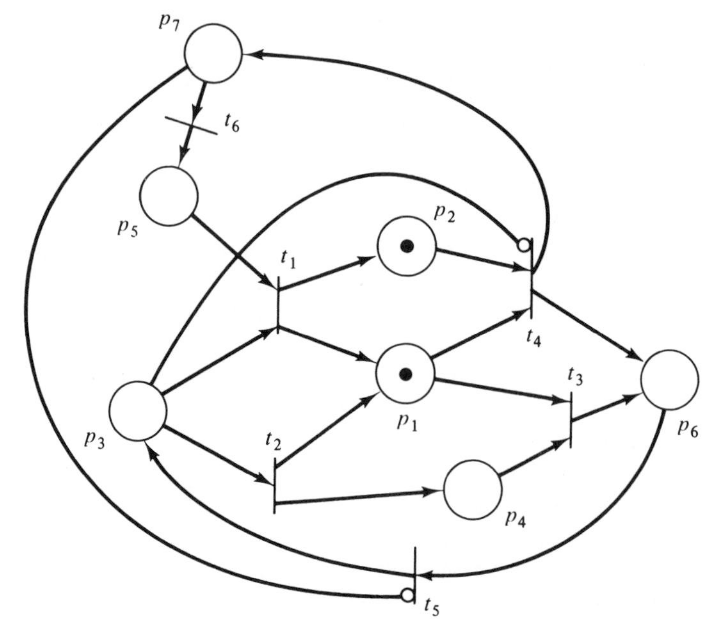

Figure 7.7 A Petri net with inhibitor arcs corresponding to the Petri net with constraints of Figure 7.6. The inhibitor arcs assure that tokens cannot simultaneously be in places p_3 and p_7.

be placed in the output by other transitions. We must concentrate on the transition. This requires splitting each transition t_j into two transitions t_j' and t_j'' and a place p_j'. We define $I(t_j') = I(t_j)$ (without the inhibitor arcs) and $O(t_j'') = O(t_j)$. The place p_j' represents the firing of t_j, so $O(t_j') = \{p_j'\} = I(t_j'')$. This is illustrated in Figure 7.8. Now, for each place p_i which is an inhibitor input to t_j, we define a constraint $\{p_i, t_j'\}$. This assures that the transition cannot fire if the marking of p_i is nonzero.

7.2.2 Exclusive-OR Transitions and Switches

An *exclusive-OR transition* t_j with input $I(t_j)$ requires that one and only one of its inputs be nonzero to enable the transition. This is equivalent to a set of transitions, one for each element in $I(t_j)$. Each transition has one (normal) input, and the rest of the inputs are inhibitor arcs. Figure 7.9 gives an example.

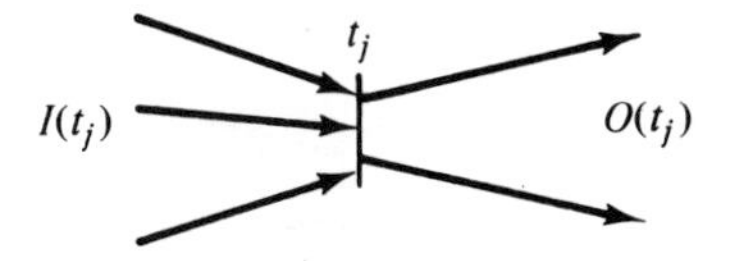

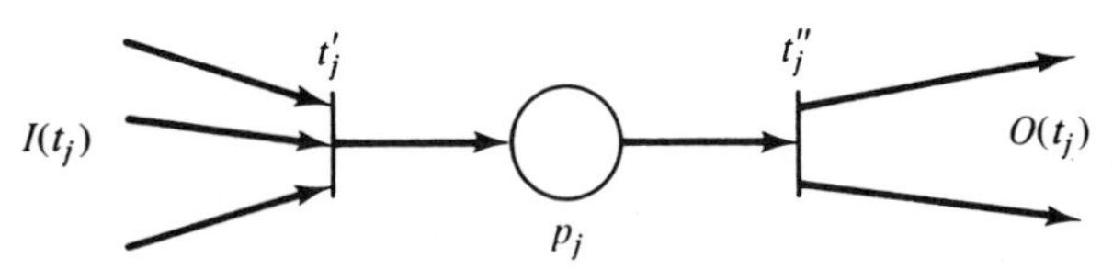

Figure 7.8 Converting a transition into a start and end transition with a place representing the transition firing.

Switches can also be easily transformed into inhibitor arcs. See Figure 7.10.

It is not immediately obvious how inhibitor arcs can be converted into either switches or exclusive-OR transitions, but it is certainly the case that they can be.

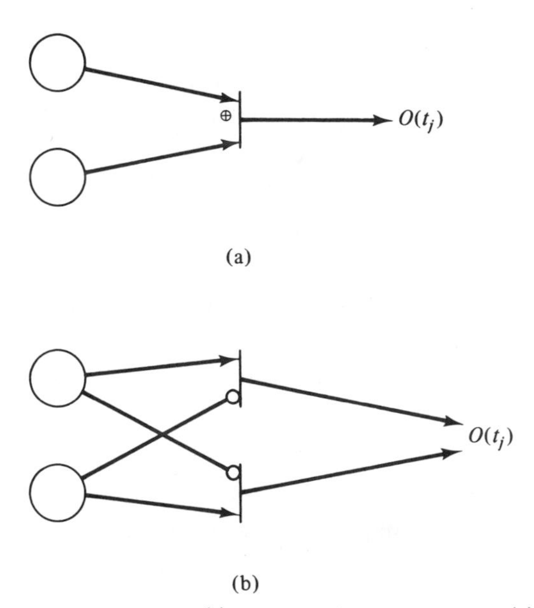

Figure 7.9 Converting from (a) exclusive-OR transitions to (b) inhibitor arcs.

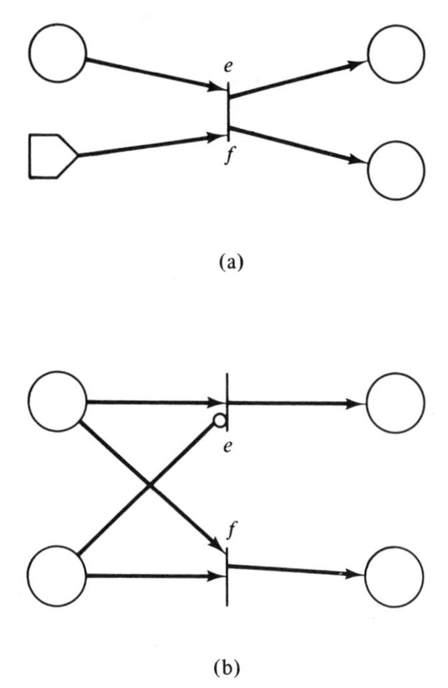

Figure 7.10 Converting (a) switches into Petri nets with (b) inhibitor arcs.

7.2.3 Other Extensions

Two other major extensions to Petri nets have been suggested. Priorities can be associated with the transition such that if t_i and t_k are both enabled, then the transition with the highest priority will fire first [Hack 1975c]. Time Petri nets [Merlin 1974] associate with each transition t_j two times $\tau_{1,j}$ and $\tau_{2,j}$. A transition t_j can fire only if it has been enabled for at least time $\tau_{1,j}$, and it *must* fire before $\tau_{2,j}$ if it is enabled. Both of these extensions can be used to test for zero.

In the case of priorities, we can easily test if a place p_i is zero. This is shown in Figure 7.11. If we put a token into place $p_i = 0$? and define the priority of transition t_1 to be higher than the priority of transition t_2, then we will get a token in one of the two places at the right depending on the marking of place p_i. This results from the fact that transition t_1 can fire only if it is enabled, and it is enabled only if place p_i has a token. If t_1 cannot fire because p_i is empty, then, and only then, will transition t_2 fire.

Hack has shown complete constructions for converting Petri nets with priorities to Petri nets with inhibitor arcs and vice versa [Hack 1975c]. Time Petri nets can also test a place for zero by simulating

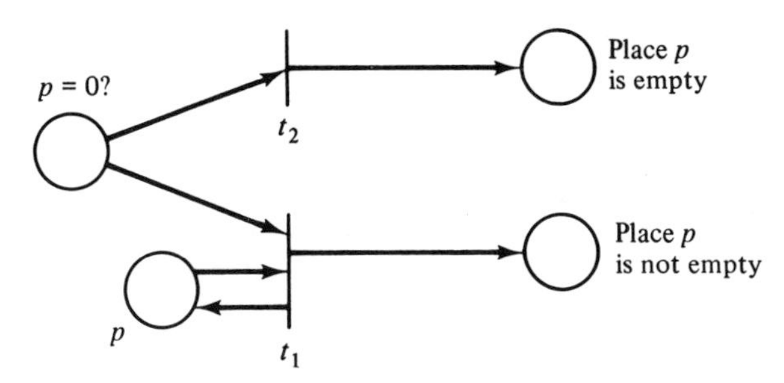

Figure 7.11 Using priorities to test if the marking of a place p is zero or nonzero. Transition t_1 has priority over transition t_2.

priorities. If we have two transitions t_j and t_k and we set $\tau_{2,j} < \tau_{1,k}$, then transition t_j has priority over transition t_k since t_j *must* fire (if it is enabled) before t_k would be allowed to fire.

7.3 Extended Petri Nets and Register Machines

We have shown that the suggested extensions to the Petri net model all allow a place to be tested for zero. How is this important to the decision power of a Petri net? Does it affect our ability to analyze Petri nets?

Zero testing decreases the decision power of Petri nets. Agerwala [1974a], Hack [1975c], Thomas [1976], and others have shown that adding the ability to test for zero to a Petri net model allows a Petri net to simulate a Turing machine. Thus a Petri net with zero testing produces a modeling scheme which can model any system. Also, however, almost all analysis questions for Petri nets become undecidable, since they are undecidable for Turing machines.

The proof of the equivalence of extended Petri nets and Turing machines is relatively simple. It is easiest to give in terms of the register machines of Shepardson and Sturgis [1963] or the program machines of Minsky [1967].

A register machine is a computer-like machine with a number of registers which are used to store arbitrarily large numbers. A program is written to manipulate the registers. The program is a sequence of instructions such as "increase register n by 1," "decrease register n by 1 (only if register n is not 0)," "jump to statement s if register n is not zero," and so on. For example, the following is a program to add the contents of register 2 to register 1.

1. If register 2 is zero, go to instruction 5.
2. Subtract 1 from register 2.
3. Add 1 to register 1.
4. Go to instruction 1.
5. Halt.

Shepardson and Sturgis have shown that a register machine with the following instructions is equivalent to a Turing machine.

1. $P(n)$: Increase register n by 1.
2. $D(n)$: Decrease register n by 1 (register n not zero).
3. $J(n)[s]$: Jump to statement s if register n is zero.

Thus, if a register machine can be converted into an equivalent extended Petri net, we see that extended Petri nets are equivalent to register machines. This conversion is relatively straightforward.

To represent a register machine as an extended Petri net, we represent the n registers used in a program by n places, $p_1', p_2', \ldots, p_n'$. We also use $s + 1$ places to represent the position of the program counter either before statement 1 (the initial marking) or after statement i, for $i = 1, \ldots, s$, in a program with s statements. Each instruction in the program is represented by a transition. Figure 7.12 shows how each of the three instructions above would be represented as a transition in an extended Petri net. This shows that a register machine can be converted into an extended Petri net and therefore that an extended Petri net is equivalent to a Turing machine. This equivalence to Turing machines destroys any hope of being able to analyze extended Petri nets. However, it does prove that extended Petri nets can model any system (or at least any computable system). Thus we see that an increase in modeling power, in this case, results in a definite decrease in decision power.

Also notice that the key point in the proof of equivalence of extended Petri nets and register machines and Turing machines is the ability to test a single place for zero. Thus, all of the extensions which have been suggested — constraints, exclusive-OR transitions, switches, priorities, timings, and inhibitor arcs — extend the Petri net model to the level of Turing machines.

There have been other suggestions for extensions which do not

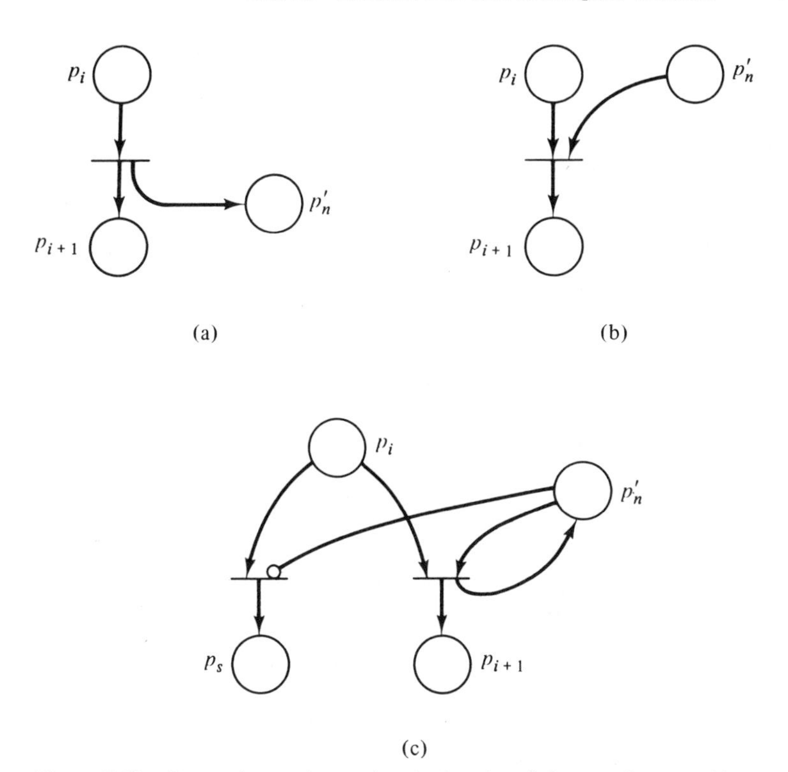

Figure 7.12 Converting an instruction (at location i) for a register machine to a transition in an extended Petri net using inhibitor arcs. (a) $P(n)$: Increase register n by 1. (b) $D(n)$: Decrease register n by 1 (register n must be nonzero). (c) $J(n)[s]$: Jump to statement s if register n is zero.

raise Petri nets to the level of Turing machines. Self-loops and multiple input and output arcs were first suggested as extensions but, as seen in Section 5.3, are in fact equivalent to restricted Petri nets. Similarly, allowing inclusive-OR inputs, inclusive-OR outputs, or exclusive-OR outputs would not increase the modeling power of Petri nets.

In general, it seems that any extension which does not allow zero testing will not actually increase the modeling power (or decrease the decision power) of Petri nets but merely result in another equivalent formulation of the basic Petri net model. (Modeling *convenience* may be increased.) At the same time, any extension which does allow zero testing will increase the modeling power to the level of Turing machines and decrease decision power to zero. Thus Petri net extensions would seem to have few practical advantages for analysis.

7.4 Subclasses of Petri Nets

The objective of extending Petri nets is to increase their modeling power; an unfortunate side effect is that the decision power of extended Petri nets is greatly reduced. The decision power of normal Petri nets is of questionable value anyway due to its complexity and expense. (Remember the results of Section 5.8 on the complexity of the reachability and boundedness problems.) This has resulted in some investigations of subclasses of Petri nets. The objective of these studies is to determine reasonable structural restrictions on Petri nets which will increase the decision power of the restricted Petri net models while not overly restricting modeling power.

Much can be done with Petri net subclasses. The goal of this part of Petri net research is quite simple: Define a subclass of Petri nets which can model a large class of systems (all or almost all interesting ones) and yet still has simple analysis procedures (at least for the interesting problems). It is also necessary that there exist a simple test for determining if a Petri net is a member of the defined subclass. The subclasses which have been defined are all *syntactic* or *structural* subclasses; one can easily examine a Petri net structure to determine if it is a member of the specified subclass. This is in distinction to subclasses which might be defined according to dynamic properties, such as, for example, persistent Petri nets [Landweber and Robertson 1975] or bounded Petri nets. These subclasses might have very nice properties, but it may be quite difficult to determine if an arbitrary given Petri net is a persistent or bounded Petri net.

Only two major subclasses of the Petri net model have been widely studied: *state machines* and *marked graphs*. In addition, Hack [1972] has studied a subclass called *free-choice Petri nets* and has suggested that another subclass, *simple Petri nets*, might have good decision properties. We present each of these classes and indicate their major properties, advantages, and disadvantages.

7.4.1 State Machines

A state machine is a Petri net in which each transition is restricted to having exactly one input and one output.

DEFINITION 7-1 A *state machine* is a Petri net $C = (P, T, I, O)$ such that for all $t_j \in T$, $|I(t_j)| = 1$ and $|O(t_j)| = 1$.

Several properties of state machines are immediately obvious. First, a state machine is strictly conservative. This means that the number of tokens in the state machine never changes, resulting in a finite system. This implies that the reachability tree for a state machine is finite, and hence all analysis questions are decidable for state machines. In fact, a state machine is equivalent to the finite state machine automata of automata and formal language theory (see Section 3.3.1). Thus, these models are of limited interest despite their high decision power, due to the limited modeling power of finite state machines.

7.4.2 Marked Graphs

Another subclass of Petri nets which is often mentioned in the literature is the class of *marked graphs.* A marked graph is a Petri net in which each place is an input for exactly one transition and an output of exactly one transition. Alternatively, we can say that each place has exactly one input and one output.

DEFINITION 7-2 A *marked graph* is a Petri net $C = (P,T,I,O)$ such that for each $p_i \in P$, $|I(p_i)| = |\{t_j | p_i \in O(t_j)\}| = 1$ and $|O(p_i)| = |\{t_j | p_i \in I(t_j)\}| = 1$.

Marked graphs are duals of state machines, in the graph-theoretic sense of the word, since transitions have one input and one output in state machines, while places have one input and one output in marked graphs. They are also duals from the point of view of modeling. A state machine can easily represent conflicts by a place with several outputs but cannot model the creation and destruction of tokens needed to model concurrency or the waiting which characterizes synchronization. Marked graphs, on the other hand, can model concurrency and synchronization but cannot model conflict or data-dependent decisions.

The properties which have been investigated for marked graphs have been liveness, safeness, and reachability. In the investigation of these properties, the major structural parts of a marked graph of interest are its *cycles.* A cycle in a marked graph is a sequence of transitions $t_{j_1}t_{j_2}t_{j_3} \cdots t_{j_k}$ such that for each t_{j_r} and $t_{j_{r+1}}$ in the sequence there is a place p_{i_r} with $p_{i_r} \in O(t_{j_r})$ and $p_{i_r} \in I(t_{j_{r+1}})$ and $t_{j_1} = t_{j_k}$. A cycle is thus a closed path from a transition back to that same transition.

For example, in the marked graph of Figure 7.13, the sequence $t_1t_2t_1$ is a cycle, as are $t_4t_3t_4$ and $t_2t_4t_3t_1t_2$.

The importance of cycles for marked graphs derives from the following theorem.

THEOREM 7-1 The number of tokens on a cycle of a marked graph does not change as a result of transition firings.

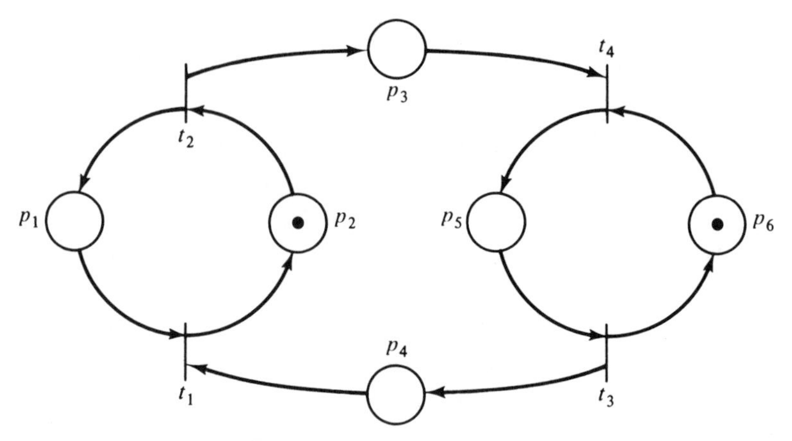

Figure 7.13 A marked graph.

From this theorem, it is easy to show the following.

THEOREM 7-2 A marking is live if and only if the number of tokens on each cycle of the marked graph is at least one.

THEOREM 7-3 A live marking is safe if and only if every place in the marked graph is in a cycle with a token count of one.

These theorems provide a simple and easy way to inspect the structure of a marked graph and from its structure and its initial marking to determine if the marked graph is live and safe. It is also possible to show that the reachability problem for markings of marked graphs is decidable. For example, note the following.

THEOREM 7-4 A marking μ' is reachable from a live marking μ in a strongly connected marked graph if and only if the total number of tokens in each cycle of the marked graph is the same in both μ and μ'.

The high decision power of marked graphs is evident from the above theorems and the work on marked graphs in [Holt and Commoner 1970; Commoner et al. 1971; Genrich and Lautenbach 1973; Izbicki 1973; Murata 1977b]. However, there is a trade-off between decision power and modeling power, and the high decision power of marked graphs results in part from low modeling power. Thus, investigators have tried to develop other subclasses of Petri nets which retain the high decision power of marked graphs while increasing their modeling power.

7.4.3 Free-choice Petri Nets

Hack, in his Master's thesis at M.I.T. [Hack 1972], defined and investigated one such subclass of Petri nets, the *free-choice Petri nets.* This subclass allows both the conflicts of state machines and the concurrency of marked graphs but in a more restricted manner than the general Petri net model.

DEFINITION 7-3 A *free-choice Petri net* is a Petri net $C = (P,T,I,O)$ such that for all $t_j \in T$ and $p_i \in I(t_j)$, either $I(t_j) = \{p_i\}$ or $O(p_i) = \{t_j\}$.

The importance of this definition is the way in which it allows controlled conflict. Conflict occurs only when one place is an input to several transitions. By the definition of free-choice Petri nets, if a place is an input to several transitions (potential conflict), then it is the *only* input for all of these transitions. Hence either all of these conflicting transitions are simultaneously enabled, or none of them are. This allows the choice (conflict resolution) as to which transition is to fire to be made freely; the presence of other tokens in other places is not involved in the choice as to which transition fires.

This restricted form of conflict allowed Hack [1972] to prove necessary and sufficient conditions for a marked free-choice Petri net to be live and safe. The liveness condition is related to the markings of traps and deadlocks in the net. A *trap* is a set of places such that every transition which inputs from one of these places also outputs to one of these places. This means that once any of the places in a trap has a token there will always be a token in one of the places of the trap. Firing transitions may move the token between places but cannot remove a token from the trap. A *deadlock* is a set of places such that every transition which outputs to one of the places in the deadlock also inputs from one of these places. This means that once all of the places in the deadlock become unmarked, the entire set of places will always be unmarked; no transition can place a token in the deadlock because there is no token in the deadlock to enable a transition which outputs to a place in the deadlock.

Hack proved that a necessary and sufficient condition for liveness in a marked free-choice Petri net is that every deadlock contain a marked trap. This theorem was based on work by Commoner [Commoner 1972; Hack 1972]. Necessary and sufficient conditions for safeness involve showing that the free-choice Petri net is covered by a union of state machines. Details can be found in [Hack 1972].

Unfortunately, no further work has been done on free-choice Petri nets, and so properties of free-choice Petri nets with respect to reachability, equivalence, containment, languages, and so on have not been considered.

7.4.4 Simple Petri Nets

Hack has also defined one other subclass of Petri nets called *simple Petri nets* [Hack 1972]. Simple nets require that each transition have at most one input place which is shared with another transition and so also serve to restrict the manner in which conflicts can occur. No investigation has been made into the properties of this subclass of Petri nets.

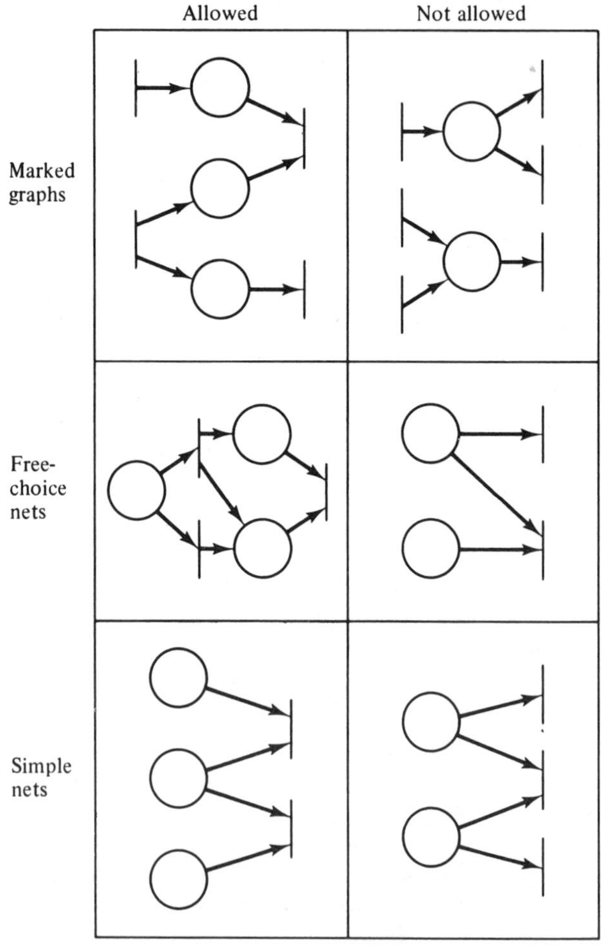

Figure 7.14 A simple chart showing some of the structural configurations which are allowed or not allowed for some subclasses of Petri nets.

7.5 Further Reading

The proof by Patil [1971] that P/V systems cannot solve all synchronization problems and the counterproof by Parnas [1972] are short and interesting. They led to the proofs of Kosaraju [1973] and Agerwala and Flynn [1973] that Petri nets cannot model all concurrent systems. These results led Agerwala to investigate what is needed in a model which can model all systems [Agerwala 1974a; Agerwala 1974b].

The major work on restricted Petri net models is in the early [Holt et al. 1968] and [Commoner et al. 1971] and the later [Hack 1972]. Some work has continued on marked graphs [Izbicki 1973; Murata 1977b], but little has been done on the other models. Some encouraging results have been obtained in [Crespi-Reghizzi and Mandrioli 1975] and [Landweber and Robertson 1975] for *conflict-free* Petri nets in which $|O(t_j)| = 1$. Reachability and liveness are decidable for conflict-free nets.

7.6 Topics for Further Study

1. One suggested extension is to associate information with tokens. This can be phrased as a Petri net with colored tokens. Define a Petri net model with colored tokens. Use this extended model to test the conjecture that either (a) it is equivalent to normal Petri nets, or (b) it is equivalent to Turing machines (by allowing zero testing). The major problem will be defining the actions of transitions with colored inputs.
2. Investigate further the properties of conflict-free, free-choice, and simple Petri nets.
3. Characterize the class of Petri nets which are both marked graphs and state machines.
4. What are the properties of the class of Petri nets whose transition inputs are either disjoint $[I(t_j) \cap I(t_k) = \emptyset]$ or identical $[I(t_j) = I(t_k)]$? This class of Petri nets properly includes free-choice Petri nets, but we would expect the properties of this new class to be very similar to those for free-choice nets.

8

Related Models of Parallel Computation

Petri nets have been defined as models of systems with concurrency. As we have seen, Petri nets have good modeling power, being able to model a large number of systems. However, Petri nets are not the only model of parallel computation; a number of other models have been defined, investigated, and used. In this chapter, we present some of these models and examine their relationship to Petri nets. Our purpose in this chapter is to give you a feeling for the variety of models which may be of use in modeling systems and the relative modeling power of these models.

A major problem raised by our intent to relate the various models to each other is the problem of establishing an appropriate method for comparing models of parallel computation. We wish to be able to prove that a model A is "less powerful" than a model B or is "equivalent." The notions of equivalence and containment are of critical importance here.

Several studies have been performed on the relationships between various models. The surveys of Bredt [1970a], Baer [1973a], and Miller [1973] helped to bring together in one place the descriptions of several models. In particular, Bredt provided a general definition of a control structure which allows the various models to be defined in a uniform way. This led to the work in [Peterson 1973] and [Peterson and Bredt 1974] which compared various models to arrive at a hierarchy of models, related by their modeling power. Independently Agerwala [1974b] compared a larger set of models and produced another hierarchy with a similar structure.

The results of both Peterson and Bredt [1974] and Agerwala [1974b] were obtained by using the *languages* of the models to compare them. A class of models A defines a class of languages $L(A)$. For two classes of models A and B, the classes were defined to be equivalent if $L(A) = L(B)$. This means that for any specific instance a of a class of models A with language $L(a)$ there would exist an instance b of class B with an identical language $L(a) = L(b)$. If the languages truly characterize the models, then they are an appropriate means for comparing two classes of models.

However, as we have seen, it is not completely obvious how to define a language for a model of parallel computation. Research into Petri net languages has resulted in at least 12 different definitions for languages, most of them apparently distinct. These different languages may result in different equivalence and inclusion relationships between models. On the other hand, if the differences between the models are truly significant, they may well be insensitive to (minor) variations in the definition of equivalence and inclusion. Thus, the similar results of Agerwala [1974b] and Peterson and Bredt [1974] are more significant because of the different definitions for equivalence and inclusion used.

However, this is not to say that these results are beyond dispute. Lipton et al. [1974] also compared a large number of models of parallel computation and arrived at different results. Their comparison is based on a quite detailed analysis of the structure and *state space* of the specific instances of a class of models, and so their results differ significantly from those of Agerwala [1974b] and Peterson and Bredt [1974].

So, given that there exists some difference of opinion among researchers as to how models should be compared, how shall we compare models? It would be best to take a conservative approach, and so we base our comparisons in this chapter on both structural and behavioral characteristics. We say that a class of models A is *less than or equal in modeling power to* (included in) a class of models B if given any instance a of the class A there is an algorithm to create an instance b of the class B such that

1. Each structural component of the model a is represented by a (small) identifiable set of components of the model b. The size of the model b (number of parts) is at worst a constant multiple of the size of the model a, with the constant being determined by the classes of models A and B and not by the specific instances a and b.
2. Any sequence of actions in a can be simulated by a sequence in b, with the length of the sequence in b being at most a constant multiple of the length of the sequence of a.
3. The model b deadlocks only when the model a does. A model deadlocks if all actions become impossible.

The motivation for these constraints should be relatively obvious. The first constraint merely tries to establish that the two models are structurally similar; the second constraint assures that the two models behave the same. However, we do not require exactness between the two models; it is permissible to represent an action in one model by a (short) sequence of actions in another model or a component (like a place or transition) by a (small) collection of components. Thus, an action in one model can be modeled by a sequence of two actions in another model. The last constraint requires that the more powerful model cannot make a mistake when the less powerful model would not. This prevents constructing a model which nondeterministically chooses one of several actions and aborts itself if it later finds that it made the wrong choice.

Two models are *equivalent* if each includes the other. This allows any specific instance of either model to be converted to an instance of the other.

With this in mind we now consider the relationships among the following models of parallel computation:

1. Finite state machines [Hopcroft and Ullman 1969; Bredt 1970b; Gilbert and Chandler 1972]
2. Marked graphs [Commoner et al. 1971]
3. Computation graphs [Karp and Miller 1966]
4. P/V systems [Dijkstra 1968; Bruno et al. 1972]
5. Message systems [Riddle 1972]
6. UCLA graphs [Gostelow 1971; Cerf et al. 1971; Cerf 1972; Cerf et al. 1972]
7. Vector addition systems [Karp and Miller 1968]
8. Vector replacement systems [Keller 1972]
9. Extended Petri nets (Chapter 7)

For each class of models, we first define the model and give an example. Then we discuss its relationship to other models of parallel computation.

8.1 Finite State Machines

We have already seen in Sections 3.3.1 and 7.4.1 that *finite state machines* can be transformed easily into Petri nets. Finite state machines have been used by several researchers as a model of parallel computation. Bredt [1970b] defined a model based on computer hardware. Each processor is modeled as a finite state machine with input and output lines which connect a processor to other processors. The state of each input and output line is either 0 or 1. Since every output line for one processor is an input line for another processor and there are a finite number of processors and a finite number of lines each with a finite state, the entire system is finite state.

Gilbert and Chandler [1972] used a model with common memory rather than communication lines. This means that their model is directed more toward modeling computer software processes with shared memory rather than the hardware model of Bredt but is nonetheless a finite state model and hence included in the Petri net model.

8.2 Marked Graphs

Marked graphs have been discussed in Section 7.4.2. As a subclass of Petri nets, marked graphs are obviously of more limited modeling power than Petri nets. Marked graphs are not directly comparable to finite state machines but rather seem to be duals. This provides us with the relationships in Figure 8.1 among Petri nets, finite state machines, and marked graphs.

8.3 Computation Graphs

One of the earliest models of parallel computation was the *computation graph model* [Karp and Miller 1966]. This model was mainly designed to represent the execution of programs evaluating arithmetic expressions in parallel.

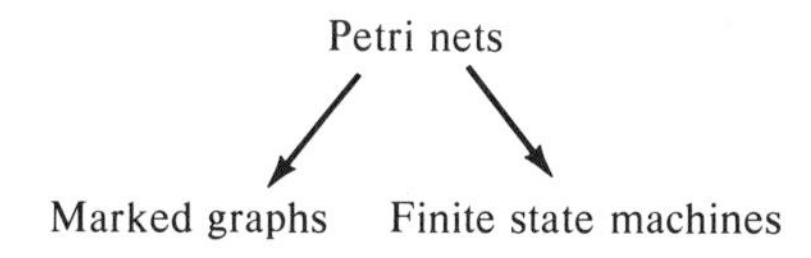

Figure 8.1 Relationship of Petri nets, marked graphs, and finite state machines.

A computation graph G is defined by a directed graph $G = (V,A)$, where

$$V = \{v_1, v_2, \ldots, v_n\} \text{ is a set of vertices}$$

$$A = \{a_1, a_2, \ldots, a_m\} \text{ is a set of arcs}$$

Each arc $a_i \in A$ is an ordered pair of vertices (v_j, v_k) representing an arc from v_j to v_k. Associated with each arc $a_i = (v_j, v_k)$ is a quadruple $(I_{j,k}, V_{j,k}, W_{j,k}, T_{j,k})$. Each arc represents a *queue* of data items produced by the processor for node v_j and to be used by the processor for node v_k. $I_{j,k}$ gives the number of data items *initially* in the queue for the arc from v_j to v_k. A node v_k is enabled if there are at least $T_{j,k}$ data items on each arc directed into node v_k from each node v_j; $T_{j,k}$ is a *threshold.* The operation associated with node v_k executes by removing $W_{j,k}$ data items from the queue ($W_{j,k} \leqslant T_{j,k}$) for each arc directed into v_k. When the operation for v_k terminates, it places $V_{k,r}$ data items on the queue associated with each arc (v_k, v_r) directed from node v_k to a node v_r.

Figure 8.2 is an example computation graph. In the initial state,

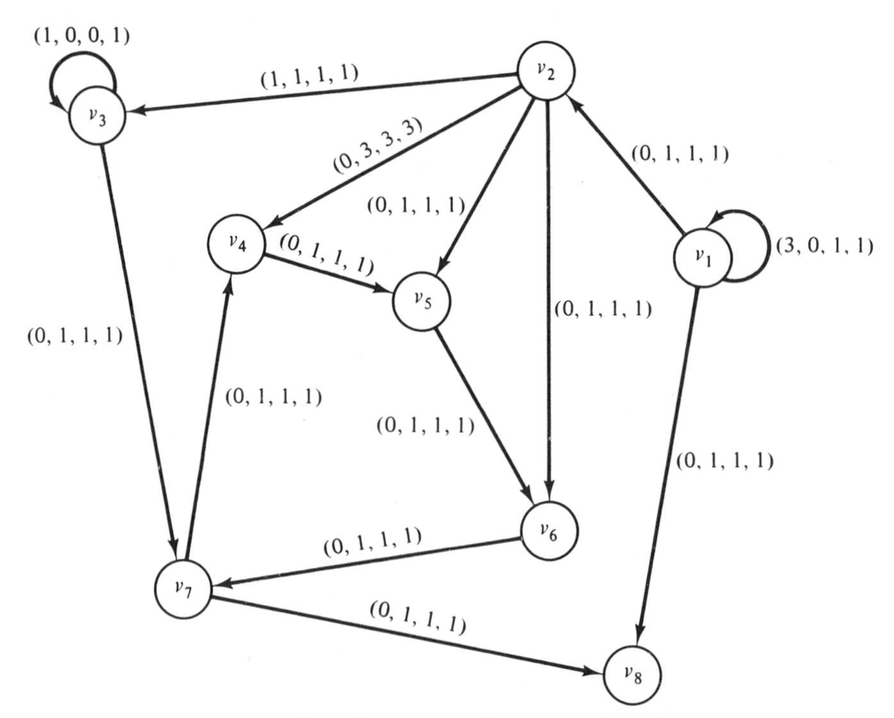

Figure 8.2 A computation graph.

the node v_1 is enabled, since it has one input and this input has three data items in its queue. When v_1 executes it removes one data item from this queue and upon completion puts one data item on the arc from v_1 to v_2 and one on the arc from v_1 to v_8. In this new state either v_1 or v_2 can execute since both have enough data items in their input queues to satisfy their thresholds.

A computation graph is easily modeled as a Petri net. Each arc is represented by a place, and each node of the computation graph becomes a transition. The transition corresponding to node v_k has $T_{j,k}$ input arcs from the place representing an arc from v_j to v_k. This assures that the transition is enabled only if the threshold is met. However, when the transition fires it can only remove $W_{j,k}$ of these tokens, so $T_{j,k} - W_{j,k}$ arcs are directed back from transition v_k to the place representing the arc from v_j to v_k. In addition $V_{k,r}$ tokens are put into the places representing arcs from node v_k to node v_r. The initial marking is determined from the $I_{j,k}$ as would be expected.

Figure 8.3 shows the Petri net constructed by this means from the computation graph of Figure 8.2.

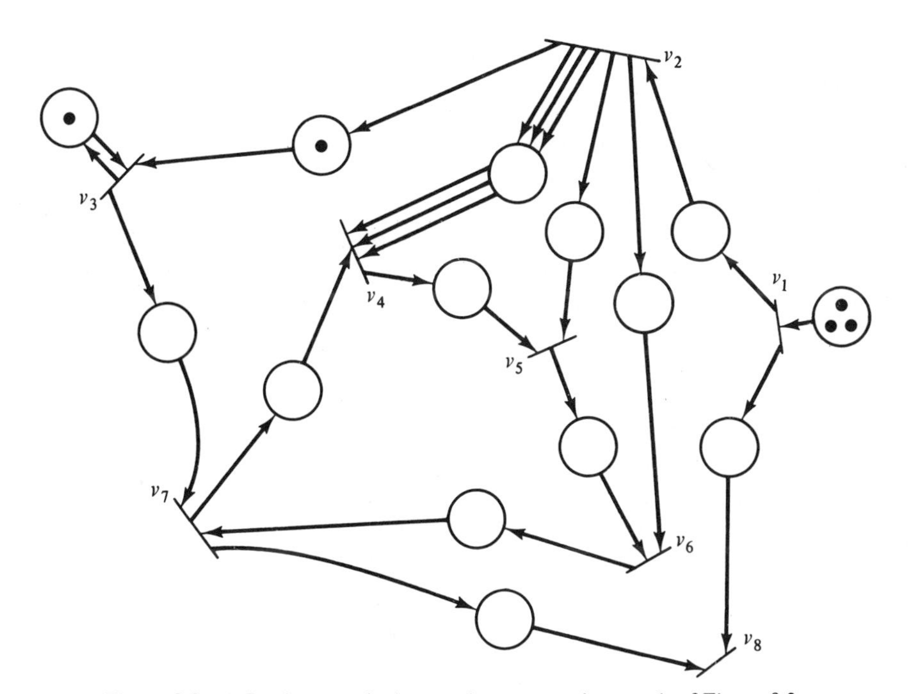

Figure 8.3 A Petri net equivalent to the computation graph of Figure 8.2.

A little thought shows that marked graphs can be modeled as computation graphs with $T_{j,k} = W_{j,k}$ for all nodes v_j and v_k. However, computation graphs are more powerful than marked graphs because of the ability to have $T_{j,k} > W_{j,k}$.

On the other hand, computation graphs and finite state machines are incomparable, as are marked graphs and finite state machines. Computation graphs cannot model decisions or conditional execution — the same limitation that marked graphs have. Thus, our hierarchy of models at this point is as shown in Figure 8.4.

Karp and Miller [1966] thoroughly investigated computation graphs, especially concerning the problems of liveness and safeness. Actually, Karp and Miller were interested in assuring that a computation graph would terminate and so determined the conditions for a computation graph to terminate (i.e., to not be live). Since the arcs (places) represented queues of data, Karp and Miller's investigation of boundedness was aimed at determining the maximum queue length. These differences in notation and motivation as well as the difference in model definition between computation graphs and marked graphs has meant that no one has tried to relate the results and algorithms of Karp and Miller [1966] on computation graphs with the work on marked graphs [Commoner et al. 1971].

8.4 P/V Systems

P and V operations on semaphores were first introduced by Dijkstra [1968] to aid in solving synchronization problems in systems of parallel processes. As such they can be used to model synchronization and communication in the same way as Petri nets. Patil used this approach when he defined the cigarette smokers' problem to show the limitations of systems which can use only P and V operations between processes. P/V systems are nonetheless quite popular, and the computing science literature abounds with discussions or applications of these operations,

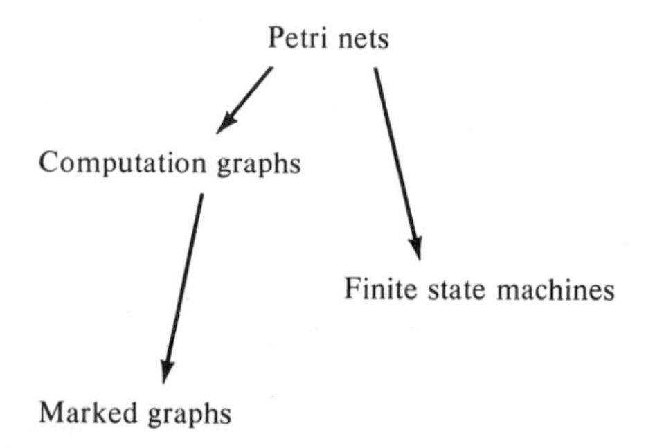

Figure 8.4 Adding computation graphs to the hierarchy.

for example, [Liskov 1972; Brinch Hansen 1973].

Section 3.4.8 showed that P and V operations can be modeled using Petri nets, and Patil's proof [Patil 1971] shows that the inclusion is proper: There are problems (the cigarette smokers' problem, for example) which can be solved with Petri nets but not with P and V operations only. However, P/V systems are powerful enough to include both computation graph models [Lipton and Snyder 1974] and finite state machine models.

To convert a finite state machine into a P/V system, we use a separate process to model each state of the finite state machine. One semaphore is associated with each state. Let $Q = \{q_1, q_2, \ldots, q_n\}$ be the set of states and $\delta: Q \times \Sigma \rightarrow Q$ be the next state function, with a set of actions Σ. Associate with state q_i a semaphore S_i and a process. The process first executes a $P(S_i)$. Generally it will wait here until the state of the finite state machine becomes q_i. After the $P(S_i)$, the process arbitrarily picks any $\sigma \in \Sigma$ for which $\delta(q_i, \sigma)$ is defined and executes a $V(S_j)$, where $q_j = \delta(q_i, \sigma)$. Then this process loops back to its $P(S_i)$. Figure 8.5 illustrates this conversion on the finite state machine of Figure 8.6. The semaphores are initially zero, except for the semaphore for the start state, which is initialized to one.

To convert a computation graph into a P/V system, we associate a semaphore $S_{j,k}$ with each arc (v_j, v_k) in the graph. The value of the semaphore will be the number of data items waiting on the queue for that arc. Thus, initially the value of semaphore $S_{j,k}$ is $I_{j,k}$. A process is created for every node in the computation graph. The process for node v_k first executes $T_{j,k}$ P operations on semaphore $S_{j,k}$ for all arcs (v_j, v_k) directed into v_k. This assures that each queue has at least $T_{j,k}$ data items in it. Then, since each P operation decremented the semaphore and the correct effect is only to decrease $S_{j,k}$ by $W_{j,k}$, we execute

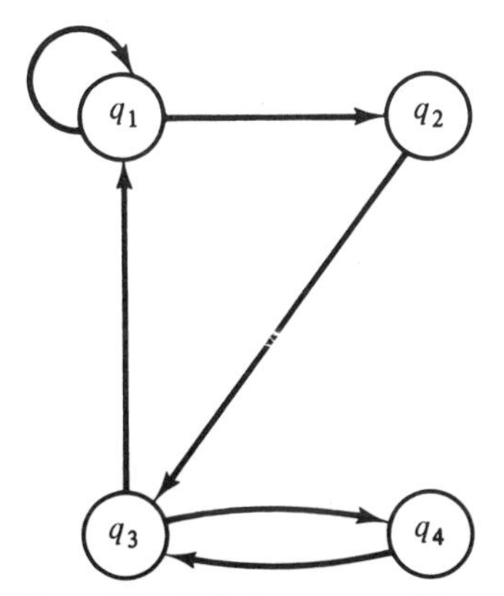

Figure 8.5 A finite state machine.

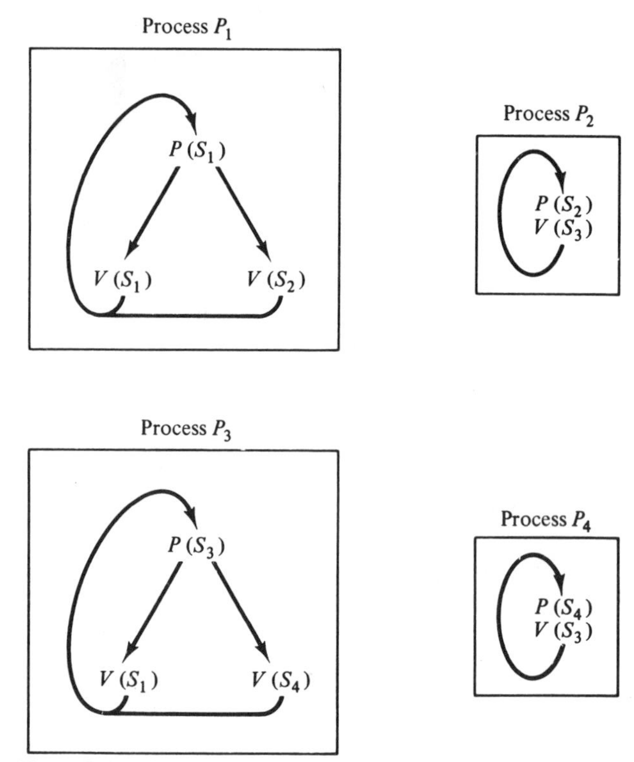

Figure 8.6 P/V system for the finite state machine of Figure 8.5.

$T_{j,k} - W_{j,k}$ V operations on $S_{j,k}$ to restore the correct value of $S_{j,k}$. Now we complete the process for node v_k by executing $V_{k,r}$ V operations on semaphore $S_{k,r}$ for each arc (v_k, v_r) directed out of node v_k.

This conversion is illustrated in Figure 8.7 for the nodes v_3 and v_4 of the computation graph of Figure 8.2.

Notice that the computation graph could test and input from several sources at once, while P/V systems can only test and input from several sources by a sequence of tests and inputs from single sources. The inability to test and input from several sources simultaneously is the key to Patil's proof of the limitations of P/V systems, the problem being that another process may grab the second source while you grab the first, leading to a deadlock. This is not a problem with computation graphs because the sources are not shared between processors — no two nodes ever share an input arc. This point is crucial to constructing a P/V system which will not deadlock (terminate) unless the corresponding computation graph also terminates.

The addition of P/V systems to our hierarchy of models gives us Figure 8.8.

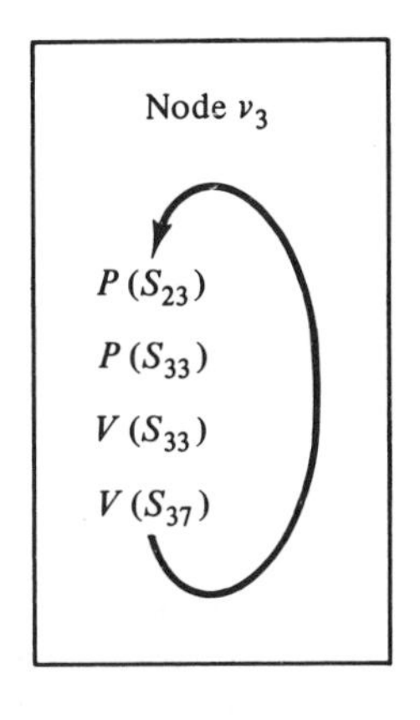

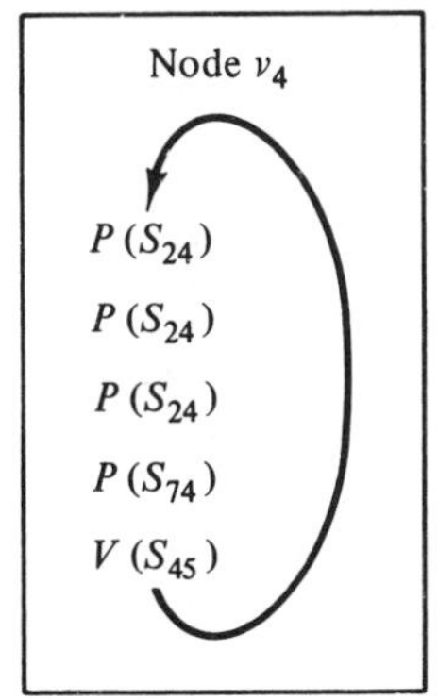

Figure 8.7 P/V system processes for two of the nodes of the computation graph of Figure 8.2.

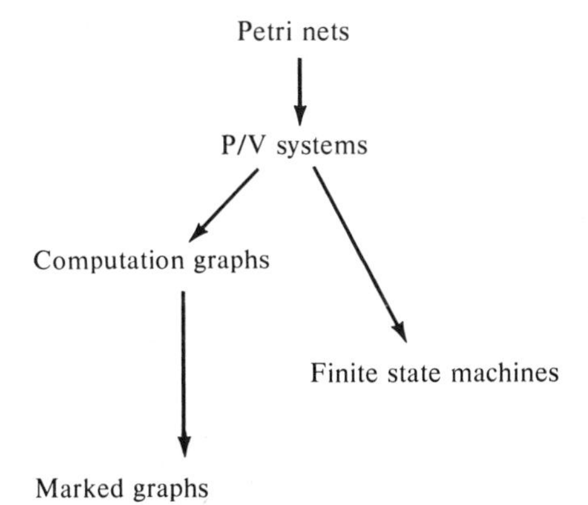

Figure 8.8 Adding P/V systems to the hierarchy of models.

8.5 Message Transmission Systems

Systems in which processes communicate via P and V operations on semaphores may be doing so for lack of a better communication mechanism. One suggestion for a better mechanism is *messages.* A message system is a collection of processes which communicate by messages. Two operations on messages are possible: *send* and *receive.* Sending a message is similar to a V operation; receiving a message is similar to a P operation. If no message is present when a *receive* is executed, the receiver waits until a message is sent.

This mechanism has been used as the basis of a modeling scheme by Riddle [1972]. This model would seem most appropriate for modeling computer network protocols. Riddle considers a (finite) set of *processes* which communicate by means of *messages.* Messages are sent to and requested from special processes called *link* processes (mailboxes). Link processes provide what is essentially a bag of messages which have been sent and not yet received or requests for messages from receives which have been made but not yet satisfied. The other processes of the system are called *program* processes and are represented in the *program process modeling language* (PPML).

An example system of three processes is given in Figure 8.9. As can be seen, the PPML description of the processes is essentially a schema. Only the transmission activity of messages in the system is of interest. Messages are abstract items whose only characteristic is a type. There may only be a finite number of types of messages in a system. Messages are sent from and received into a message buffer in each process. There is only one buffer per process. The statements in PPML are

set t: Place a message of type *t* into the message buffer.

send l: Send the message in the message buffer to link process *l*.

receive l: Request a message from link process *l*. Wait (if necessary) until one is returned. The message is placed in the message buffer.

unless t s: Test the type of the message in the message buffer and branch to statement *s* unless the message is of type *t*.

if-internal-test s: Model an internal, data-dependent test. Either continue with the next statement or branch to the statement labeled *s*.

go-to s: Transfer control to the statement *s*.

end: Terminate the process.

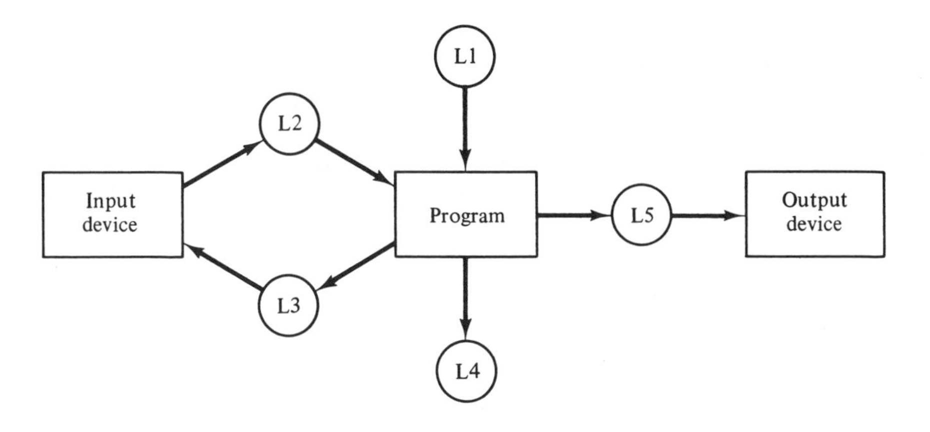

Program

```
        Receive L1;
A1:     Set READ;
        Send L3;
        Receive L2;
        Unless INPUT A2;
        Set OUTPUT
        Send L5;
        Go-to A1;
A2:     Send L4;
        End;
```

Input device

```
A1:     Receive L3;
        If-internal-test A2;
        Set INPUT;
        Go-to A3;
A2:     Set EOF;
A3:     Send L2;
        Go-to A1;
        End;
```

Output Device

```
A1:     Receive L5;
        Go-to A1;
        End;
```

Figure 8.9 A system of Processes described in the Program Process Modeling Language (PPML).

The PPML system models a set of parallel processes. Each process is started at the beginning of its program and executes its program until it encounters an *end* statement. Riddle shows how to construct a *message transfer expression* which represents the possible flow of messages in the system and uses this expression to examine the structure of the system for correct operation. This message transfer expression is used for the same purpose as the language of a Petri net. Therefore, we show how a PPML description of a system of processes can be converted into a Petri net whose language is equal to the message transfer expression of Riddle's analysis. This conversion ignores the execution of the individual statements of the PPML descriptions, although with only minor modification these could also be represented in the Petri net language.

To model a process as a Petri net, we use one token per process as a program counter. The presence of a message in a link process is also represented by a token. Since the messages are identified by type, it is necessary to model each type of message in a link process by a separate place. A very important property of PPML systems is that the number of message types is finite. Each program process is also finite. Only the queueing of messages involves a potentially unbounded amount of storage. Thus the ability to model the link processes and to represent the *send* and *receive* statements correctly is the most important aspect of the transformation from a PPML description to a Petri net. By modeling the link processes by a set of places (one for each message type), we can represent a *send* statement by a transition which places a token in the place representing the appropriate link process and message type. The *receive* statement merely removes a token from any one of the places of a link process. The particular place which supplies the token determines the type of the message which was received. This information can be used in any subsequent *unless* statement.

The only symbols in a message transfer expression are the message types for the messages which are sent to or received from a link process. Since every transition in a Petri net results in a symbol in the Petri net language for that Petri net, only the *send* and *receive* statements in a PPML system can be modeled. Thus there are two kinds of places in the Petri net. One kind of place, labeled p_{l_i,m_j}, acts as a counter of the number of messages of type m_j in link process l_i. The other kind of place represents the *send* and *receive* statements of the PPML programs. Let the statements be uniquely labeled $s_1, \ldots, s_r$. We label the place representing a statement s_i with a message of type m_j in the message buffer by p_{s_i,m_j}. A token in a place associated with a statement s_i means that statement s_i has just been executed. Figure 8.10 illustrates how a *send* and *receive* statement at s_k would be modeled in the Petri net. In Figure 8.10, place $p_{s_i,m}$ represents the place associated with any statement which precedes the statement s_k.

It remains only to show that it is possible to determine which statements can precede other statements in the PPML program. Notice that we must consider each statement as a pair consisting of a message type and a statement number, since the same statement with different message types in the message buffer will be modeled differently in the Petri net. The most obvious method to determine the predecessors of a statement is to start at the beginning of each PPML program with a

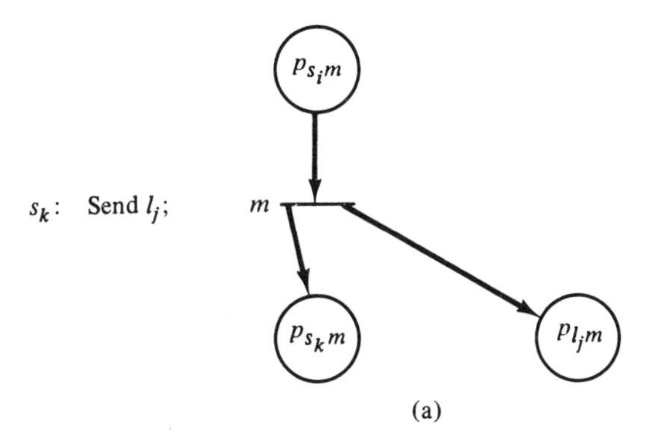

(a)

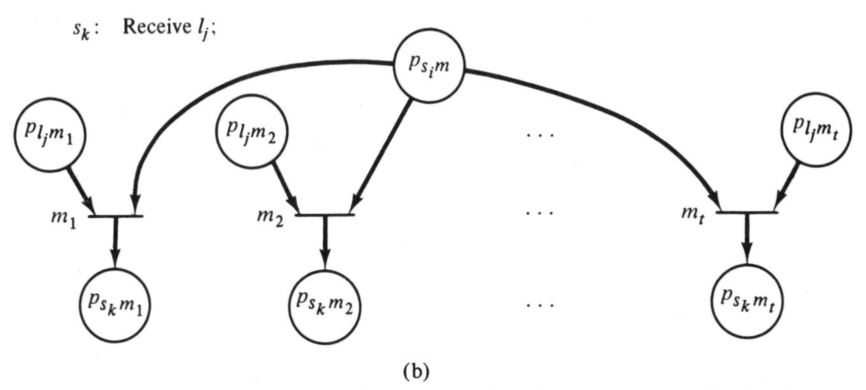

(b)

Figure 8.10 Transformation from *send* and *receive* statements to Petri net transitions. (a) Model of a *send* statement at s_k with a message of type m in the message buffer. The link process is l_j. (b) Model of a *receive* statement at s_k from link process l_j. The possible message types in l_j are $m_1, m_2, \ldots, m_t$.

special start statement (which will become the start place) and follow the program description generating all possible *send* and *receive* statements which can follow this statement with their corresponding message buffer contents. This process is repeated for all newly reachable statements until all such *send* and *receive* statements have been generated and their successors identified. Since the number of statements in a PPML description and the number of message types are finite, only a finite number of statement/message-type pairs will be generated. This procedure is similar to the characteristic equations used by Riddle [1972] to construct the message transfer expression. Figure 8.11 lists

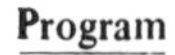

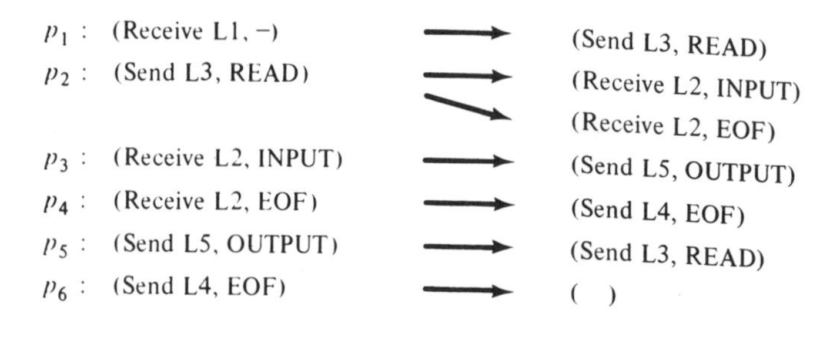

Input device

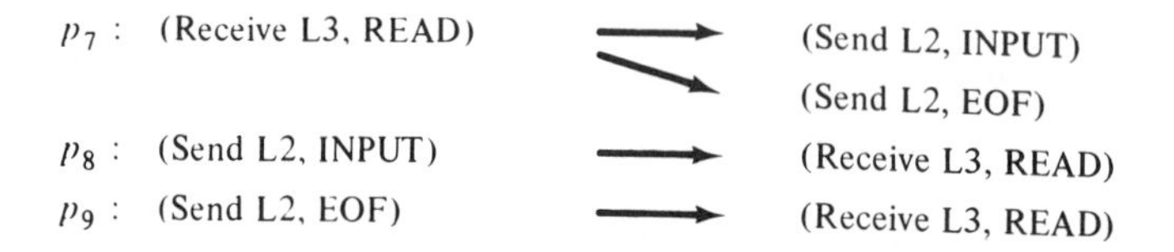

Output device

p_{10} : (Receive L5, OUTPUT) → (Receive L5, OUTPUT)

Figure 8.11 Successor statements for the PPML system of Figure 8.9.

the statements and their possible successors for the PPML system of Figure 8.9.

Once the successors of a statement have been determined, this information can be used to identify the possible predecessors of a statement and hence to construct a Petri net which is equivalent to the PPML system by use of transitions such as given in Figure 8.10. A special start place is the predecessor of the first statements of each of the processes of the system. Figure 8.12 transforms the PPML system of Figure 8.9 into an equivalent Petri net.

This brief description of the transformation from message transmission systems to Petri nets shows that this model is included in the modeling power of the Petri net. It also shows that the set of message transfer expressions, considered as a class of languages, is a subset of the class of Petri net languages.

Since P/V systems can be modeled as message transmission systems with all messages being the same type, P/V systems are included

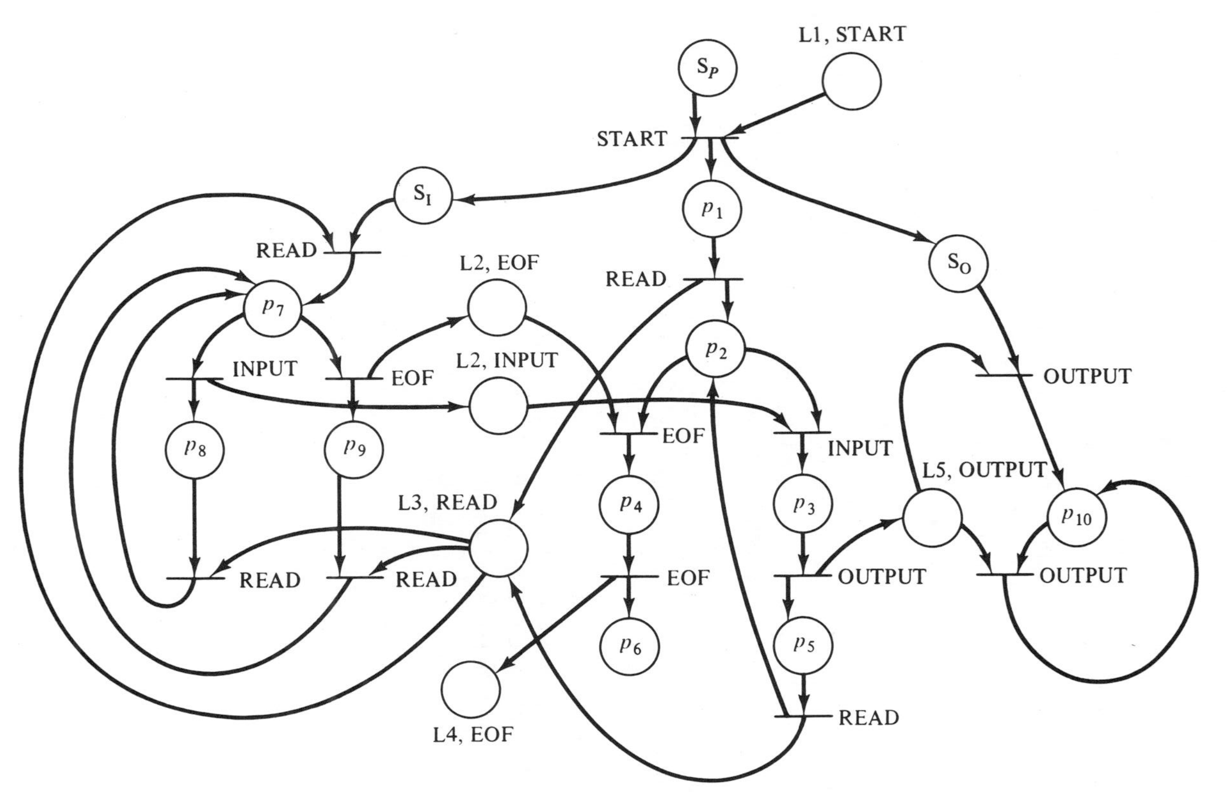

Figure 8.12 A Petri net equivalent to the PPML system of Figure 8.9.

in message transmission systems. It is fairly easy to construct a message system to solve the cigarette smokers' problem, however, so the inclusion of P/V systems in message systems is proper. Message systems, on the other hand, lack the ability to input from several sources simultaneously and so are not equivalent to Petri nets. One of two cases will occur in an attempt to model a transition with multiple inputs:

1. A process will attempt to receive tokens (messages) from all of its inputs but be unable and so will block while holding onto tokens which are needed to allow some other transition to continue. This will lead to deadlocks in the message system which do not correspond to deadlocks in the Petri net, violating the third constraint.
2. The process will avoid creating false deadlocks by determining that the remaining tokens which are needed are not present and returning the tokens it has received to the places (link processes) from which they were received. This activity could occur arbitrarily often and will mean that there is no bound on the length of the sequence of actions in the message system corresponding to a bounded sequence of transition firings in the Petri net. This violates our second constraint.

Riddle [1974] presents a transformation which suffers from case 1, leading to spurious deadlocks. In either case, we see that message systems cannot model arbitrary Petri nets (under our constraints). This results in the hierarchy of Figure 8.13.

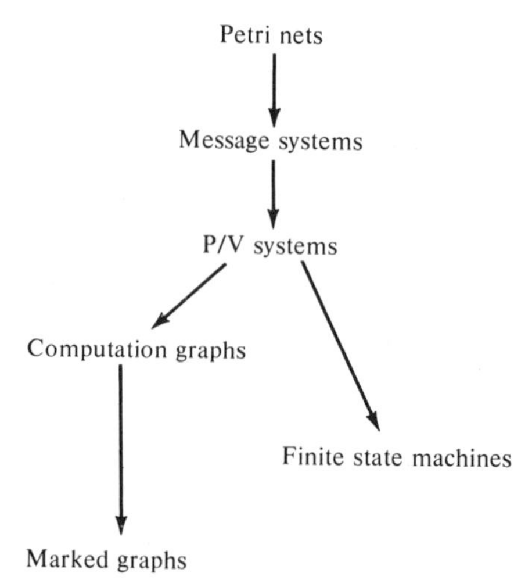

Figure 8.13 Adding message systems to the hierarchy.

8.6 UCLA Graphs

Petri nets are a graph model of parallel computation. Another graph model has been developed at the University of California at Los Angeles under the direction of Professor Estrin. This model is the *complex bilogic graph model of computation* (or UCLA graph model) [Baer 1968; Baer et al. 1970; Volansky 1970; Gostelow 1971; Cerf 1972]. In this model, systems are represented by a graph with complex directed arcs. A complex arc is an arc with (potentially) multiple sources and destinations.

Combinational logic controls the sequencing of operations at the nodes. If the input logic of a node is *AND* (*), tokens are needed on each input arc to enable an operation. For *OR* (+) logic, tokens are only needed on some one input arc. Execution of the node removes the enabling tokens on the input arcs and places tokens on the output arcs according to the output logic. For AND output logic, tokens are placed on all output arcs, while for OR logic, tokens are placed on any one output arc. The number of tokens involved for a given node-arc pair is the *degree* (or multiplicity) of that pair and may be any nonnegative integer.

Figure 8.14 is an example UCLA graph. Notice that some arcs have multiple sources (tails) and destinations (heads). Also notice that the logic of each arc-node pair is marked on the graph as either * for AND logic or + for OR logic. The degree of an arc is indicated by a small number where the arc meets the node. The degree is omitted if it is one, as is the logic when only one arc is the input to a node. In the example node a can fire whenever arc S has a token. When node a fires, it removes the token from arc S and puts tokens on both arc A and arc B (AND logic). Node g, on the other hand, will place a token on either arc K or arc G (OR logic). Node i is enabled whenever there are two tokens on arc J or one token on arc K.

DEFINITION 8-1 A *UCLA graph* is a six-tuple $C = (V,A,L,Q,S,F)$, where

$V = \{v_1,v_2, \ldots ,v_r\}$ is a set of vertices

$A = \{a_1,a_2, \ldots ,a_s\}$ is a set of arcs

$L = (L^-,L^+)\colon V \rightarrow \{*,+\}$ is the input (L^-) and output (L^+) logic mapping for each vertex

$Q = (Q^-,Q^+)\colon V \times A \rightarrow N$ is the input (Q^-) and output (Q^+) degree of each arc-vertex pair

$S \in A$ is the start arc

$F \in A$ is the final arc

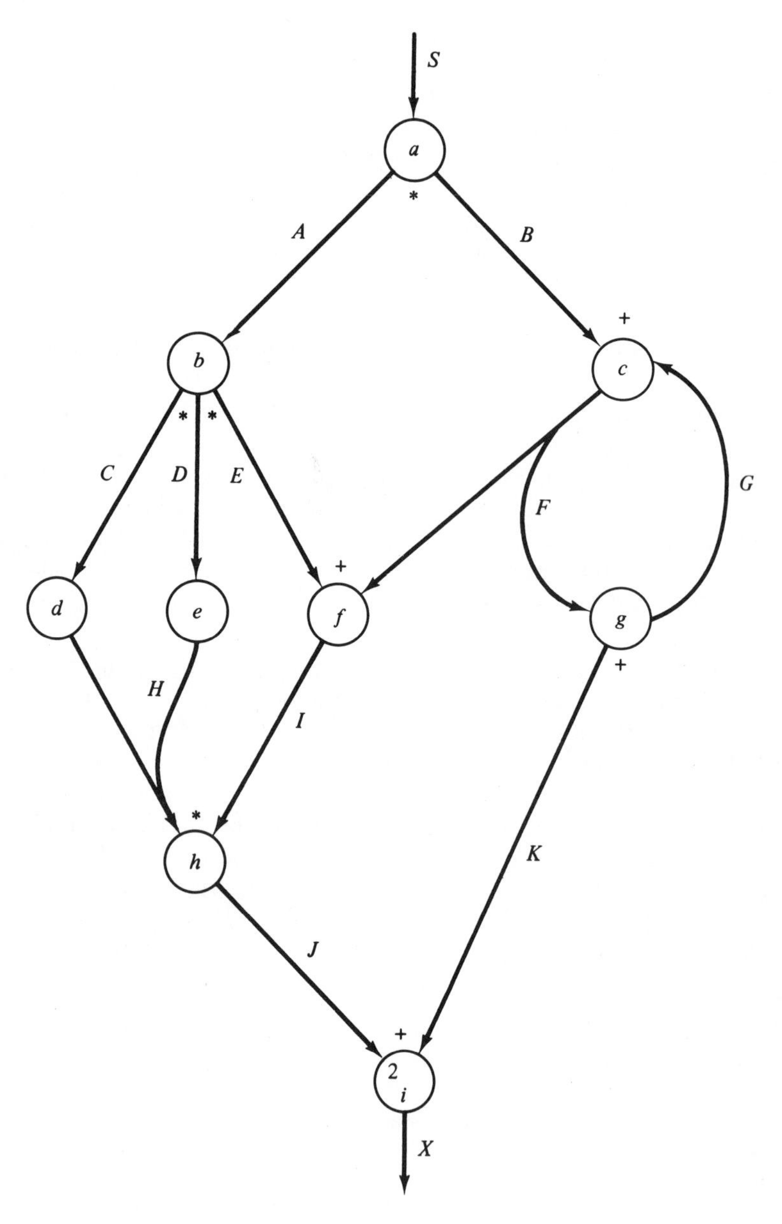

Figure 8.14 A complex bilogic directed graph (UCLA graph).

The arcs of the graph are defined as ordered pairs of sets of vertices. The first component of the pair is the set of input vertices, and the second component is the set of output vertices. The start arc has an empty set of input vertices, and the final arc has an empty set of output vertices.

The transformation from a UCLA graph to a Petri net is straightforward due to the similarity of the two systems. Every arc in a UCLA graph is represented by a place in a Petri net. In addition, we represent a node v by a place p_v and two transitions t_v' and t_v''. The first transition t_v' represents the *initiation* of the operation associated with node v; the second transition represents the *termination* of the operation. This is sketched in Figure 8.15. (The modeling of UCLA graph nodes by initiation and termination transitions is not strictly necessary but is convenient.) Figure 8.16 indicates how the input and output logic for UCLA graphs is represented in an equivalent Petri net. Degrees greater than 1 are modeled by multiple arcs between places and transitions in the Petri net.

Figure 8.17 transforms the UCLA graph of Figure 8.14 into an equivalent Petri net.

This transformation shows that the modeling power of UCLA graphs is included in the modeling power of Petri nets. It should be obvious that a Petri net can be converted into an equivalent UCLA graph by representing places as UCLA graph arcs and transitions as nodes with AND input and output logic. Thus these two models are equivalent in modeling power. Figure 8.18 shows the modified hierarchy of models.

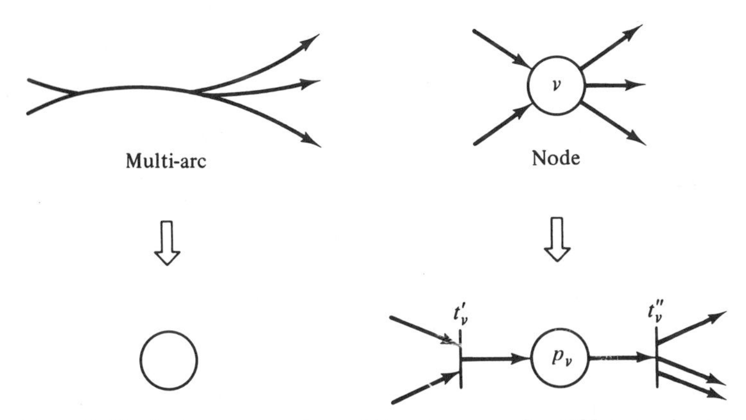

Figure 8.15 Basic representation of the elements of a UCLA graph as a Petri net.

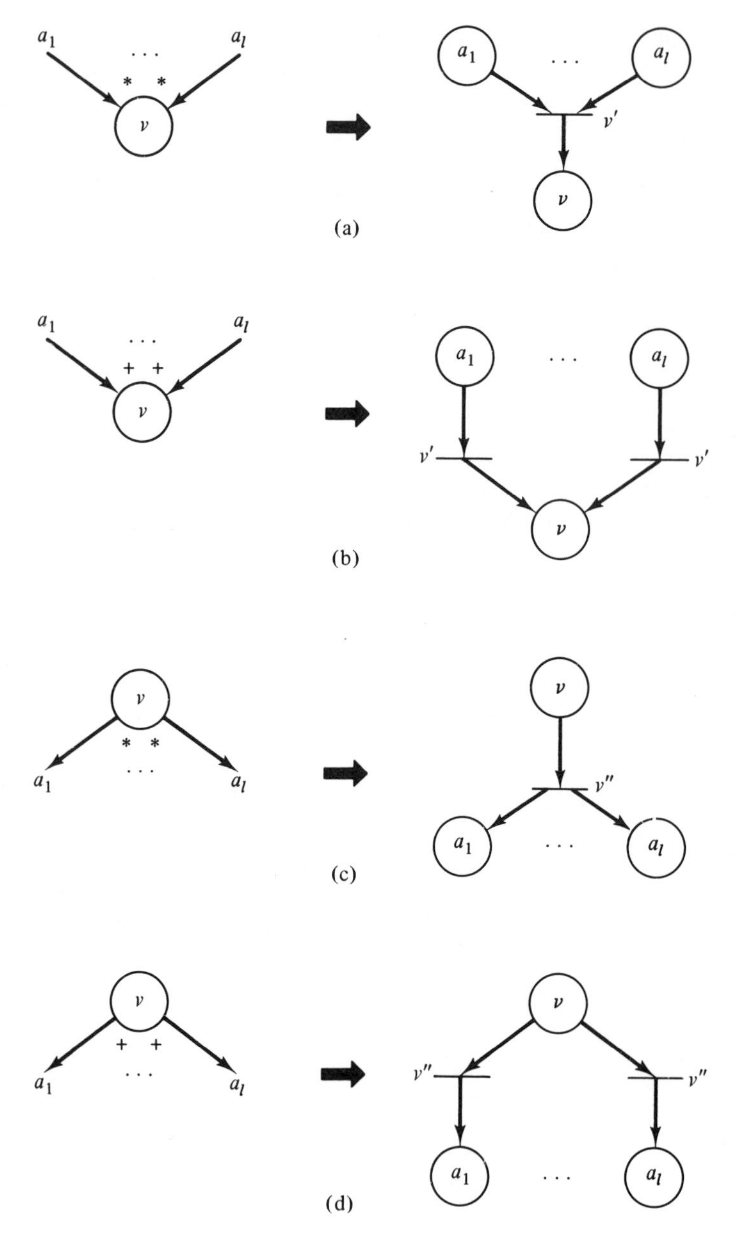

Figure 8.16 Transformation of UCLA graph parts into Petri nets.

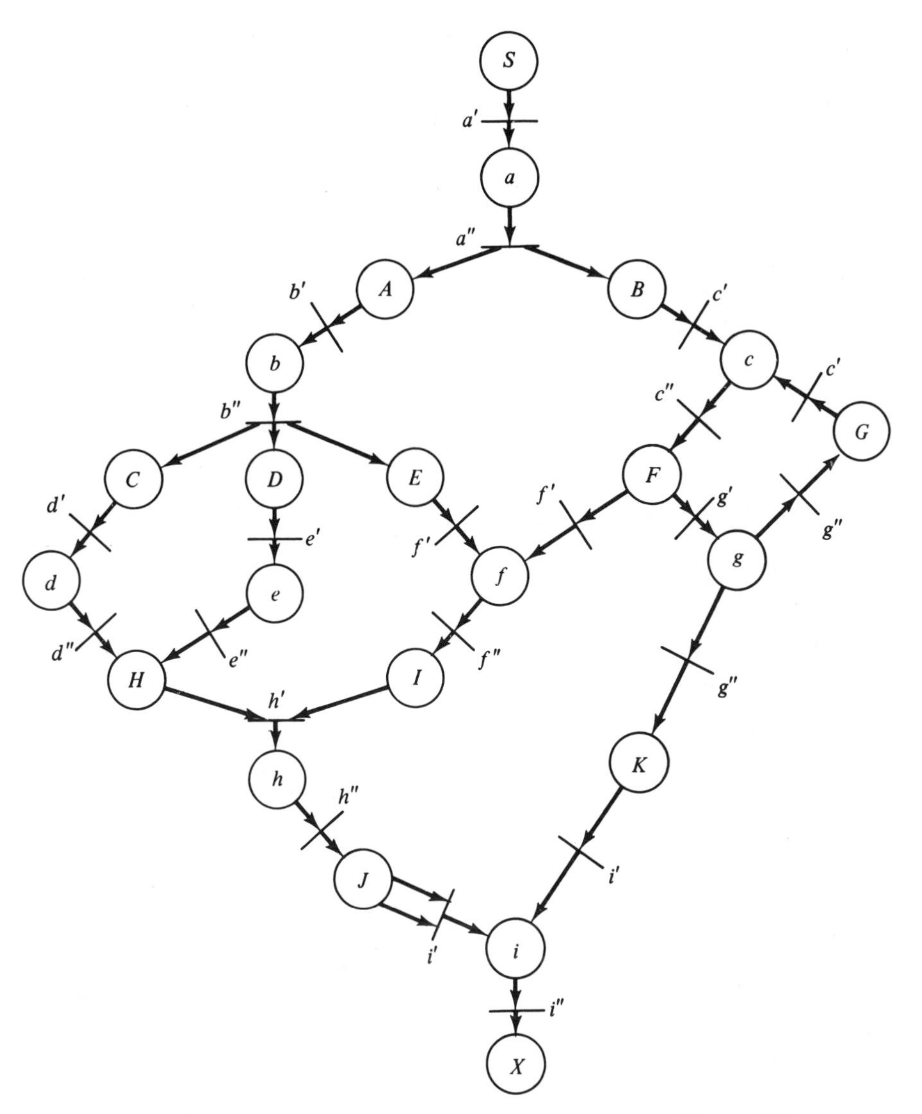

Figure 8.17 The Petri net equivalent of the UCLA graph of Figure 8.14.

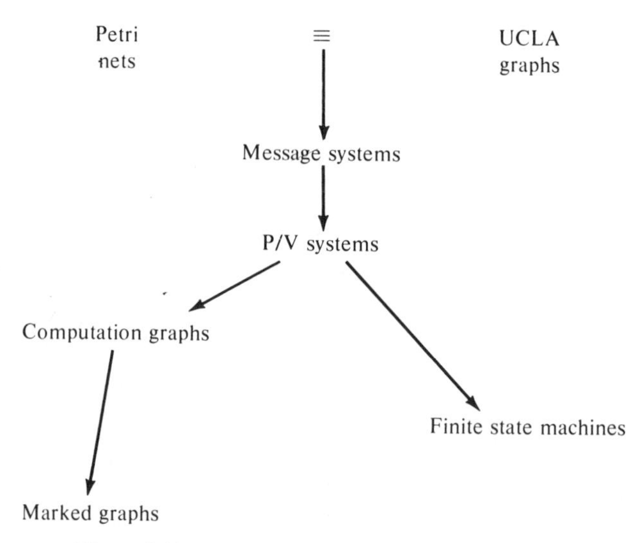

Figure 8.18 Adding UCLA graphs to the hierarchy.

8.7 Vector Addition and Replacement Systems

If you have glanced at the bibliography you may have noticed that many of the references have titles referring not to Petri nets but to *vector addition systems.* Vector addition systems were introduced by Karp and Miller [1968] as a mathematical tool for analyzing systems of parallel processes. Because of their simple mathematical formulation, vector addition systems are typically used for formal proofs of properties of Petri nets or similar systems.

DEFINITION 8-2 A *vector addition system* V is a pair $V = (B,s)$, where $B = \{b_1,b_2, \ldots ,b_m\}$ is a set of m vectors, called *basis* or *displacement* vectors. The vector s is the *start* vector. All vectors are composed of n integer values. The elements of s are nonnegative.

The reachability set of a vector addition system V is denoted $R(V)$ and can be defined either recursively by the following definition

DEFINITION 8-3 The reachability set $R(V)$ for a vector addition system $V = (B,s)$ is the smallest set such that

1. $s \in R(V)$.
2. If $x \in R(V)$ and $(x + b_j) \geqslant 0$, then $(x + b_j) \in R(V)$.

or by

DEFINITION 8-4 $x \in R(V)$ if there exists a sequence $b_{i_1}, b_{i_2}, \ldots, b_{i_k}$ of basis vectors such that

$$x = s + \sum_{j=1}^{k} b_{i_j}$$

$$s + \sum_{j=1}^{r} b_{i_j} \geqslant 0 \text{ for all } r,\ 0 \leqslant r \leqslant k$$

With these definitions, it is easy to see that vector addition systems and Petri nets are equivalent. Given a Petri net, we can construct a vector addition system whose start vector s is the initial marking, with one basis vector for each transition. The n components of the vectors of the vector addition system correspond to the markings of the n places of the Petri net, or in the case of the basis vectors to the change in marking resulting from firing the associated transition.

Similarly, a vector addition system can be converted into an equivalent Petri net, using places for the components of the vectors and transitions to represent basis vectors. Figure 8.19 illustrates the equivalence of these two models.

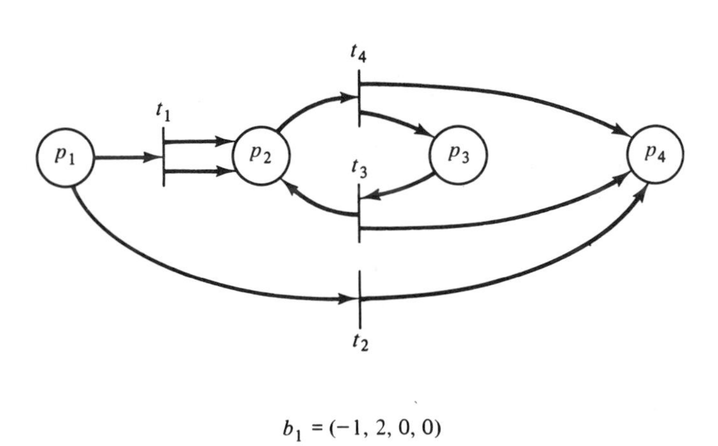

$b_1 = (-1, 2, 0, 0)$

$b_2 = (-1, 0, 0, 1)$

$b_3 = (0, 1, -1, 1)$

$b_4 = (0, -1, 1, 1)$

Figure 8.19 A Petri net and its equivalent vector addition system.

In fact, vector addition systems are equivalent to self-loop-free Petri nets. Remember that with a self-loop, the change is zero, but the number of tokens in the self-loop place must be nonzero. This does not diminish the power of vector addition systems, since we have seen (in Section 5.3) that self-loop-free Petri nets are equivalent to general Petri nets. However, to more directly model Petri nets with self-loops in a vector addition system-like model, Keller has defined *vector replacement systems* [Keller 1972].

DEFINITION 8-3 A *vector replacement system* consists of a start vector $s \geqslant 0$ and m pairs of vectors (U_i, V_i) such that $U_i \leqslant V_i$.

The U_i vectors are called *test* vectors. The reachability set is redefined such that s is in the reachability set, and if x is in the reachability set and $x + U_i \geqslant 0$, then $x + V_i$ is in the reachability set.

The vector replacement system model explicitly separates the test for enabling a transition from the action of firing the transition. The equivalence of vector replacement systems to (general) Petri nets is obvious.

Adding vector addition systems and vector replacement systems to our hierarchy yields Figure 8.20. The importance of vector addition

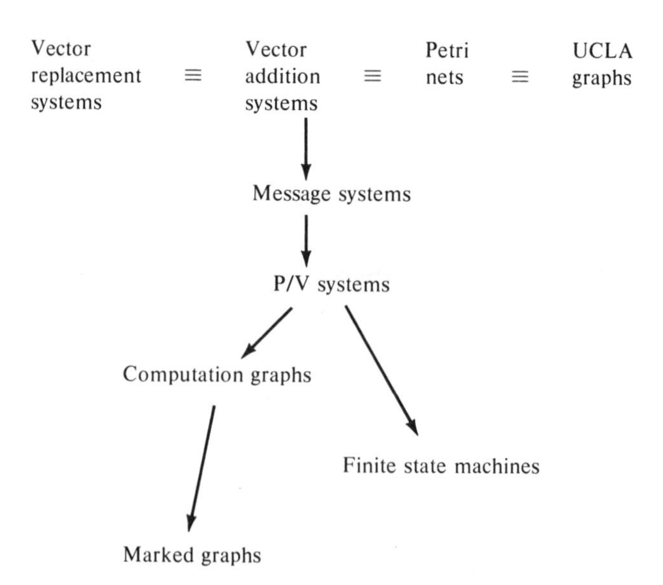

Figure 8.20 Adding vector addition systems and vector replacement systems to the hierarchy.

and replacement systems stems from their concise mathematical definition and the usefulness of this definition in proofs of the mathematical properties of systems.

8.8 Extended Petri Net Models

As a last addition to our hierarchy, we again mention the extended Petri net models studied in Chapter 7: Petri nets with constraints, exclusive-OR transitions, switches, inhibitor arcs, priorities, or times. We have seen that all of these models are equivalent to Turing machines. Thus these models properly include Petri net models. Our final hierarchy of models is shown in Figure 8.21.

8.9 Further Reading

The studies by Peterson and Bredt [1974], Agerwala [1974b], and Lipton et al. [1974] should be read first, since these are the most directly related works. Also read the surveys by Bredt [1970a] and Baer [1973a]

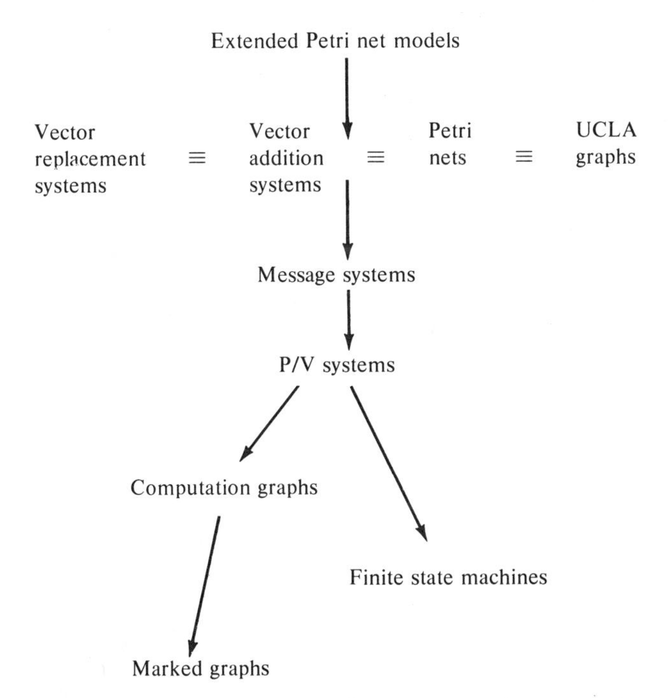

Figure 8.21 The complete hierarchy of models of parallel computation.

and the work of Miller [1973; 1974]. The references in these papers will lead to the original work on the individual models.

Riddle's model [Riddle 1972] would appear to be the best for modeling large software systems and deserves detailed study.

8.10 Topics for Further Study

1. Expand the hierarchy of Figure 8.21 by including the restricted Petri net models discussed in Chapter 7: free-choice Petri nets and simple Petri nets.
2. Investigate the properties of the languages defined by the classes of models of this chapter and relate them to regular, context-free, and context-sensitive languages.
3. Determine the decidability of the reachability problem for each of the classes of models discussed in this chapter.
4. Extend the work of this chapter to include the models of Adams [1968; 1970], Luconi [1968], Rodriguez [1967], Seitz [1971], and Slutz [1968].

A

A Brief Theory of Bags

Set theory has long been useful in mathematics and computer science. *Bag theory* is a natural extension of set theory. A bag, like a set, is a collection of elements over some domain. However, unlike a set, bags allow *multiple* occurrences of elements. In set theory, an element is either a member of a set or not a member of a set. In bag theory, an element may be in a bag zero times (not in the bag) or one time, two times, three times, or any specified number of times.

Bag theory has been developed in [Cerf et al. 1971] and [Peterson 1976].

As examples, consider the following bags over the domain $\{a,b,c,d\}$:

$$B_1 = \{a,b,c\}$$

$$B_2 = \{a\}$$

$$B_3 = \{a,b,c,c\}$$

$$B_4 = \{a,a,a\}$$

$$B_5 = \{b,c,b,c\}$$

$$B_6 = \{c,c,b,b\}$$

$$B_7 = \{a,a,a,a,a,b,b,c,d,d,d,d,d,d,d\}$$

Some bags are sets (B_1 and B_2, for example). Also, as with sets, the order of the elements is not important, so B_5 and B_6 are the same bag. (Ordered bags are sequences.)

In set theory, the basic concept is the *is a member of* relationship.

This relationship is between elements and sets and defines which elements are members of which sets. The basic concept of bag theory is the *number of occurrences* function. This function defines the number of occurrences of an element in a bag. For an element x and a bag B, we denote the number of occurrences of x in B by $\#(x,B)$ (pronounced "the number of x in B").

With this conceptual basis, we can define the fundamentals of bag theory. Most of the concepts and notation are borrowed from set theory in the obvious way. If we restrict the number of elements in a bag such that $0 \leqslant \#(x,B) \leqslant 1$, then set theory results.

A.1 Membership

The $\#(x,B)$ function defines the number of occurrences of an element x in a bag B. From this it follows that $\#(x,B) \geqslant 0$ for all x and B. We distinguish the zero and the nonzero cases. An element x is a *member* of a bag B if $\#(x,B) > 0$. This is denoted $x \in B$. Similarly, if $\#(x,B) = 0$, then $x \notin B$.

We define the empty bag $\varnothing$ with no members: For all x, $\#(x, \varnothing) = 0$.

A.2 Cardinality

The *cardinality* $|B|$ of a bag B is the total number of occurrences of elements in the bag:

$$|B| = \sum_{x} \#(x,B)$$

A.3 Bag Inclusion and Equality

A bag A is a *subbag* of a bag B (denoted $A \subseteq B$) if every element of A is also an element of B at least as many times:

$$A \subseteq B \text{ iff } \#(x,A) \leqslant \#(x,B) \text{ for all } x$$

Two bags are *equal* ($A = B$) if $\#(x,A) = \#(x,B)$ for all x.

From these definitions, we can immediately show that

$$A = B \text{ iff } A \subseteq B \text{ and } B \subseteq A$$

$$\varnothing \subseteq B \text{ for all bags } B$$

$$A = B \text{ implies } |A| = |B|$$

$$A \subseteq B \text{ implies } |A| \leqslant |B|$$

A bag A is *strictly contained* in a bag B ($A \subset B$) if $A \subseteq B$ and $A \neq B$. Note that $\#(x,A) < \#(x,B)$ does not follow from $A \subset B$, although we do have that $|A| < |B|$.

A.4 Operations

Four operations are defined on bags. For two bags A and B, we define

Bag union	$A \bigcup B$	$\#(x,A \bigcup B) = \max(\#(x,A),\#(x,B))$
Bag intersection	$A \bigcap B$	$\#(x,A \bigcap B) = \min(\#(x,A),\#(x,B))$
Bag sum	$A + B$	$\#(x,A + B) = \#(x,A) + \#(x,B)$
Bag difference	$A - B$	$\#(x,A - B) = \#(x,A) - \#(x,A \bigcap B)$

These operators have most of the properties that would be expected. Union, intersection, and sum are commutative and associative. In addition, the expected inclusions hold:

$$A \bigcap B \subseteq A \subseteq A \bigcup B$$

$$A - B \subseteq A \subseteq A + B$$

The distinction between union and sum is clearly stated by

$$|A \bigcup B| \leqslant |A| + |B|$$

$$|A + B| = |A| + |B|$$

Unfortunately no such simple statement distinguishes $A \bigcap B$ from $A - B$. The latter is complicated by the impossibility of removing elements from a bag which are not there.

A.5 Bag Spaces

We define a *domain* D as a set of elements from which bags are constructed. The bag space D^n is the set of all bags whose elements are in D such that no element occurs more than n times. That is, for all $B \in D^n$,

1. $x \in B$ implies $x \in D$.
2. $\#(x,B) \leqslant n$ for all x.

The set D^∞ is the set of all bags over a domain D; there is no limit on the number of occurrences of an element in a bag.

A.6 Parikh Mappings

For a finite domain $D = \{d_1, d_2, \ldots, d_n\}$, there is a natural correspondence between each bag B over D and the n-vector $f = (f_1, f_2, \ldots, f_n)$ defined by

$$f_i = \#(d_i, B)$$

This vector is known as the Parikh mapping [Parikh 1966].

A.7 Examples

Let $D = \{a,b,c,d\}$ be a domain. Then for the following bags,

$$A = \{a,b\}$$

$$B = \{a,a,b,c\}$$

$$C = \{a,a,a,c,c\}$$

we have

$$|A| = 2$$

$$|C| = 5$$

$$A \cup B = \{a,a,b,c\} = B$$

$$A \cup C = \{a,a,a,b,c,c\} = B \cup C$$

$$A \cap C = \{a\}$$

$$B \cap C = \{a,a,c\}$$

$$A + B = \{a,a,a,b,b,c\}$$

$$A - B = \varnothing$$

$$C - A = \{a,a,c,c\}$$

$$C - B = \{a,c\}$$

Annotated Bibliography

This annotated bibliography brings to a close our presentation on the theory of Petri nets. As can be expected, we have not been able to present all of the work which has been done on Petri nets. More complete information is available in the scientific literature.

Unfortunately much of the research on Petri nets is available only as technical reports from the institutions where the research has been done; only a fraction appears in journals with wide circulation. Even the Petri net literature which has appeared in journals is scattered. Thus, there may be some problems in finding these works.

There are three main sources for literature on Petri nets: journals, reports, and conferences. Theoretical journals in computer science are the most readily available source. These journals would include *Theoretical Computer Science*, *Journal of Computer and System Sciences*, *Information and Control*, and *Journal of the ACM*.

The papers published in these journals are often available first as technical reports. These reports are generally available only from the author(s) or the issuing agency, typically a department of a university. Many departments issue Ph.D. dissertations as technical reports; alternatively these are available from University Microfilms in Ann Arbor, Michigan.

A major source of Petri net research has been the work performed at the Massachusetts Institute of Technology (M.I.T.). Originally these reports were issued by Project MAC; Project MAC has since changed its name to the Laboratory for Computer Science. Another important source of research reports is the Institut fur Informationssystemforschung of the Gesellschaft fur Mathematik und Datenverarbeitung in Bonn, West Germany.

A final source of research results in Petri net theory is the proceedings of conferences. Work on Petri nets is very scattered among conferences with few predictable forums for presenting new results. The Annual Symposium on Switching and Automata Theory (now called the Symposium on the Foundations of Computer Science) has often had some presentations related to Petri nets. Petri net research is also reported at the Symposium on Mathematical Foundations of Computer Science, the Allerton Conference on Communication, Control and Computing, the Design Automation Conference, the ACM Symposium on Theory of Computing, the Sagamore Computer Conference on Parallel Processing, the Conference on Information Sciences and Systems, the International Computing Symposium, and the IFIP Congress. The proceedings of these conferences are often available from ACM, IEEE, North-Holland Publishing Company, or Springer-Verlag.

This bibliography is mainly the result of my extensive research on Petri nets. An extensive search was made of journals and conference proceedings for papers on Petri nets. In addition, any reference listed by a new paper which was about Petri nets was also searched for until no new references were produced. I have read most of these references, but there are some which I have not been able to find (their existence is known only by being referenced in some paper which I have found) and others which I cannot read because they are not in English. I hope that this bibliography will be of use to you in any further research you may do on Petri nets.

[Abraham 1970] S. Abraham, "On Matrix Grammars," Technical Report 3, Department of Computer Sciences, Technion — Israel Institute of Technology, Haifa, Israel, (April 1970), 12 pages.

[Adams 1968] D. Adams, "A Computational Model with Data Flow Sequencing," Ph.D. dissertation, Computer Science Department, Stanford University, Stanford, California, (December 1968), 134 pages; Also Technical Report 117, Computer Science Department, Stanford University, Stanford, California, (December 1968), 130 pages.

[Adams 1970] D. Adams, "A Model for Parallel Computations," *Parallel Processor Systems, Technologies and Applications*, New York: Spartan Books, (1970), pages 311-334.

[Agerwala 1974a] T. Agerwala, "A Complete Model for Representing the Coordination of Asynchronous Processes," Hopkins Computer Research Report Number 32, Computer Science Program, Johns Hopkins University, Baltimore, Maryland, (July 1974), 58 pages.

An extended Petri net model is defined which allows inhibitor arcs. It is shown that this extended model is equivalent to a Turing machine model of computation. Since a Turing machine can model any algorithmic scheme for coordinating processes, it is called complete. The extended Petri net model is also complete.

[Agerwala 1974b] T. Agerwala, "An Analysis of Controlling Agents for Asynchronous Processes," Hopkins Computer Research Report Number 35, Computer Science Program, Johns Hopkins University, Baltimore, Maryland, (August 1974), 85 pages.

Eighteen models of parallel computation as well as nine variants of Petri nets are analyzed and compared to create a lattice of models, similar to the lattice of Chapter 8. These results are compatible with the results of [Peterson and Bredt 1974] but independently derived and more ambitious.

[Agerwala 1978] T. Agerwala, "Some Applications of Petri Nets," *Proceedings of the 1978 National Electronics Conference*, Volume 23, (October 1978), pages 149-154; also "Putting Petri Nets to Work," *Computer*, Volume 12, Number 12, (December 1979), pages 85-94.

An excellent, although short, paper on how Petri nets can be used to model systems with concurrency. A reasonable alternative to Chapter 3 of this text. A variety of applications are presented including computer software, hardware, asynchronous circuits, design languages, and some novel applications.

[Agerwala and Flynn 1973] T. Agerwala, and M. Flynn, "Comments on Capabilities, Limitations and 'Correctness' of Petri Nets," Hopkins Computer Research Report Number 26, Computer Science Program, Johns Hopkins University, Baltimore, Maryland, (July 1973), 58 pages; Also *Proceedings of the First Annual Symposium on Computer Architecture*, New York: ACM, (1973), pages 81-86.

[Agerwala and Flynn 1976a] T. Agerwala, and M. Flynn, "On the Completeness of Representation Schemes for Concurrent Systems," Unpublished, (1976), 16 pages.

[Agerwala and Flynn 1976b] T. Agerwala, and M. Flynn, "Modeling with Extended Petri Nets," Unpublished, (1976), 18 pages.

[Anderson et al. 1967] D. Anderson, F. Sparacio, and R. Tomasulo, "The IBM System/360 Model 91: Machine Philosophy and Instruction Handling," *IBM Journal of Research and Development*, Volume 11, Number 1, (January 1967), pages 8-24.

The 360/91 computer was designed to utilize a great deal of parallelism to provide high performance. The control unit should therefore be able to be modeled by a Petri net if Petri nets are good for modeling parallel hardware. This paper describes the basic operation of the control unit of the 360/91.

[Andre et al. 1979] C. Andre, M. Diaz, C. Girault, and J. Sifakis, "Survey of French Researches and Applications Based on Petri Nets," Advanced Course on General Net Theory of Processes and Systems, Hamburg, (October 1979); also Lecture Notes in Computer Science, Berlin: Springer-Verlag, (1980).

[Anshel 1976] M. Anshel, "Decision Problems for HNN Groups and Vector Addition Systems," *Mathematics of Computation*, Volume 30, Number 133, (January 1976), pages 154-156.

[Araki and Kasami 1976] T. Araki, and T. Kasami, "Some Decision Problems

Related to the Reachability Problem for Petri Nets," *Theoretical Computer Science*, Volume 3, Number 1, (October 1976), pages 85-104.

[Araki and Kasami 1977] T. Araki, and T. Kasami, "Decidable Problems on the Strong Connectivity of Petri Net Reachability Sets," *Theoretical Computer Science*, Volume 4, Number 1, (February 1977), pages 99-119.

[Ayache et al. 1979] J. Ayache, M. Diaz, and R. Valette, "A Methodology for Specifying Control in Electronic Switching Systems," *Proceedings of the International Switching Symposium*, Paris, (May 1979), pages 1049-1056.

[Azema et al. 1975] P. Azema, M. Diaz, and J. Doucet, "Multilevel Description Using Petri Nets," *Proceedings of the 1975 International Symposium on Computer Hardware Description Languages and Their Applications*, New York: IEEE, (September 1975), pages 188-190.

[Azema et al. 1976] P. Azema, R. Valette, and M. Diaz, "Petri Nets as a Common Tool for Design Verification and Hardware Simulation," *Proceedings 13th Design Automation Conference*, New York: IEEE, (June 1976), pages 109-116.

[Azema et al. 1977] P. Azema, R. Valette, and M. Renalier, "Programme de Simulation et d'Analyse des Schemas a Reseaux de Petri, en Langage APL," AFCET Journees sur les Reseaux de Petri (AFCET Workshop on Petri Nets), Paris, France, (March 1977), pages 73-88.

[Baer 1968] J. Baer, "Graph Models of Computations in Computer Systems," Ph.D. dissertation, Department of Electrical Engineering, University of California, Los Angeles, California, (1968), 223 pages.

[Baer 1973a] J. Baer, "A Survey of Some Theoretical Aspects of Multiprocessing," *Computing Surveys*, Volume 5, Number 1, (March 1973), pages 31-80.

A survey of some of the theory which has been developed for parallel computation. First the problem of representing parallelism in a program, or automatically detecting hidden parallelism if the programming language does not allow explicit representation of parallelism, is considered. Then several models of parallel computation, including Petri nets, UCLA graphs, and parallel program schemata, are presented. Finally some techniques for attempting to predict the performance of a multiprocessor system are described. An appendix describes different types of multiprocessor hardware systems.

[Baer 1973b] J. Baer, "Modeling for Parallel Computation: A Case Study," *Proceedings of the 1973 Sagamore Computer Conference on Parallel Processing*, New York: IEEE, (August 1973), pages 13-22.

A report on early work in modeling a compiler with an extended Petri net is presented here. Only modeling is of concern. The Petri net model is extended by the addition of disjunctive (OR) logic, switches, and token absorbers. Then this model is used to describe the lexical analysis phase of a compiler. This work appears to have led to the results in [Baer and Ellis 1977].

[Baer and Ellis 1977] J. Baer, and C. Ellis, "Model, Design and Evaluation of a Compiler for a Parallel Processing Environment," *IEEE Transactions on Software Engineering*, Volume SE-3, Number 6, (November 1977), pages 394-405.

This is a continuation of the work reported in [Baer 1973b]. An XPL compiler was modeled as a Petri net. This provided insight into the structure of the compiler which allowed it to be restructured as three separate stages, suitable for a three-processor pipelined compiler with a resulting speed up by a factor of two. A practical application of Petri nets.

[Baer et al. 1970] J. Baer, D. Bovet, and G. Estrin, "Legality and Other Properties of Graph Models of Computations," *Journal of the ACM*, Volume 17, Number 3, (July 1970), pages 543-554.

[Baker 1972] H. Baker, Jr., "Petri Nets and Languages," Computation Structures Group Memo 68, Project MAC, Massachusetts Institute of Technology, Cambridge, Massachusetts, (May 1972), 6 pages.

A short note describing the basic idea of associating a language with a Petri net and using this language to describe the behavior of the Petri net.

[Baker 1973a] H. Baker, Jr., "Equivalence Problems of Petri Nets," Master's thesis, Department of Electrical Engineering, Massachusetts Institute of Technology, Cambridge, Massachusetts, (May 1973), 53 pages.

The title of this thesis is somewhat misleading. Some equivalence problems are mentioned, but the major emphasis and importance concern Baker's early thoughts on Petri net languages. This is one of the first works to consider Petri nets in the context of formal language theory. Baker works with a prefix language of a Petri net and derives a canonical form for prefix languages of marked graphs.

[Baker 1973b] H. Baker, Jr., "Rabin's Proof of the Undecidability of the Reachability Set Inclusion Problem of Vector Addition Systems," Computation Structures Group Memo 79, Project MAC, Massachusetts Institute of Technology, Cambridge, Massachusetts, (July 1973), 18 pages.

Rabin was misquoted in [Karp and Miller 1968] as having proved that the equality problem for reachability sets is undecidable when in fact he showed only that the inclusion problem was undecidable. This proof was never published, but in 1972 a new proof was presented at a talk at M.I.T.. This proof is presented here. It uses Hilbert's tenth problem to show that the inclusion problem is undecidable. Hilbert's tenth problem is reduced to nondeterministic register machines, which are in turn equivalent to Petri nets. This proof is the basis of the proof given in Chapter 5 for the undecidability of the equality and subset problems for Petri net reachability sets.

[Berlin 1979] F. Berlin, "Time-Extended Petri Nets," Master's Thesis, Department of Computer Sciences, University of Texas, Austin, Texas, (August 1979), 152 pages.

[Bernstein 1966] A. Bernstein, "Program Analysis for Parallel Processing," *IEEE Transactions on Electronic Computers*, Volume EC-15, Number 5, (October 1966), pages 757-762.

An early paper on automatic detection of parallelism in a program, this paper contains the definition of Bernstein's conditions: Two activities can be executed concurrently if the output of each activity is neither an input nor an output of the other activity.

[Bernstein 1973] P. Bernstein, "Description Problems in the Modeling of Asynchronous Computer Systems," Technical Report 48, Department of

Computer Science, University of Toronto, Toronto, Ontario, (January 1973), 116 pages.

This report mentions Petri nets as a model of parallel computation.

[Bernstein and Tsichritzis 1974] P. Bernstein, and D. Tsichritzis, "Models for Description of Computer Systems," *Proceedings of the Eighth Annual Princeton Conference on Information Sciences and Systems*, Princeton University, Princeton, New Jersey, (March 1974), pages 340-343.

[Berthelot 1977a] G. Berthelot, "Checking Liveness of Petri-Nets," *Parallel Computers — Parallel Mathematics*, Amsterdam: North-Holland, (March 1977), pages 217-220.

[Berthelot 1977b] G. Berthelot, "Une Methode de Verification des Reseaux de Petri," AFCET Journees sur les Reseaux de Petri (AFCET Workshop on Petri Nets), Paris, France, (March 1977), pages 33-54, (In French).

[Berthelot 1978] G. Berthelot, "Verification de Reseaux de Petri," These Doctorat Troisieme Cycle, Universite Pierre et Marie Curie, Paris, France, (January 1978), 220 pages, (In French).

[Berthelot and Roucairol 1976] G. Berthelot, and G. Roucairol, "Reduction of Petri Nets," *Proceedings of the Fifth Symposium on Mathematical Foundations of Computer Science*, Lecture Notes in Computer Science, Volume 45, Berlin: Springer-Verlag, (September 1976), pages 202-209.

[Best 1974] E. Best, "Beitraege zur Petrinetz-Theorie," Thesis for diploma, Institut fur Informatik, Universitat Karlsruhe, Karlsruhe, West Germany, (September 1974), (In German).

[Best 1976a] E. Best, "On the Liveness Problem of Petri Net Theory," ASM/6, Computing Laboratory, University of Newcastle upon Tyne, Newcastle upon Tyne, England, (June 1976).

[Best 1976b] E. Best, "The SOLO Operating System Described by Petri Nets," ASM/8, Computing Laboratory, University of Newcastle upon Tyne, Newcastle upon Tyne, England, (August 1976).

[Best 1979a] E. Best, "Path Programs," Advanced Course on General Net Theory of Processes and Systems, Hamburg, (October 1979); Also Lecture Notes in Computer Science, Berlin: Springer-Verlag, (1980).

[Best 1979b] E. Best, "Aspects of Occurrence Nets," Advanced Course on General Net Theory of Processes and Systems, Hamburg, (October 1979); Also Lecture Notes in Computer Science, Berlin: Springer-Verlag, (1980).

[Best and Schmid 1975] E. Best, and H. Schmid, "Systems of Open Paths in Petri Nets," *Proceedings of the Fourth Symposium on Mathematical Foundations of Computer Science*, Lecture Notes in Computer Science, Volume 32, Berlin: Springer-Verlag, (September 1975), pages 186-193.

[Bredt 1970a] T. Bredt, "A Survey of Models for Parallel Computing," Technical Report 8, Digital Systems Laboratory, Stanford University, Stanford, California, (August 1970), 58 pages; Also Technical Report STAN-CS-70-171, Computer Science Department, Stanford University, Stanford, California, (August 1970), 58 pages.

[Bredt 1970b] T. Bredt, "Analysis of Parallel Systems," Technical Report 7, Digital Systems Laboratory, Stanford University, Stanford, California, (August 1970), 59 pages; Also *IEEE Transactions on Computers*, Volume C-20, Number 11, (November 1971), pages 1403-1407.

[Bredt and McCluskey 1970] T. Bredt, and E. McCluskey, "A Model for Parallel Computer Systems," Technical Report 5, Digital Systems Laboratory, Stanford University, Stanford, California, (April 1970), 62 pages; Also Technical Report STAN-CS-70-160, Department of Computer Sciences, Stanford University, Stanford, California, (April 1970), 62 pages.

[Brinch Hansen 1973] P. Brinch Hansen, "Concurrent Programming Concepts," *Computing Surveys*, Volume 5, Number 4, (December 1973), pages 223-245.

[Brinsfield and Miller 1971] W. Brinsfield, and R. Miller, "On the Composition of Parallel Program Schemata," *Conference Record 1971 12th Annual Symposium on Switching and Automata Theory*, New York: IEEE, (October 1971), pages 20-23.

[Bruno et al. 1972] J. Bruno, E. Coffman, and W. Hosken, "Consistency of Synchronization Nets Using P and V Operations," Technical Report 117, Computer Science Department, Pennsylvania State University, University Park, Pennsylvania, (June 1972); Also *Proceedings of the 13th Annual Symposium on Switching and Automata Theory*, New York: IEEE, (October 1972), pages 71-77.

[Cardoza 1975] E. Cardoza, "Computational Complexity of the Word Problem for Commutative Semigroups," Master's thesis, Department of Electrical Engineering and Computer Science, Massachusetts Institute of Technology, Cambridge, Massachusetts, (August 1975), 67 pages; Also Technical Memorandum 67, Project MAC, Massachusetts Institute of Technology, Cambridge, Massachusetts, (October 1975), 67 pages.

Comments on the close relation of commutative semigroups to vector addition systems and Petri nets.

[Cardoza et al. 1976] E. Cardoza, R. Lipton, and A. Meyer, "Exponential Space Complete Problems for Petri Nets and Commutative Subgroups," *Proceedings of the Eighth Annual ACM Symposium on Theory of Computing*, New York: ACM, (May 1976), pages 50-54.

[Cerf 1972] V. Cerf, "Multiprocessors, Semaphores, and a Graph Model of Computation," Ph.D. dissertation, Computer Science Department, University of California, Los Angeles, California, (April 1972), 337 pages.

One of a series of reports and dissertations developing the UCLA graph model of computation.

[Cerf et al. 1971] V. Cerf, E. Fernandez, K. Gostelow, and S. Volansky, "Formal Control Flow Properties of a Model of Computation," Report ENG-7178, Computer Science Department, University of California, Los Angeles, California, (December 1971), 81 pages.

[Cerf et al. 1972] V. Cerf, K. Gostelow, G. Estrin, and S. Volansky, "Proper Termination of Flow of Control in Programs Involving Concurrent

Processes,'' *SIGPLAN Notices*, Volume 7, Number 11, (November 1972), pages 15-27; Also *Proceedings of the ACM National Conference*, ACM, New York, (August 1972), pages 742-754.

[Chen 1971] T. Chen, ''Overlap and Pipeline Processing,'' in H. Stone, (Editor), *Introduction to Computer Architecture*, Chicago: Science Research Associates, (1975), pages 375-431.

This article is a good introduction to pipelined computer architectures which can be modeled by Petri nets, as in Chapter 3.

[Commoner 1972] F. Commoner, ''Deadlocks in Petri Nets,'' Report CA-7206-2311, Massachusetts Computer Associates, Wakefield, Massachusetts, (June 1972), 50 pages.

A deadlock is a situation where two activities are each waiting for the other to proceed before they proceed themselves. Liveness is (in some sense) the opposite of deadlock. Commoner examines various degrees of deadlock, or liveness, and obtains a sufficient condition for liveness.

[Commoner et al. 1971] F. Commoner, A. Holt, S. Even, and A. Pnueli, ''Marked Directed Graphs,'' *Journal of Computer and System Sciences*, Volume 5, Number 5, (October 1971), pages 511-523.

Marked graphs are a subclass of Petri nets which exhibit concurrency but not conflict. This paper is the basic presentation of marked graphs and results. Algorithms are given to solve the safeness, liveness, and reachability problems among other things.

[Cooper 1971] D. Cooper, ''Programs for Mechanical Program Verification,'' in B. Meltzer and D. Michie (Editors), *Machine Intelligence 6*, New York: Halstead Press, (1971), pages 43-59.

Pages 44 to 47 describe a Presburger arithmetic algorithm which is used to prove simple theorems about algebraic formulas.

[Cooprider 1976] L. Cooprider, ''Petri Nets and the Representation of Standard Synchronizations,'' Department of Computer Science, Carnegie-Mellon University, Pittsburgh, Pennsylvania, (January 1976), 30 pages.

Some of the presentation in Chapter 3 on the use of Petri nets to model computer software is based on this draft. The presentation is somewhat mechanical, but the basic contents present and analyze many of the classical problems in process synchronization.

[Cotronis and Lauer 1977] J. Cotronis, and P. Lauer, ''Verification of Concurrent Systems of Processes,'' Technical Report 97, Computing Laboratory, University of Newcastle upon Tyne, Newcastle upon Tyne, England, (January 1977), 11 pages; Also *Proceedings of the International Computing Symposium 1977*, Amsterdam: North-Holland, (April 1977), pages 197-207.

[Courtois et al. 1971] P. Courtois, F. Heymans, and D. Parnas, ''Concurrent Control with 'Readers' and 'Writers',,'' *Communications of the ACM*, Volume 14, Number 10, (October 1971), pages 667-668.

[Courvoisier 1977] M. Courvoisier, ''Une Architecture Parallele Asynchrone pour les Systemes de Commande,'' AFCET Journees sur les Reseaux de Petri (AFCET Workshop on Petri Nets), Paris, France, (March 1977), pages 56-72, (In French).

[Cox 1978] L. Cox, Jr., "Predicting Concurrent Computer System Performance Using Petri Net Models," *Proceedings of the 1978 ACM National Conference*, New York: ACM, (December 1978), pages 901-913.

[Crespi-Reghizzi and Mandrioli 1974] S. Crespi-Reghizzi, and D. Mandrioli, "Petri Nets and Commutative Grammars," Internal Report 74-5, Laboratorio di Calcolatori, Instituto di Electtrotecnica ed Elettronica del Politecnico di Milano, Italy, (March 1974), 80 pages.

An attempt is made to characterize Petri net systems. First an algebraic characterization, based on the matrix representation of Petri nets, is investigated; then a language approach is used to try to characterize Petri nets by their sequences of transition firings. Petri nets are shown to be equivalent to commutative phrase structure grammars, grammars in which the ordering of symbols is irrelevant. The report is a well written and understandable treatment of basic Petri net properties.

[Crespi-Reghizzi and Mandrioli 1975] S. Crespi-Reghizzi, and D. Mandrioli, "A Decidability Theorem for a Class of Vector Addition Systems," *Information Processing Letters*, Volume 3, Number 3, (January 1975), pages 78-80.

It is shown that the reachability problem for conflict-free Petri nets is decidable, and a brief description of the algorithm for finding a firing sequence leading from one marking to the other (if the marking is reachable) is given.

[Crespi-Reghizzi and Mandrioli 1976] S. Crespi-Reghizzi, and D. Mandrioli, "Some Algebraic Properties of Petri Nets," *Alta Frequenza*, Volume 45, Number 2, (February 1976), pages 130-137.

[Crespi-Reghizzi and Mandrioli 1977] S. Crespi-Reghizzi, and D. Mandrioli, "Petri Nets and Szilard Languages," *Information and Control*, Volume 33, Number 2, (February 1977), pages 177-192.

[Crespi-Reghizzi and Schreiber 1972] S. Crespi-Reghizzi, and F. Schreiber, "Le Reti di Petri: Un Modello per la Descrizione dei Sistemi Asincroni," *Rivista di Informatica*, Volume 3, (1972), pages 3-20, (In Italian).

[Crowley and Noe 1975] C. Crowley, and J. Noe, "Interactive Graphical Simulation Using Modified Petri Nets," *Proceedings of Symposium on Simulation of Computer Systems*, New York: ACM, (August 1975), pages 177-184.

[Dadda 1976] L. Dadda, "The Synthesis of Petri Nets for Controlling Purposes and the Reduction of their Complexity," *Proceedings of the Second Euromicro Symposium on Microprocessing and Microprogramming*, Amsterdam: North-Holland, (October 1976), pages 251-259.

[Davis 1957] M. Davis, "A Program for Presburger's Algorithm," *Summer Institute for Symbolic Logic*, Cornell University, Ithaca, New York, (1957), pages 215-233.

A machine language program for the computer at the Institute for Advanced Study is described. The program will determine the validity of a formula in Presburger arithmetic (subject to a limit of 96 symbols). The value of this paper is its explanation of Presburger's algorithm before the program.

[Davis 1958] M. Davis, *Computability and Unsolvability*, New York: McGraw-Hill, (1958), 210 pages.

A classic work on computability theory.

[Davis 1973] M. Davis, "Hilbert's Tenth Problem Is Unsolvable," *The American Mathematical Monthly*, Volume 80, Number 3, (March 1973), pages 233-269.

Davis reports the complete proof by Matijasevic of the undecidability of Hilbert's tenth problem.

[Davis and Hersh 1973] M. Davis, and R. Hersh, "Hilbert's 10th Problem," *Scientific American*, Volume 229, Number 5, (November 1973), pages 84-91.

A popular and very readable explanation of the proof that Hilbert's tenth problem is undecidable.

[Dennis 1970a] J. Dennis, "Modular, Asynchronous Control Structures for a High Performance Processor," *Record of the Project MAC Conference on Concurrent Systems and Parallel Computation*, New York: ACM, (June 1970), pages 55-80.

Examples of the use of Petri nets for the description of the control mechanisms of complex computers are given. The eventual goal would be to develop automatic mechanisms for implementing the Petri net as a digital system.

[Dennis 1970b] J. Dennis (Editor), *Record of the Project MAC Conference on Concurrent Systems and Parallel Computation*, New York: ACM, (June 1970), 199 pages.

The Woods Hole Conference brought together 27 researchers on parallel computation for a few days and eventually resulted in this proceedings. Papers dealing with Petri nets include [Holt and Commoner 1970; Dennis 1970a; Seitz 1970; Patil 1970b] and there is an excellent bibliography.

[Dennis 1972] J. Dennis, "Concurrency in Software Systems," Computation Structures Group Memo 65-1, Project MAC, Massachusetts Institute of Technology, Cambridge, Massachusetts, (June 1972), 18 pages; Also *Advanced Course in Software Engineering*, Berlin: Springer-Verlag, (1973), pages 111-127; Also Lecture Notes in Computer Science, Volume 30, Berlin: Springer-Verlag, (1975), pages 111-127.

Petri nets are used as a model of parallel computation to illustrate the problems of determinacy and communication between processes.

[Dennis and Patil 1971] J. Dennis, and S. Patil, "Speed Independent Asynchronous Circuits," Computation Structures Group Memo 54, Project MAC, Massachusetts Institute of Technology, Cambridge, Massachusetts, (January 1971), 5 pages; Also *Proceedings Fourth Hawaii International Conference on Systems Sciences*, University of Hawaii, Honolulu, Hawaii, (January 1971), pages 55-58.

[Dennis and Van Horn 1966] J. Dennis, and E. Van Horn, "Programming Semantics for Multiprogrammed Computations," *Communications of the ACM*, Volume 9, Number 3, (March 1966), pages 143-155.

[DeVillers and Louchard 1973] R. DeVillers, and G. Louchard, "Realization of Petri Nets Without Conditional Statements," *Information Processing Letters*, Volume 2, Number 4, (October 1973), pages 105-107.

[Dijkstra 1965] E. Dijkstra, "Solution of a Problem in Concurrent Program Control," *Communications of the ACM*, Volume 8, Number 9, (September 1965), page 569.

[Dijkstra 1968] E. Dijkstra, "Cooperating Sequential Processes," in F. Genuys (Editor), *Programming Languages*, New York: Academic Press, (1968), pages 43-112.

This classic paper first introduced the P and V operations on semaphores.

[Ellis 1977] C. Ellis, "Consistency and Correctness of Duplicate Database Systems," Xerox Systems Science Laboratory, Palo Alto Research Center, Palo Alto, California, (May 1977), 28 pages.

[Fernandez 1975] C. Fernandez, "Net Topology I," Internal Report 75-9, Institut fur Informationssystemforschung, Gesellschaft fur Mathematik und Datenverarbeitung, Bonn, West Germany, (December 1975).

[Fernandez 1976] C. Fernandez, "Net Topology II," Internal Report 76-2, Institut fur Informationssystemforschung, Gesellschaft fur Mathematik und Datenverarbeitung, Bonn, West Germany, (April 1976).

General net theory has been under development by Petri and others since the early work which lead to the Petri net model. These two reports, "Net Topology I" and "Net Topology II," are concerned with the development of the algebraic (mathematical) properties of general net theory. Many definitions, theorems, and proofs; no intuition.

[Foo and Musgrave 1975] S. Foo, and G. Musgrave, "Comparison of Graph Models for Parallel Computation and Their Extension," *Proceedings of the 1975 International Symposium on Computer Hardware Description Languages and Their Applications*, New York: IEEE, (September 1975), pages 16-21.

[Furtek 1971] F. Furtek, "Modular Implementation of Petri Nets," Master's thesis, Department of Electrical Engineering, Massachusetts Institute of Technology, Cambridge, Massachusetts, (September 1971), 136 pages.

[Furtek 1975] F. Furtek, "A New Approach to Petri Nets," Computation Structures Group Memo 123, Project MAC, Massachusetts Institute of Technology, Cambridge, Massachusetts, (April 1975), 26 pages.

[Furtek 1976] F. Furtek, "The Logic of Systems," Ph.D. dissertation, Department of Electrical Engineering, Massachusetts Institute of Technology, Cambridge, Massachusetts, (July 1976), 176 pages; Also Technical Report 170, Laboratory for Computer Science, Massachusetts Institute of Technology, Cambridge, Massachusetts, (December 1976), 176 pages.

[Genrich 1971] H. Genrich, "Einfache nicht-sequentielle Prozesse," (Simple Nonsequential Processes), Report 37, Institut fur Informationssystemforschung, Gesellschaft fur Mathematik und Datenverarbeitung, Bonn, West Germany, (1971), (In German).

Mentioned in [Holt and Commoner 1970] as being a thorough study of marked graphs.

[Genrich 1976] H. Genrich, "Extended Simple Regular Expressions," Lecture Notes in Computer Science, Volume 32, Berlin: Springer-Verlag, (1975), pages 231-237.

[Genrich 1976] H. Genrich, "The Petri Net Representation of Mathematical Knowledge," Internal Report 76-5, Institut fur Informationssystem-forschung, Gesellschaft fur Mathematik und Datenverarbeitung, Bonn, West Germany, (May 1976), 30 pages.

Uses Petri nets to represent theorems, axioms, lemmas, and their relationships.

[Genrich and Lautenbach 1973] H. Genrich, and K. Lautenbach, "Synchronisationsgraphen," *Acta Informatica*, Volume 2, Number 2, (1973), pages 143-161, (In German).

[Genrich and Lautenbach 1978] H. Genrich, and K. Lautenbach, "Facts in Place/Transition-Nets," *Proceedings of the Seventh Symposium on Mathematical Foundations of Computer Science 1978*, Lecture Notes in Computer Science, Volume 64, Berlin: Springer-Verlag, (September 1978), pages 213-231.

[Genrich and Stankiewicz-Wiechno 1979] H. Genrich and E. Stankiewicz-Wiechno, "A Dictionary of Some Basic Notions of Net Theory," Advanced Course on General Net Theory of Processes and Systems, Hamburg, (October 1979); Also Lecture Notes in Computer Science, Berlin: Springer-Verlag, (1980).

[Genrich and Thieler-Mevissen 1976] H. Genrich, and G. Thieler-Mevissen, "The Calculus of Facts," *Proceedings of the Fifth Symposium on the Mathematical Foundations of Computer Science*, Lecture Notes in Computer Science, Volume 45, Berlin: Springer-Verlag, (September 1976), pages 588-595.

A representation of first-order predicate calculus as a Petri net is given.

[Genrich et al. 1979] H. Genrich, K. Lautenbach, and P. Thiagarajan, "An Overview of Net Theory," Advanced Course on General Net Theory of Processes and Systems, Hamburg, (October 1979); Also Lecture Notes in Computer Science, Berlin: Springer-Verlag, (1980).

[Ghosh 1977a] S. Ghosh, "Some Comments on Timed Petri Nets," AFCET Journees sur les Reseaux de Petri (AFCET Workshop on Petri Nets), Paris, France, (March 1977), pages 213-226.

[Ghosh 1977b] S. Ghosh, "Structured Petri Nets," Report 49/77, Lehrstuhl Informatik, Universitat Dortmund, West Germany, (August 1977), 27 pages.

A subclass of Petri nets is defined which can be hierarchically combined to create larger nets whose liveness is guaranteed.

[Gilbert and Chandler 1972] P. Gilbert, and W. Chandler, "Interference Between Communicating Processes," *Communications of the ACM*, Volume 15, Number 3, (March 1972), pages 171-176.

A finite state machine model of parallel computation, using shared memory for communication.

[Gill 1962] A. Gill, *Introduction to the Theory of Finite State Machines*, New York: McGraw-Hill, (1962), 207 pages.

[Ginsburg 1966] S. Ginsburg, *The Mathematical Theory of Context Free Languages*, New York: McGraw-Hill, (1966), 232 pages.

A very mathematical treatment of a part of formal language theory.

The work on bounded languages, Parikh's theorem, and semilinear sets relates to work in Petri net languages.

[Ginsburg 1975] S. Ginsburg, *Algebraic and Automata-Theoretic Properties of Formal Languages*, Amsterdam: North-Holland, (1975), 313 pages.

[Ginsburg and Spanier 1966] S. Ginsburg, and E. Spanier, "Semigroups, Presburger Formulas, and Languages," *Pacific Journal of Mathematics*, Volume 16, Number 2, (1966), pages 285-296.

The connection is made among context-free languages, semilinearity, and Presburger arithmetic.

[Girault 1978] C. Girault, "Reseaùx de Petri et Synchronisation de Processus," Technical Report 78.02, Institut de Programmation, University Pierre et Marie Curie, Paris, France, (1978), 19 pages, (In French).

[Godbersen and Meyer 1978] H. Godbersen and B. Meyer, "Function Nets as a Tool for the Simulation of Information Systems," *Proceedings of the Summer Computer Simulation Conference*, (July 1978), pages 46-53.

[Gostelow 1971] K. Gostelow, "Flow of Control, Resource Allocation and the Proper Termination of Programs," Ph.D. dissertation, Computer Science Department, University of California, Los Angeles, California, (December 1971), 301 pages.

One of a series of reports and dissertations developing the UCLA graph model of computation. This dissertation defines the property of termination and develops an algorithm to decide proper termination. An appendix shows that Petri nets and UCLA graphs are almost equivalent, but the constructions mistakenly associate the places of a Petri net with the nodes of the UCLA graph, rather than the correct association with the arcs of the graph, so the results are misleading.

[Gostelow 1975] K. Gostelow, "Computation Modules and Petri Nets," *Proceedings of the Third IEEE-ACM Milwaukee Symposium on Automatic Computation and Control*, New York: IEEE, (April 1975), pages 345-353.

The property of proper termination of a Petri net is defined and discussed. A properly terminating Petri net can be substituted for a transition in another net without creating deadlocks.

[Grabowski 1979] J. Grabowski, "The Unsolvability of Some Petri Net Language Problems," *Information Processing Letters*, Volume 9, Number 2, (August 1979), pages 60-63.

[Hack 1972] M. Hack, "Analysis of Production Schemata by Petri Nets," Master's thesis, Department of Electrical Engineering, Massachusetts Institute of Technology, Cambridge, Massachusetts, (February 1972), 119 pages; Also Technical Report 94, Project MAC, Massachusetts Institute of Technology, Cambridge, Massachusetts, (February 1972), 119 pages; Errata: "Corrections to Analysis of Production Schemata by Petri Nets," Computation Structures Group Note 17, Project MAC, Massachusetts Institute of Technology, Cambridge, Massachusetts, (June 1974), 11 pages.

This thesis introduces free-choice Petri nets and develops their properties. Necessary and sufficient conditions for liveness and safeness are the main results. There is also some work on decomposing free-choice nets into simpler subnets.

[Hack 1973a] M. Hack, "The Godelization of Petri Nets and Vector Addition Systems," Unpublished, (May 1973), 9 pages.

[Hack 1973b] M. Hack, "Extended State-Machine Allocatable Nets (ESMA), an Extension of Free Choice Petri Net Results," Computation Structures Group Memo 78, Project MAC, Massachusetts Institute of Technology, Cambridge, Massachusetts, (May 1973), 33 pages; revised as Memo 78-1, (June 1974), 38 pages.

[Hack 1973c] M. Hack, "A Petri Net Version of Rabin's Undecidability Proof for Vector Addition Systems," Computation Structures Group Memo 94, Project MAC, Massachusetts Institute of Technology, Cambridge, Massachusetts, (December 1973), 12 pages.

A Petri net version of the undecidability of the subset problem for reachability sets is given. Rabin's original proof, as reported in [Baker 1973b], was given for vector addition systems. This proof is also given in [Hack 1974a] and [Hack 1975c] and is given here in Chapter 5.

[Hack 1974a] M. Hack, "Decision Problems for Petri Nets and Vector Addition Systems," Computation Structures Group Memo 95, Project MAC, Massachusetts Institute of Technology, Cambridge, Massachusetts, (March 1974), 79 pages; revised as Memo 95-1, (August 1974); Also Technical Memo 59, Project MAC, Massachusetts Institute of Technology, Cambridge, Massachusetts, (March 1975), 79 pages.

Three major results are collected into this report: (1) the equivalence of generalized Petri nets to restricted Petri nets (no self-loops and multiplicity of either zero or one), (2) the Petri net version of the undecidability of the subset problem for reachability sets (also [Hack 1973c]), and (3) the equivalence of the liveness and reachability problems (also [Hack 1974c]).

[Hack 1974b] M. Hack, "Petri Nets and Commutative Semigroups," Unpublished, (July 1974), 5 pages.

[Hack 1974c] M. Hack, "The Recursive Equivalence of the Reachability Problem and the Liveness Problem for Petri Nets and Vector Addition Systems," Computation Structures Group Memo 107, Project MAC, Massachusetts Institute of Technology, Cambridge, Massachusetts, (August 1974), 9 pages; Also *Proceedings of the 15th Annual Symposium on Switching and Automata Theory*, New York: IEEE, (October 1974), pages 156-164.

It is shown that the reachability problem and the liveness problems are mutually reducible.

[Hack 1975a] M. Hack, "The Equality Problem for Vector Addition Systems Is Undecidable," Computation Structures Group Memo 121, Project MAC, Massachusetts Institute of Technology, Cambridge, Massachusetts, (April 1975), 32 pages; Also *Theoretical Computer Science*, Volume 2, Number 1, (June 1976), pages 77-96.

It is shown that the subset problem for reachability sets is reducible to the equality problem. Thus since the subset problem is undecidable (see [Baker 1973b], [Hack 1973c], or [Hack 1974a]), the equality problem is undecidable also. This proof is given here in Chapter 5.

[Hack 1975b] M. Hack, "Petri Net Languages," Computation Structures

Group Memo 124, Project MAC, Massachusetts Institute of Technology, Cambridge, Massachusetts, (June 1975), 128 pages; Also Technical Report 159, Laboratory for Computer Science, Massachusetts Institute of Technology, Cambridge, Massachusetts, (March 1976), 128 pages.

Following Baker's suggestion [Baker 1972], Hack carefully investigates Petri net languages. Two basic language classes are distinguished: the prefix languages of all sequences which lead to a reachable marking and the languages of sequences leading to a defined final marking. In addition, Hack considers languages which result from allowing or not allowing transitions to be labeled with λ, the empty string. For each of these four classes, simple closure properties and characterizations are obtained. In addition these languages are related to other classes of languages, including regular languages. It is shown that membership, emptiness, and finiteness are decidable for some languages and equivalent to the reachability problem for other languages. An excellent piece of research into a new subject.

[**Hack 1975c**] M. Hack, "Decidability Questions for Petri Nets," Ph.D. dissertation, Department of Electrical Engineering, Massachusetts Institute of Technology, Cambridge, Massachusetts, (December 1975), 194 pages; Also Technical Report 161, Laboratory for Computer Science, Massachusetts Institute of Technology, Cambridge, Massachusetts, (June 1976), 194 pages.

This dissertation brings together much of Hack's work into one report. The report is an excellent treatment of Petri nets covering both basic definitions and properties and advanced research. Each topic is carefully introduced and developed. Major chapters include the following: Decidability of Boundedness and Coverability (using the reachability tree); Reachability Problems; Liveness and Persistence; Undecidability of Subset and Equality Problems for Petri net Reachability sets (as first presented in [Hack 1973c; Hack 1975a]), Petri net languages (from [Hack 1975b]). The dissertation concludes with some open questions and conjectures.

[**Han 1978**] Y. Han, "Performance Evaluation of a Digital System Using a Petri Net-like Approach," *Proceedings of the National Electronics Conference*, Volume 23, (October 1978), pages 166-172.

[**Han and Kinney 1977**] Y. Han, and L. Kinney, "Petri Net Reduction and Verification," Honeywell, Minneapolis, Minnesota, (1977), 52 pages.

[**Hebalkar 1970**] P. Hebalkar, "Deadlock-free Sharing of Resources in Asynchronous Systems," Ph.D. dissertation, Department of Electrical Engineering, Massachusetts Institute of Technology, Cambridge, Massachusetts, (September 1970), 185 pages; Also Technical Report 75, Project MAC, Massachusetts Institute of Technology, Cambridge, Massachusetts, (September 1970), 185 pages.

[**Heimerdinger 1978**] W. Heimerdinger, "A Petri Net Approach to System Level Fault Tolerance Analysis," *Proceedings of the National Electronics Conference*, Volume 23, (October 1978), pages 161-165.

[**Henhapl 1971**] W. Henhapl, "A Transformation of Marked Graphs," *Information Processing Letters*, Volume 2, Number 1, (March 1973), pages 26-29.

[Herzog 1977] O. Herzog, "Automatic Deadlock Analysis of Parallel Programs," *Proceedings of the International Computing Symposium 1977*, Amsterdam: North-Holland, (April 1977), pages 209-216.

[Holt 1971] A. Holt, "Introduction to Occurrence Systems," *Associative Information Techniques*, New York: American Elsevier, (1971), pages 175-203.

A well-organized summary of the results of the Information System Theory Project. Tutorial presentation of occurrence systems with examples and excellent discussion of concurrency and conflict in terms of occurrence systems and (thus) safe Petri nets.

[Holt 1975a] A. Holt, "Role/Activity Models and Petri Nets," *Second Semi-Annual Technical Report*, Report CADD-7503-1411, Massachusetts Computer Associates, Wakefield, Massachusetts, (March 1975), pages 5-26.

[Holt 1975b] A. Holt, "Definitions Pertaining to Petri Nets, States and Events and Behavior," *Second Semi-Annual Technical Report*, Report CADD-7503-1411, Massachusetts Computer Associates, Wakefield, Massachusetts, (March 1975), pages 58-68.

[Holt 1975c] A. Holt, "Communication Mechanics," *Second Semi-Annual Technical Report*, Report CADD-7503-1411, Massachusetts Computer Associates, Wakefield, Massachusetts, (March 1975), pages 143-168.

[Holt and Commoner 1970] A. Holt, and F. Commoner, "Events and Conditions," (in three parts), Applied Data Research, New York, (1970); Also *Record of the Project MAC Conference on Concurrent Systems and Parallel Computation*, New York: ACM, (June 1970), pages 1-52.

Systemics is the name of Holt's work to develop a science of information and systems. This paper is an excellent presentation of the fundamentals of this work and the beginnings of Petri net theory. Petri nets are defined and illustrated. Marked graphs and state machines are defined and discussed at length.

This paper relaxes the constraint of safeness in one example and thus plants the seed of modern definitions. Very readable and interesting.

[Holt et al. 1968] A. Holt, H. Saint, R. Shapiro, and S. Warshall, "Final Report of the Information System Theory Project," Technical Report RADC-TR-68-305, Rome Air Development Center, Griffiss Air Force Base, New York, (September 1968), 352 pages.

A massive work, this report presents the results of the Information System Theory Project which was concerned with finding a proper descriptive means for modeling, evaluating, and implementing systems. Petri nets were a major portion of this work and are presented here in detail. The Petri net concept itself is still a safe net. This leads to a fair amount of work on occurrence graphs and occurrence systems, an approach to representation and analysis of firing sequences which is only applicable to safe nets. After some presentation in [Holt and Commoner 1970], occurrence systems have been dropped.

Most works on Petri nets refer to this report, although it is now of mainly historical interest. Section V was revised as [Shapiro and Saint 1970].

[Hopcroft and Pansiot 1976] J. Hopcroft, and J. Pansiot, "On the Reachability

Problem for 5-Dimensional Vector Addition Systems," Technical Report 76-280, Department of Computer Science, Cornell University, Ithaca, New York, (June 1976), 42 pages.

The reachability problem for vector addition systems with up to five dimensions (Petri nets with up to five places) is studied. The reachability sets are shown to be effectively semilinear, which provides an algorithm for their solution. Also containment and equivalence, not generally decidable, are decidable for these systems. Unfortunately this approach will not work for more than five places and is, therefore, of limited practical use.

[**Hopcroft and Ullman 1969**] J. Hopcroft, and J. Ullman, *Formal Languages and Their Relation to Automata*, Reading, Massachusetts: Addison-Wesley, (1969), 242 pages.

An excellent text, this book is a classic in the field. It is a concise, yet understandable, treatment of formal language theory (regular, context-free, context-sensitive, and type-0 languages) and automata theory (finite state pushdown, linear bounded, and Turing machines). Time and space complexity issues and decidability questions are well presented. Must reading and a constant reference source for work in any of these areas.

[**Hopner and Opp 1977**] M. Hopner, and M. Opp, "Renaming and Erasing in Szilard Languages," Technical Report 35, Institut fur Informatik, University of Hamburg, (May 1977), 37 pages; Also *Automata, Languages and Programming*, Lecture Notes in Computer Science, Volume 52, Berlin: Springer-Verlag, (July 1977), pages 244-257.

[**Huen and Siewiorek 1975**] W. Huen, and D. Siewiorek, "Intermodule Protocol for Register Transfer Level Modules: Representation and Analytic Tools," *Proceedings of the Second Annual Symposium on Computer Architecture*, New York: ACM, (January 1975), pages 56-62.

Petri nets, UCLA graphs, and vector addition systems are considered as models of the interconnections of hardware modules. Analysis using the reachability tree then shows some problems with the modules.

[**Hughes 1977**] C. Hughes, "The Equivalence of Vector Addition Systems to a Subclass of Post Canonical Systems," Technical Report CS-77-22, Computer Science Department, University of Tennessee, Knoxville, Tennessee, (August 1977), 6 pages; Also *Information Processing Letters*, Volume 7, Number 4, (June 1978), pages 201-204.

[**Hyman 1966**] H. Hyman, "Comments on a Problem in Concurrent Programming Control," *Communications of the ACM*, Volume 9, Number 1, (January 1966), page 45.

[**Iazeolla 1973**] G. Iazeolla, "Modelli per Elaborazione Parallela," Technical Report 73-03, Instituto di Automatica, Universita di Roma, Rome, Italy, (March 1973), 54 pages, (In Italian).

[**Izbicki 1973**] H. Izbicki, "Report on Marked Graphs," Technical Report 25-136, IBM Vienna Laboratories, Vienna, Austria, (April 1973), 37 pages.

This is the final report on the investigations at the IBM Vienna Laboratories on marked graphs. It brings together into a consistent development the work of Izbicki, Henhapl, Hansal, and Schwab. The development is very formal, overloaded with arcane notation and theorems, and concerns marked graphs. The results are minimal.

[Jaffe 1977] J. Jaffe, "Semilinear Sets and Applications," Master's thesis, Department of Electrical Engineering and Computer Science, Massachusetts Institute of Technology, Cambridge, Massachusetts, (June 1977), 79 pages; Also Technical Report 183, Laboratory for Computer Science, Massachusetts Institute of Technology, Cambridge, Massachusetts, (July 1975), 79 pages.

[Jantzen 1979a] M. Jantzen, "On the Hierarchy of Petri Net Languages," *R.A.I.R.O. Theoretical Informatics*, Volume 13, Number 1, (1979), pages 19-30.

[Jantzen 1979b] M. Jantzen, "Structured Representation of Knowledge by Petri Nets as an Aid for Teaching and Research," Advanced Course on General Net Theory of Processes and Systems, Hamburg, (October 1979); Also Lecture Notes in Computer Science, Berlin: Springer-Verlag, (1980).

[Jantzen and Valk 1979] M. Jantzen and R. Valk, "Formal Properties of Place Transition Nets," Advanced Course on General Net Theory of Processes and Systems, Hamburg, (October 1979); Also Lecture Notes in Computer Science, Berlin: Springer-Verlag, (1980).

[Jensen 1978] K. Jensen, "Extended and Hyper Petri Nets," Technical Report DAIMI TR-5, Computer Science Department, Aarhus University, Aarhus, Denmark, (August 1978).

[Jensen 1979] K. Jensen, "Coloured Petri Nets and the Invariant Method," Technical Report DAIMI PB-104, Computer Science Department, Aarhus University, Aarhus, Denmark, (October 1979), 27 pages.

[Johnson 1970] R. Johnson, "Needed: A Measure for Measure," *Datamation*, Volume 16, Number 12, (December 1970), pages 22-70.

An introduction to Petri nets, among other subjects.

[Jones et al. 1976] N. Jones, L. Landweber, and Y. Lien, "Complexity of Some Problems in Petri Nets," Technical Report 276, Computer Science Department, University of Wisconsin, Madison, Wisconsin, (September 1976), 43 pages; Also *Theoretical Computer Science*, Volume 4, Number 3, (June 1977), pages 277-299.

Complexity theory is applied to Petri nets in this paper. Time and space bounds are given for several Petri net problems, including reachability, liveness, and safeness. It is shown that persistence is reducible to reachability. The complexity results are given both for Petri nets and for state machines, marked graphs, and free-choice Petri nets.

[Jump and Thiagarajan 1972] J. Jump, and P. Thiagarajan, "On the Equivalence of Asynchronous Control Structures," *Proceedings of the 13th Annual Symposium on Switching and Automata Theory*, New York: IEEE, (October 1972), pages 212-223.

[Karp 1972] R. Karp, "Reducibility Among Combinatorial Problems," in R. Miller and J. Thatcher (Editors), *Complexity of Computer Computations*, New York: Plenum Press, (March 1972), pages 85-103.

[Karp and Miller 1966] R. Karp, and R. Miller, "Properties of a Model for

Parallel Computation: Determinacy, Termination and Queueing," *SIAM Journal of Applied Math*, Volume 14, Number 6, (November 1966), pages 1390-1411.

The computation graph model of computation is defined and its properties investigated. This model influenced many of the later models, such as the parallel program schemata of [Karp and Miller 1968].

[Karp and Miller 1968] R. Karp, and R. Miller, "Parallel Program Schemata," RC-2053, IBM T. J. Watson Research Center, Yorktown Heights, New York, (April 1968), 54 pages; Also *Journal of Computer and System Science*, Volume 3, Number 4, (May 1969), pages 167-195; preliminary draft, *IEEE Conference Record of the 1967 Eighth Annual Symposium on Switching and Automata Theory*, New York: IEEE, (October 1967), pages 55-61.

Much of the early work on modeling parallelism in computer systems was based on the model of a computer system with memory and multiple processors. Processors executed programs which were modeled more or less as flowcharts. This paper by Karp and Miller is one of the high points of this approach. It presents the parallel program schemata model and then proceeds to determine various properties of the model, such as algorithms for equivalence, determinacy, and boundedness. Of particular interest to Petri nets is Section V, which introduces vector addition systems and the reachability tree to decide boundedness and coverability. A classic paper.

[Kasami 1974] T. Kasami, "Vector Addition Systems and Synchronization Problems of Concurrent Processes," Technical Report ARO-23, University of Hawaii, Honolulu, Hawaii, (September 1974).

[Keller 1972] R. Keller, "Vector Replacement Systems: A Formalism for Modeling Asynchronous Systems," Technical Report 117, Computer Science Laboratory, Princeton University, Princeton, New Jersey, (December 1972), 38 pages; revised (January 1974), 57 pages.

An important paper which is commonly referenced. Keller defines a "transition system" as a general model for parallel computation and then narrows this to vector replacement systems specifically. Vector replacement systems are the natural generalization of vector addition systems and an equivalent formalization to Petri nets. The liveness problem is the major property investigated.

[Keller 1974] R. Keller, "A Fundamental Theorem of Asynchronous Parallel Computation," *Proceedings of the Sagamore Computer Conference*, Lecture Notes in Computer Science, Volume 24, Berlin: Springer-Verlag, (August 1974), pages 102-112.

[Keller 1975a] R. Keller, "Generalized Petri Nets as Models for System Verification," Technical Report 202, Department of Electrical Engineering, Princeton University, Princeton, New Jersey, (August 1975), 50 pages.

Keller is interested here in defining the modeling power (systems which can be modeled) and decision power (questions which are decidable) of Petri nets. The modeling power is characterized by *additive monotone systems.* Decision properties discussed here center on the reachability tree for showing correctness.

[Keller 1975b] R. Keller, "Look-Ahead Processors," *Computing Surveys*, Volume 7, Number 4, (December 1975), pages 177-196.

[Keller 1976] R. Keller, "Formal Verification of Parallel Programs," *Communications of the ACM*, Volume 19, Number 7, (July 1976), pages 371-384.

The model used here is more powerful than Petri nets but has been developed from Keller's work with vector replacement systems and Petri nets.

[Kinney and Han 1976] L. Kinney, and Y. Han, "Reduction of Petri Nets," *Proceedings of the 14th Allerton Conference on Circuits and Systems Theory*, (September 1976).

[Knuth 1973] D. Knuth, *The Art of Computer Programming, Volume 3: Sorting and Searching*, Reading, Massachusetts: Addison-Wesley, (1973), 772 pages.

[Knuth 1978a] E. Knuth, "Petri Nets and Regular Trace Languages," Report ASM/47, Computing Laboratory, University of Newcastle upon Tyne, Newcastle upon Tyne, England, (April 1978).

[Knuth 1978b] E. Knuth, "Cycles of Partial Orders," *Proceedings of the Seventh Symposium on Mathematical Foundations of Computer Science 1978*, Lecture Notes in Computer Science, Volume 64, Berlin: Springer-Verlag, (September 1978), pages 315-325.

[Kosaraju 1973] S. Kosaraju, "Limitations of Dijkstra's Semaphore Primitives and Petri Nets," Technical Report 25, Computer Science Program, Johns Hopkins University, Baltimore, Maryland, (May 1973), 20 pages; Also *Operating Systems Review*, Volume 7, Number 4, (October 1973), pages 122-126.

This was the first paper to define a synchronization problem which Petri nets could not model. Since Petri nets can model P and V operations on semaphores, these were also shown to be incomplete. The problem simply encoded the state by one, two, or three tokens in a place. Since Petri nets are permissive, it is not possible to prevent transitions meant for one or two tokens from firing when there are three tokens.

[Kotov 1978] V. Kotov, "An Algebra for Parallelism Based on Petri Nets," *Proceedings of the Seventh Symposium on Mathematical Foundations of Computer Science 1978*, Lecture Notes in Computer Science, Volume 64, Berlin: Springer-Verlag, (September 1978), pages 39-55.

[Landweber and Robertson 1975] L. Landweber, and E. Robertson, "Properties of Conflict-free and Persistent Petri Nets," Technical Report 264, Computer Sciences Department, University of Wisconsin, Madison, Wisconsin, (December 1975), 30 pages; Also *Journal of the ACM*, Volume 25, Number 3, (July 1978), pages 352-364.

Conflict-free Petri nets allow shared input places only if they are on self-loops; persistent Petri nets cannot disable a transition except by firing it. The reachability sets of both types of Petri nets are characterized as semilinear. Complexity and decidability are the main subjects of this work.

[Lauer 1974] P. Lauer, "Path Expressions as Petri Nets, or Petri Nets with Fewer Tears," MRM 70, Computing Laboratory, University of Newcastle upon Tyne, Newcastle upon Tyne, England, (January 1974), 61 pages.

A very informal transcript of the development of the relationship between path expressions and Petri nets. This led to the writing of [Lauer and Campbell 1974].

[Lauer and Campbell 1974] P. Lauer, and R. Campbell, "A Description of Path Expressions by Petri Nets," Technical Report 64, Computing Laboratory, University of Newcastle upon Tyne, Newcastle upon Tyne, England, (May 1974), 39 pages; Also *Proceedings of the Second ACM Symposium on Principles of Programming Languages*, New York: ACM, (January 1975), pages 95-105.

A constructive approach is taken to showing how path expressions can be represented by Petri nets. Path expressions are a means of describing the sequences of allowable interactions between concurrent processes.

[Lauer and Campbell 1975] P. Lauer, and R. Campbell, "Formal Semantics of a Class of High-Level Primitives for Coordinating Concurrent Processes," *Acta Informatica*, Volume 5, Number 4, (1975), pages 297-332.

[Lauer and Shields 1977] P. Lauer, and M. Shields, "Abstract Specification of Resource Accessing Disciplines: Adequacy, Starvation, Priority and Interrupts," Technical Report 117, Computing Laboratory, University of Newcastle upon Tyne, Newcastle upon Tyne, England, (December 1977), 40 pages.

[Lauer et al. 1977] P. Lauer, E. Best, and M. Shields, "On the Problem of Achieving Adequacy of Concurrent Programs," Technical Report 113, Computing Laboratory, University of Newcastle upon Tyne, Newcastle upon Tyne, England, (June 1977), 36 pages.

[Lautenbach 1975] K. Lautenbach, "Liveness in Petri Nets," Internal Report ISF-75-02.1, Institut fur Informationssystemforschung, Gesellschaft fur Mathematik und Datenverarbeitung, Bonn, West Germany, (July 1975), 33 pages.

A review of the state of knowledge about liveness and deadlock in Petri nets, this paper brings together most of the results in one place.

[Lautenbach and Schmid 1974] K. Lautenbach, and H. Schmid, "Use of Petri Nets for Proving Correctness of Concurrent Process Systems," *Information Processing 74, Proceedings of the 1974 IFIP Congress*, Amsterdam: North-Holland, (August 1974), pages 187-191.

Systems of concurrent processes are more difficult to analyze and prove correct than sequential programs. Petri nets can be used to model such a system and prove properties of the modeled system. This paper treats semaphores, bounded buffers, and the five philosophers to rederive in Petri net terms results that have been developed separately.

[Lautenbach and Thiagarajan 1978] K. Lautenbach and P. Thiagarajan, "Analysis of a Resource Allocation Problem Using Petri Nets," Internal Report ISF-78-05, Institut fur Informationssystemforschung, Gesellschaft fur Mathematik und Datenverarbeitung, Bonn, West Germany, (June 1978), 17 pages; also *Proceedings of the First European Conference on Parallel and Distributed Processing*, (February 1979).

[Leung et al. 1977] K. Leung, C. Michel, and P. Le Beux, "Logical Systems Design Using PLAs and Petri Nets — Programmable Hardwired Systems,"

Information Processing 77, Proceedings of the 1977 IFIP Congress, Amsterdam: North-Holland, (August 1977), pages 607-611.

[Lien 1972] Y. Lien, "A Study of the Theoretical and Practical Aspects of Transition Systems," Ph.D. dissertation, Department of Computer Sciences, University of California, Berkeley, California, (September 1972).

[Lien 1974] Y. Lien, "Termination and Finiteness Properties of Transition Systems," Technical Report 74-4, Department of Computer Science, University of Kansas, Lawrence, Kansas, (September 1974), 32 pages.

[Lien 1976a] Y. Lien, "Termination Properties of Generalized Petri Nets," *SIAM Journal of Computing*, Volume 5, Number 2, (June 1976), pages 251-265.

Termination properties mean deadlock and liveness properties. Since the problem for general Petri nets is still open, Lien investigates the problem for subclasses which are conflict-free or concurrent-free. Some graph theory concepts such as strongly connected are used as are Petri net concepts such as conservation. Lots of formal mathematics.

[Lien 1976b] Y. Lien, "A Note on Transition Systems," *Information Sciences*, Volume 10, Number 4, (1976), pages 347-362.

[Lien and Margrave 1974] Y. Lien, and M. Margrave, "Transition Systems — A Model for Concurrent Activities," Technical Report 74-5, Department of Computer Science, University of Kansas, Lawrence, Kansas, (October 1974), 43 pages.

A tutorial paper showing the basic concepts of Petri nets and the related transition system model. Applications of the modeling techniques to transportation systems, genetic systems, and computations are given.

[Lipton 1976] R. Lipton, "The Reachability Problem Requires Exponential Space," Research Report 62, Department of Computer Science, Yale University, New Haven, Connecticut, (January 1976), 15 pages.

The results of this paper were first presented at the Conference on Petri Nets and Related Methods in July 1975. This paper is the most commonly cited result in complexity for Petri nets. Five lemmas are used to prove that the reachability problem for Petri nets and equivalent models such as vector addition systems require at least $2^{c \quad n}$ space. The result is based on a construction showing that it is possible to construct a counter which can effectively count from 0 to 2^{2^k} with a number of transitions proportional to k. The operations of adding, subtracting, and testing for zero are possible on the counter. (It is the test for zero that is hard.) The proofs are constructive in a parallel program language which is easily shown to be equivalent to Petri nets. A rather tricky proof that requires several readings.

[Lipton and Snyder 1974] R. Lipton, and L. Snyder, "PV Simulation of Computation Graphs," Unpublished, (1974), 8 pages.

It is shown that computation graphs can be simulated by sets of processes using P and V operations.

[Lipton et al. 1974] R. Lipton, L. Snyder, and Y. Zalcstein, "A Comparative Study of Models of Parallel Computation," *Proceedings of the 15th Annual Symposium on Switching and Automata Theory*, New York: IEEE, (October 1974), pages 145-155.

Lipton, Snyder and Zalcstein compare the major models of parallel computation, similar to the work in [Agerwala 1974b] and [Peterson and Bredt 1974], but their results are substantially different. The differences are caused by the definition of equivalence and containment. Their research is more formal and harder to follow than [Agerwala 1974b] or [Peterson and Bredt 1974] but of equal validity and interest.

[Liskov 1972] B. Liskov, "The Design of the Venus Operating System," *Communications of the ACM*, Volume 15, Number 3, (March 1972), pages 144-149.

[Luconi 1968] F. Luconi, "Asynchronous Computational Structures," Ph.D. dissertation, Department of Electrical Engineering, Massachusetts Institute of Technology, Cambridge, Massachusetts, (January 1968), 139 pages; Also Technical Report 49, Project MAC, Massachusetts Institute of Technology, Cambridge, Massachusetts, (February 1968), 139 pages.

[Mandrioli 1976] D. Mandrioli, "A Note on Petri Net Languages," Internal Report 76-5, Laboratori di Calcolatori, Instituto di Electtrotecnica ed Elettronica del Politecnico di Milano, Milan, Italy, (1976); Also *Information and Control*, Volume 34, Number 2, (June 1977), pages 169-171.

It is shown that Petri net languages which allow λ transitions are not closed under complement.

[Marin et al. 1976] J. Marin, C. Andre, and F. Boeri, "Conception de Systemes Sequentiels Totalement Autotestables a partir des Reseaux de Petri," *Revue Francaise d'Automatique, Informatique et Recherche Operationnelle: Automatique*, Volume 10, Number 11, (November 1976), pages 5-22, (In French).

[Mayr 1977] E. Mayr, "The Complexity of the Finite Containment Problem for Petri Nets," Master's thesis, Department of Electrical Engineering and Computer Science, Massachusetts Institute of Technology, Cambridge, Massachusetts, (May 1977), 65 pages; Also Technical Report 181, Laboratory for Computer Science, Massachusetts Institute of Technology, Cambridge, Massachusetts, (June 1977), 65 pages.

[Meldman 1977] J. Meldman, "A New Technique for Modeling the Behavior of Man-Machine Information Systems," *Sloan Management Review*, Volume 18, Number 3, (Spring 1977), pages 29-46.

[Meldman 1978] J. Meldman, "A Petri-Net Representation of Civil Procedure," *IDEA — The Journal of Law and Technology*, Volume 19, Number 2, (1978), pages 123-148.

[Meldman and Holt 1971] J. Meldman, and A. Holt, "Petri Nets and Legal Systems," *Jurimetrics Journal*, Volume 12, Number 2, (December 1971), pages 65-75.

A curious presentation of Petri nets for the legal profession. Legal systems are social systems including lawyers, judges, plaintiffs, defendants, clerks, and so on. The interactions among these participants are described by such volumes as the Federal Rules of Civil Procedure. This presentation shows that at least some of these rules can be represented by an interpreted Petri net. It is suggested that modeling the legal system as a Petri net could lead to better understanding through analysis and then to a better legal system.

[Memmi 1977] G. Memmi, "Appliations de la Notion de Semi-Flots aux Reseaux de Petri," AFCET Journees sur les Reseaux de Petri (AFCET Workshop on Petri Nets), Paris, France, (March 1977), pages 207-212, (In French).

[Memmi 1978a] G. Memmi, "Applications of the Semiflow Notion to the Boundedness and Liveness Problems in Petri Net Theory," *Proceedings of the 1978 Conference on Information Sciences and Systems*, Johns Hopkins University, Baltimore, Maryland, (March 1978), pages 505-509; Also Technical Report 78-3, Institut de Programmation, Universite Pierre et Marie Curie, Paris, France, (March 1978), 5 pages.

[Memmi 1978b] G. Memmi, "Fuites dans les Reseaux de Petri," *R.A.I.R.O. Informatique/Theoretical Computer Science*, Volume 12, Number 2, (1978), pages 125-144, (In French).

[Memmi 1978c] G. Memmi, "Fuites et Semi-Flots dans les Reseaux de Petri," These de Docteur Ingenieur, Universite Pierre et Marie Curie, Paris, France, (December 1978), 117 pages, (In French).

[Memmi 1979] G. Memmi, "Notion de Dualite dans les Reseaux de Petri," *International Symposium on the Semantics of Concurrent Computation*, Lecture Notes in Computer Science, Berlin: Springer-Verlag, (July 1979), (In French).

[Memmi and Roucairol 1979] G. Memmi and G. Roucairol, "Linear Algebra in Net Theory," Advanced Course on General Net Theory of Processes and Systems, Hamburg, (October 1979); Also Lecture Notes in Computer Science, Berlin: Springer-Verlag, (1980).

[Merlin 1974] P. Merlin, "A Study of the Recoverability of Computing Systems," Ph.D. dissertation, Department of Information and Computer Science, University of California, Irvine, California, (1974), 181 pages; Also Technical Report 58, Department of Information and Computer Science, University of California, Irvine, California, (1974), 181 pages.

This dissertation is a strange combination of many topics. Petri nets are used to model systems. Then faults are introduced into the model by considering the behavior of the system when tokens are added or removed with no transition firing. "Good" Petri nets will eventually get back into an acceptable marking, while "bad" nets will either deadlock or go off into new markings where they shouldn't be. This work was strongly influenced by the work on the UCLA model and its token machine concepts.

Another part of the dissertation presents the timed Petri net model in which transitions must wait a certain time after they are enabled before they can fire and then must fire within a certain amount of time. These nets can represent more systems since they are equivalent to Turing machines. However, Merlin's work is basically limited to finite systems, so no problems arise.

[Merlin 1975] P. Merlin, "A Methodology for the Design and Implementation of Communication Protocols," Report RC-5541, IBM T. J. Watson Research Center, Yorktown Heights, New York, (June 1975); Also *IEEE Transactions on Communications*, Volume COM-24, Number 6, (June 1976), pages 614-621.

Time Petri nets are used to model communication protocols. An example is presented of the development and modeling of a protocol by a time Petri net followed by the validation of the protocol and an indication of how this protocol could then be implemented directly from the Petri net by a table-driven microprogrammed Petri net interpreter. The time Petri net model is, of course, derived from Merlin's dissertation [Merlin 1974].

[**Merlin and Farber 1976**] P. Merlin, and D. Farber, "Recoverability of Communication Protocols — Implications of a Theoretical Study," *IEEE Transactions on Communications*, Volume COM-24, Number 9, (September 1976), pages 1036-1043.

[**Merlin and Randell 1977**] P. Merlin, and B. Randell, "Consistent State Restoration in Distributed Systems," Technical Report 113, Computing Laboratory, University of Newcastle upon Tyne, Newcastle upon Tyne, England, (October 1977), 41 pages.

[**Miller 1973**] R. Miller, "A Comparison of Some Theoretical Models of Parallel Computation," Report RC-4230, IBM T. J. Watson Research Center, Yorktown Heights, New York; Also *IEEE Transactions on Computers*, Volume C-22, Number 8, (August 1973), pages 710-717.

Petri nets, computation graphs, parallel program schemata, and marked graphs are compared.

[**Miller 1974**] R. Miller, "Some Relationships Between Various Models of Parallelism and Synchronization," Report RC-5074, IBM T. J. Watson Research Center, Yorktown Heights, New York, (October 1974), 43 pages.

Reducibility between Petri nets, computation graphs, semaphore systems, vector addition systems, and vector replacement systems are considered.

[**Miller 1977**] R. Miller, "Theoretical Studies of Asynchronous and Parallel Processing," *Proceedings of the 1977 Conference on Information Sciences and Systems*, Johns Hopkins University, Baltimore, Maryland, (March 1977), pages 333-339.

[**Minsky 1967**] M. Minsky, *Computation: Finite and Infinite Machines*, Englewood Cliffs, New Jersey: Prentice-Hall, (1967), 317 pages.

A textbook, somewhat dated by its treatment of McCullouch-Pitts nets, which is still a fine treatment of computability theory. Of most interest to us here is Chapter 11 on register machines, but the chapters on Turing machines are related also.

[**Misunas 1973**] D. Misunas, "Petri Nets and Speed Independent Design," *Communications of the ACM*, Volume 16, Number 8, (August 1973), pages 474-481.

[**Moalla et al. 1978a**] M. Moalla, J. Pulou, and J. Sifakis, "Reseaux de Petri Synchronises," *R.A.I.R.O. Automatique/Systems Analysis and Control*, Volume 12, Number 2, (1978), pages 103-130, (In French).

[Moalla et al. 1978b] M. Moalla, J. Pulou, and J. Sifakis, "Synchronized Petri Nets: A Model for the Description of Non-Autonomous Systems," *Proceedings of the Seventh Symposium on Mathematical Foundations of Computer Science 1978*, Lecture Notes in Computer Science, Volume 64, Berlin: Springer-Verlag, (September 1978), pages 374-384.

[Moalla et al. 1978c] M. Moalla, J. Sifakis, and M. Silva, "A la Recherche d'une Methodologie de Conception sure des Automatismes Logiques Basee sur l'Utilisation des Reseaux de Petri," Technical Report 138, Institute of Applied Mathematics and Computer Science, Grenoble, France, (October 1978), 43 pages, (In French).

[Morcrette 1977] M. Morcrette, "Validation d'Algorithmes Paralleles, Application a l'Analyse Syntaxique," AFCET Journees sur les Reseaux de Petri (AFCET Workshop on Petri Nets), Paris, France, (March 1977), pages 13-32, (In French).

[Murata 1975] T. Murata, "State Equations, Controllability and Maximal Matchings of Petri Nets," Research Report MDC 1.1.10, Department of Information Engineering, University of Illinois, Chicago, Illinois, (December 1975); Also *IEEE Transactions on Automatic Control*, Volume AC-22, Number 3, (June 1977), pages 412-416.

An introduction to Petri nets for those in control theory, this paper concentrates on presenting and analyzing Petri nets as discrete time systems. Controllability and reachability are examined in terms of the matrix representation of a Petri net and maximal matchings. This is a first step toward an interchange between Petri net theory and optimal control theory.

[Murata 1976] T. Murata, "A Method for Synthesizing Marked Graphs from Given Markings," *Proceedings of the Tenth Annual Asilomar Conference on Circuits, Systems, and Computers*, (November 1976), pages 202-206.

[Murata 1977a] T. Murata, "Petri Nets, Marked Graphs, and Circuit-System Theory," *IEEE Circuits and Systems Society Newsletter*, Volume 11, Number 3, (June 1977), pages 2-12.

A very nice introduction to Petri nets and marked graphs, stressing the circuit theory viewpoint. The matrix approach to representing Petri nets is used throughout and such concepts as controllability, in the systems theoretic sense, are considered. The presentation is tutorial and is a fine example of how Petri nets can be widely used in other fields of study. The informal treatment here is an interesting contrast from [Murata 1977b].

[Murata 1977b] T. Murata, "Circuit Theoretic Analysis and Synthesis of Marked Graphs," *IEEE Transactions on Circuits and Systems*, Volume CAS-24, Number 7, (July 1977), pages 400-405.

[Murata and Church 1975] T. Murata, and R. Church, "Analysis of Marked Graphs and Petri Nets by Matrix Equations," Research Report MDC 1.1.8, Department of Information Engineering, University of Illinois, Chicago, Illinois, (November 1975).

A matrix representation of Petri nets can lead to some useful analysis techniques. This approach is developed in this report, mainly for marked graphs but to some degree for Petri nets. This is the most complete investigation of Petri nets and matrix equations.

[Murata and Shah 1976] T. Murata, and T. Shah, "On Liveness, Deadlock, and Reachability of E-Nets," *Proceedings of the 14th Annual Allerton Conference on Circuits and Systems Theory*, (September 1976), pages 597-605.

Uses the incidence matrix of an E-net to study the properties of liveness, deadlock, and reachability.

[Murata et al. 1975] T. Murata, R. Church, and A. Amin, "Matrix Equations for Petri Nets and Marked Graphs," *Proceedings of the Ninth Annual Asilomar Conference on Circuits, Systems, and Computers*, (November 1975), pages 36-41.

[Nash 1973] B. Nash, "Reachability Problems in Vector Addition Systems," *The American Mathematical Monthly*, Volume 80, Number 3, (March 1973), pages 292-295.

The reachability problem for vector addition systems is defined as a research problem. It is shown equivalent to the zero-reachability problem and the reachable-from-zero problems.

[Noe 1971] J. Noe, "A Petri Net Model of the CDC 6400," Technical Report 71-04-03, Department of Computer Science, University of Washington, Seattle, Washington, (April 1971), 16 pages; Also *Proceedings ACM SIGOPS Workshop on System Performance Evaluation*, New York: ACM, (April 1971), pages 362-378.

The title is rather misleading since not the CDC 6400 but the SCOPE 3.2 operating system is modeled. Also the model used is not strictly Petri nets but seems to have been extended to include inclusive- and exclusive-OR logic. It is not obvious whether this is necessary, but it is not relevant; this paper has mainly served as an example to show that Petri nets can model real systems (even if the details don't quite agree). This was also the start of the work which led to E-nets (see [Nutt 1972a] and [Noe and Nutt 1973]).

[Noe 1975] J. Noe, "Pro-Nets: For Modeling Processes and Processors," Technical Report 75-07-15, Department of Computer Science, University of Washington, Seattle, Washington, (July 1975).

[Noe 1977a] J. Noe, "Machine Aided Modeling Using Modified Petri Nets," AFCET Journees sur les Reseaux de Petri (AFCET Workshop on Petri Nets), Paris, France, (March 1977), pages 89-114.

[Noe 1977b] J. Noe, "Abstraction and Refinement with Modified Petri Nets," AFCET Journees sur les Reseaux de Petri (AFCET Workshop on Petri Nets), Paris, France, (March 1977), pages 157-160.

[Noe 1977c] J. Noe, "Abstraction Levels with Pro-Nets: An Algorithm and Examples," Technical Report 77-03-01, Department of Computer Science, University of Washington, Seattle, Washington, (June 1977), 16 pages.

[Noe 1978] J. Noe, "Hierarchical Modeling with Pro-Nets," *Proceedings of the National Electronics Conference*, Volume 23, (October 1978), pages 155-160.

[Noe 1979a] J. Noe, "Nets in Modeling and Simulation," Advanced Course on General Net Theory of Processes and Systems, Hamburg, (October

1979); also Lecture Notes in Computer Science, Berlin: Springer-Verlag, (1980).

[Noe 1979b] J. Noe, "Abstraction of Net Models," Advanced Course on General Net Theory of Processes and Systems, Hamburg, (October 1979); also Lecture Notes in Computer Science, Berlin: Springer-Verlag, (1980).

[Noe 1979c] J. Noe, "Applications of Net-Based Models," Advanced Course on General Net Theory of Processes and Systems, Hamburg, (October 1979); also Lecture Notes in Computer Science, Berlin: Springer-Verlag, (1980).

[Noe and Kehl 1975] J. Noe, and T. Kehl, "A Petri Net Model of A Modular Microprogrammable Computer (LM^2)," Technical Report 75-09-01, Computer Science Department, University of Washington, Seattle, Washington, (September 1975), 23 pages.

[Noe and Nutt 1973] J. Noe, and G. Nutt, "Macro E-Nets for Representation of Parallel Systems," *IEEE Transactions on Computers*, Volume C-22, Number 8, (August 1973), pages 718-727.

E-nets (evaluation nets) are an extended, interpreted model for parallel computation (derived from Petri nets) for performance measurement, evaluation, and simulation. E-nets represent one approach toward adding timing information to a Petri net.

[Noe et al. 1974] J. Noe, C. Crowley, and T. Anderson, "The Design of a Interactive Graphical Net Editor," Technical Report 74-07-30, Department of Computer Science, University of Washington, Seattle, Washington, (July 1974), 31 pages; Also *Proceedings CIPS-ACM Pacific Regional Conference*, (May 1974), pages 386-402;

[Nutt 1972a] G. Nutt, "The Formulation and Application of Evaluation Nets," Ph.D. dissertation, Computer Science Group, University of Washington, Seattle, Washington, (July 1972), 181 pages; Also Technical Report 72-07-02, Computer Science Group, University of Washington, Seattle, Washington, (July 1972), 170 pages.

The complete work on E-nets, an extension of Petri nets for performance modeling and evaluation.

[Nutt 1972b] G. Nutt, "Evaluation Nets for Computer Systems Performance Analysis," Technical Report 72-04-03, Computer Science Group, University of Washington, Seattle, Washington, (April 1972); Also *Proceedings of the 1972 Fall Joint Computer Conference*, Montvale, New Jersey: AFIPS Press, (December 1972), pages 279-286.

[Oberquelle 1979] H. Oberquelle, "Nets as a Tool in Teaching and in Terminology Work," Advanced Course on General Net Theory of Processes and Systems, Hamburg, (October 1979); Also Lecture Notes in Computer Science, Berlin: Springer-Verlag, (1980).

[Parikh 1966] R. Parikh, "On Context-Free Languages," *Journal of the ACM*, Volume 13, Number 4, (October 1966), pages 570-581.

The mapping from a string to a vector whose ith component is the number of occurrences in the string of the ith symbol is defined. It is shown that this mapping (the Parikh mapping) applied to a context-free language leads to a semilinear set.

[Parnas 1972] D. Parnas, "On a Solution to the Cigarette Smokers' Problem (Without Conditional Statements)," Department of Computer Science, Carnegie-Mellon University, Pittsburgh, Pennsylvania, (July 1972), 11 pages; Also *Communications of the ACM*, Volume 18, Number 3, (March 1975), pages 181-183.

A solution to the cigarette smokers' problem posed by Patil [1971] is presented. The solution uses an array of semaphores and a global variable, which were implicitly not allowed by Patil.

[Patil 1970a] S. Patil, "Coordination of Asynchronous Events," Ph.D. dissertation, Department of Electrical Engineering, Massachusetts Institute of Technology, Cambridge, Massachusetts, (May 1970), 234 pages; Also Technical Report 72, Project MAC, Massachusetts Institute of Technology, Cambridge, Massachusetts, (June 1970), 234 pages.

One of the early works on Petri nets, this dissertation touches a lot of bases. It has a reasonable introduction to the problems of controlling concurrent processes and to Petri nets. Then it moves on to an extended model, called coordination nets, and their properties. In particular, Patil is concerned with how his coordination nets can be implemented in hardware. This shows up in his later work also.

[Patil 1970b] S. Patil, "Closure Properties of Interconnections of Determinate Systems," *Record of the Project MAC Conference on Concurrent Systems and Parallel Computation*, New York: ACM, (June 1970), pages 107-116.

A system is determinate if it produces the same output for the same inputs. This will not always be the case if concurrent operations may happen in arbitrary order, and this order affects the output. Determinacy was a big problem when concurrent systems were first being defined, but at least the term is no longer broadly used. In this paper, Patil was concerned with how to connect smaller, determinate systems together to form a larger determinate system.

[Patil 1971] S. Patil, "Limitations and Capabilities of Dijkstra's Semaphore Primitives for Coordination Among Processes," Computation Structures Group Memo 57, Project MAC, Massachusetts Institute of Technology, Cambridge, Massachusetts, (February 1971), 18 pages.

This short note defines the cigarette smokers' problem and shows that it cannot be solved by using semaphores. Petri nets are used to model the problem. It is shown that any solution must result in a form of Petri net which does not correspond to those resulting from P and V operations. This research led to [Kosaraju 1973] and [Agerwala and Flynn 1973], but also see the note by [Parnas 1972].

[Patil 1972] S. Patil, "Circuit Implementation of Petri Nets," Computation Structures Group Memo 73, Project MAC, Massachusetts Institute of Technology, Cambridge, Massachusetts, (December 1972), 14 pages.

Gate-level circuits for implementing Petri nets in hardware are given. These circuits differ from those in [Patil 1970a] by their detail (down to gate level) and by the assumption of a known bound on transmission delays.

[Patil and Dennis 1972] S. Patil, and J. Dennis, "The Description and Realization of Digital Systems," Computation Structures Group Memo 71, Project MAC, Massachusetts Institute of Technology, Cambridge, Massachusetts, (October 1972), 4 pages; Also *COMPCON 72: Sixth Annual IEEE Computer Society International Conference Digest of Papers*, New York: IEEE, (September 1972), pages 223-226.

A short presentation is given of the work on describing digital systems by Petri nets and then implementing them as speed-independent (asynchronous) circuits.

[Peterson 1973] J. Peterson, "Modeling of Parallel Systems," Ph.D. dissertation, Department of Electrical Engineering, Stanford University, Stanford, California, (December 1973), 255 pages; Also Technical Report 46, Digital Systems Laboratory, Stanford University, Stanford, California, (February 1974), 241 pages; Also Technical Report STAN-CS-74-410, Computer Science Department, Stanford University, Stanford, California (February 1974), 241 pages.

Two major results of this dissertation were the comparison of models later summarized in [Peterson and Bredt 1974] and the Petri net languages theory later printed in [Peterson 1976].

[Peterson 1976] J. Peterson, "Computation Sequence Sets," *Journal of Computer and System Sciences*, Volume 13, Number 1, (August 1976), pages 1-24.

Chapter 4 of [Peterson 1973], revised, is published. A specific class of Petri net languages, those languages defined by executions from an initial state to a final state, are defined and investigated. Closure properties and the place of these languages in the hierarchy of languages (regular, context-free, context-sensitive, type-0) are the main points considered.

[Peterson 1977] J. Peterson, "Petri Nets," *Computing Surveys*, Volume 9, Number 3, (September 1977), pages 223-252.

This is the first widely circulated survey and tutorial on Petri nets. It touches briefly on modeling with Petri nets, basic definitions, analysis problems and techniques, Petri net languages, and related models of computation. A good introduction which should be readable by any technically minded college student or graduate.

[Peterson 1978] J. Peterson, "An Introduction to Petri Nets," *Proceedings of the National Electronics Conference*, Volume 32, (October 1978), pages 144-148.

[Peterson and Bredt 1974] J. Peterson, and T. Bredt, "A Comparison of Models of Parallel Computation," *Information Processing 74, Proceedings of the 1974 IFIP Congress*, Amsterdam: North-Holland, (August 1974), pages 466-470.

The control structures of several models of parallel computation are described and compared. The models include finite state machines, Petri nets, UCLA graphs, message systems, computation graphs, parallel program schemata, marked graphs, and coordination nets. Also see the related papers [Agerwala 1974b], [Lipton et al. 1974], and [Miller 1973].

[Petri 1962a] C. Petri, "Kommunikation mit Automaten," Ph.D. dissertation, University of Bonn, Bonn, West Germany, (1962), (In German); Also

M.I.T. Memorandum MAC-M-212, Project MAC, Massachusetts Institute of Technology, Cambridge, Massachusetts; Also Clifford F. Greene, Jr. (translator), "Communication with Automata," Supplement 1 to Technical Report RADC-TR-65-377, Volume 1, Rome Air Development Center, Griffiss Air Force Base, New York, (January 1966), 89 pages.

This dissertation is the starting point for Petri net theory. Petri starts by discussing finite automata and shows (by reference to work by Davis, Turing, and others) that regular automata cannot recognize context-free or context-sensitive grammars. Petri then claims that if these automata were not finite but unbounded, then signal delay times are also unbounded and so synchronization problems result.

Petri then proceeds to construct a net for simulating the tape of a Turing machine. This net is completely asynchronous and constructed of identical cells which pass information to their neighbors (left or right) as needed. Since the tape is infinite, an infinite number of cells are needed. However, the importance of this work is its stress on the asynchronous nature of the interactions.

In modern terms, all of the nets in Petri's dissertation are safe by definition. Petri's dissertation is of only historical interest today.

[Petri 1962b] C. Petri, "Fundamentals of a Theory of Asynchronous Information Flow," *Information Processing 62, Proceedings of the 1962 IFIP Congress*, Amsterdam: North-Holland, (August 1962), pages 386-390.

[Petri 1973] C. Petri, "Concepts of Net Theory," *Proceedings of the Symposium and Summer School on Mathematical Foundations of Computer Sciences, High Tatras*, Mathematics Institute of the Slovak Academy of Science, (September 1973), pages 137-146.

The development of a general net theory, related to general systems theory and topology.

[Petri 1975] C. Petri "Interpretations of Net Theory," Internal Report 75-07, Institut fur Informationssystemforschung, Gesellschaft fur Mathematik und Datenverarbeitung, Bonn, West Germany, (July 1975), 34 pages; revised (December 1976).

Further development of net theory.

[Petri 1976] C. Petri, "General Net Theory," *Proceedings of the Joint IBM/University of Newcastle upon Tyne Seminar on Computing System Design*, Computing Laboratory, University of Newcastle upon Tyne, Newcastle upon Tyne, England, (September 1976), pages 131-169.

[Petri 1978] C. Petri, "Concurrency as a Basis of Systems Thinking," Internal Report ISF-78-06, Institut fur Informationssystemforschung, Gesellschaft fur Mathematik und Datenverarbeitung, Bonn, West Germany, (September 1978), 20 pages.

[Petri 1979a] C. Petri, "Introduction to General Net Theory," Advanced Course on General Net Theory of Processes and Systems, Hamburg, (October 1979); also Lecture Notes in Computer Science, Berlin: Springer-Verlag, (1980).

[Petri 1979b] C. Petri, "Concurrency," Advanced Course on General Net

Theory of Processes and Systems, Hamburg, (October 1979); also Lecture Notes in Computer Science, Berlin: Springer-Verlag, (1980).

[Pless and Plunnecke 1978] E. Pless and H. Plunnecke, "A Bibliography of Net Theory," Internal Report ISF-78-08, Institut fur Informationssystemforschung, Gesellschaft fur Mathematik und Datenverarbeitung, Bonn, West Germany, (November 1978), 57 pages.

A compilation of most of the known works on Petri nets.

[Postel 1974] J. Postel, "A Graph Model Analysis of Computer Communications Protocols," Ph.D. dissertation, Computer Science Department, University of California, Los Angeles, California, (1974), 191 pages.

[Postel and Farber 1976] J. Postel, and D. Farber, "Graph Modeling of Computer Communications Protocols," *Proceedings of the Fifth Texas Conference on Computing Systems*, University of Texas, Austin, Texas, (October 1976), pages 66-77.

This paper uses the UCLA graph model rather than Petri nets (but the two are equivalent) to model communications protocols. Similar in intent to [Merlin 1975].

[Presburger 1929] M. Presburger, "Uber die Vollstandigkeit eines gewissen Systems der Arithmetic ganzer Zahlen, in Welchem die Addition als einzige Operation hervortritt," *Proceedings of the First Congress of Slavic Mathematicians*, Warsaw, (1929), pages 92-101, 395, (In German).

The basic algorithm for deciding the validity of a statement which uses only addition over integer numbers with quantifiers and logical connectives (Presburger arithmetic) is given.

[Rackoff 1976] C. Rackoff, "The Covering and Boundedness Problems for Vector Addition Systems," Technical Report Number 97, Department of Computer Sciences, University of Toronto, Toronto, Ontario, (July 1976), 14 pages; Also *Theoretical Computer Science*, Volume 6, Number 2, (April 1978), pages 223-231.

New algorithms for the covering and boundedness problems are given. These results are presented for vector addition systems but are, of course, equally applicable to Petri nets. The algorithms use $2^{c \cdot n \log n}$ space and so are close to optimal. Lipton's lower bound is $2^{c \cdot n}$. The main emphasis of this paper is establishing complexity results.

[Ramchandani 1973] C. Ramchandani, "Analysis of Asynchronous Concurrent Systems by Petri Nets," Ph.D. dissertation, Department of Electrical Engineering, Massachusetts Institute of Technology, Cambridge, Massachusetts, (July 1973), 219 pages; Also Technical Report 120, Project MAC, Massachusetts Institute of Technology, Cambridge, Massachusetts, (February 1974), 219 pages.

This dissertation is concerned with the analysis of a Petri net model to determine its performance. The Petri net model is extended to associate a time with each transition. Analysis of some subclasses of these timed Petri nets then leads to calculation of throughput rates for these systems.

[Rammig 1977] F. Rammig, "Petri-Net Based Description, Analysis and Simulation of Concurrent Processes," *Proceedings 14th Design Automation*

Conference, New York: IEEE, (June 1977).

[Reddi 1978] S. Reddi, "A Modular Computer with Petri Net Array Control," *Proceedings of the 1978 ACM National Conference*, New York: ACM, (December 1978), pages 79-85.

[Reif 1978] J. Reif, "Analysis of Communicating Processes," Technical Report 30, Computer Science Department, University of Rochester, Rochester, New York, (May 1978), 45 pages.

Uses reducibility to Petri net reachability problem to show difficulty of analyzing programs of a distributed environment.

[Riddle 1972] W. Riddle, "The Modeling and Analysis of Supervisory Systems," Ph.D. dissertation, Computer Science Department, Stanford University, Stanford, California, (March 1972), 181 pages; Also Technical Report STAN-CS-72-271, Computer Science Department, Stanford University, Stanford, California, (March 1972), 174 pages.

Riddle's Ph.D. dissertation defined a model for parallel computation which assumes that a system is composed of independent asynchronously operating sequential processes. Communication between processes is by way of messages. The objective is to develop a method for analysis of computer systems. In Riddle's model, each process is described by a pseudoprogram. These are used to produce message transfer expressions (which are languages), and then the message transfer expressions are used for analysis. Decision problems were left open, but several examples show the power and usefulness of this approach. Peterson showed that these systems could be converted to Petri nets [Peterson 1973], and Riddle proved the reverse [Riddle 1974], showing the essential equivalence of these two modeling schemes.

[Riddle 1974] W. Riddle, "The Equivalence of Petri Nets and Message Transmission Models," SRM/97, Computing Laboratory, University of Newcastle upon Tyne, Newcastle upon Tyne, England, (August 1974), 11 pages.

The work reported in [Peterson 1973] and [Peterson and Bredt 1974] states that message systems are a proper subset of Petri nets. Riddle shows that if equivalence does not consider false deadlocks, this is incorrect: They are equivalent models. The proof shows how to construct a message system with the same (nondeadlock) behavior as a Petri net. The equivalence of these two models is comforting, since the message systems are much easier to use for modeling systems of parallel processes.

[Rodriguez 1967] J. Rodriguez, "A Graph Model for Parallel Computations," Ph.D. dissertation, Department of Electrical Engineering, Massachusetts Institute of Technology, Cambridge, Massachusetts, (September 1967), 130 pages; Also Technical Report 64, Project MAC, Massachusetts Institute of Technology, Cambridge, Massachusetts, (September 1969), 121 pages.

[Roucairol 1974] G. Roucairol, "Une Transformation de Programmes Sequentiels en Programmes Paralleles," *Programming Symposium: Proceedings*, Lecture Notes in Computer Science, Volume 19, Berlin: Springer-Verlag, (April 1974), pages 327-349, (In French).

Uses a Petri net to transform sequential programs to parallel programs by modeling the precedence relations which exist in the algorithm rather than the ones introduced by writing the program in a linear programming language.

[Roucairol 1978] G. Roucairol, "Contribution a l'Etude des Equivalences Syntaxiques et Transformations de Programmes Paralleles," These Doctorat d'Etat es Sciences, Universite Pierre èt Marie Curie, Paris, France, (November 1978), 228 pages, (In French).

[Roucairol and Valk 1979] G. Roucairol and R. Valk, "Reductions of Nets and Parallel Programs," Advanced Course on General Net Theory of Processes and Systems, Hamburg, (October 1979); also Lecture Notes in Computer Science, Berlin: Springer-Verlag, (1980).

[Sacerdote and Tenney 1977] S. Sacerdote, and R. Tenney, "The Decidability of the Reachability Problem for Vector Addition Systems," COINS Technical Report 77-3, Computer and Information Sciences Department, University of Massachusetts, Amherst, Massachusetts, (1977), 38 pages; Also *Proceedings of the Ninth Annual ACM Symposium on Theory of Computing*, New York: ACM, (May 1977), pages 61-76.

An extended abstract of this paper surfaced in November 1976. A constructive "proof" is given of the decidability of the reachability problem for vector addition systems and Petri nets. Unfortunately, the "proof" is in error. From the preface to the Proceedings of the Tenth Symposium: "... in [Sacerdote and Tenney 1977], a number of difficulties have been pointed out. A new version, simpler and more accurate, is in preparation and will be available shortly from the authors."

[Salomaa 1973] A. Salomaa, *Formal Languages*, New York: Academic Press, (1973), 322 pages.

A comprehensive treatment of most of modern formal language theory, this reference text is particularly useful for its presentation of matrix grammars and Lindenmayer systems; both of these classes of languages have been linked to Petri net languages.

[Schiffers and Wedde 1978] M. Schiffers, and H. Wedde, "Analyzing Program Solutions of Coordination Problems by CP-Nets," *Proceedings of the Seventh Symposium on Mathematical Foundations of Computer Science 1978*, Lecture Notes in Computer Science, Volume 64, Berlin: Springer-Verlag, (September 1978), pages 462-473.

[Schmid 1973] H. Schmid, "An Approach to the Communication and Synchronization of Processes," *Proceedings of the International Computing Symposium 1973*, Amsterdam: North-Holland, (September 1973), pages 163-171.

[Schmid and Best 1976] H. Schmid, and E. Best, "Towards a Constructive Solution of the Liveness Problem in Petri Nets," Technical Report 4/76, Institut fur Informatik, Universitat Stuttgart, West Germany, (April 1976), 56 pages.

This report unifies the previous work of [Commoner 1972] and [Lautenbach 1975] with respect to the liveness problem.

[Schmid and Best 1978] H. Schmid, and E. Best, "A Step Towards a Solution of the Liveness Problem in Petri Nets," Technical Report 114, Computing

Laboratory, University of Newcastle upon Tyne, Newcastle upon Tyne, England, (February 1978), 40 pages.

[Seitz 1970] C. Seitz, "Asynchronous Machines Exhibiting Concurrency," *Record of the Project MAC Conference on Concurrent Systems and Parallel Computation*, New York: ACM, (June 1970), pages 93-106.

This paper follows an example through to present a tentative methodology for constructing an asynchronous machine from a Petri net.

[Seitz 1971] C. Seitz, "Graph Representations for Logical Machines," Ph.D. dissertation, Department of Electrical Engineering, Massachusetts Institute of Technology, Cambridge, Massachusetts, (January 1971), 136 pages.

[Shapiro 1975] R. Shapiro, "System Modeling with Net Structures," *Second Semi-Annual Technical Report*, Report CADD-7503-1411, Massachusetts Computer Associates, Wakefield, Massachusetts, (March 1975), pages 87-135.

[Shapiro 1979] R. Shapiro, "The Application of General Net Theory — A Personal History," Advanced Course on General Net Theory of Processes and Systems, Hamburg, (October 1979); also Lecture Notes in Computer Science, Berlin: Springer-Verlag, (1980).

[Shapiro and Saint 1970] R. Shapiro, and H. Saint, "A New Approach to Optimization of Sequencing Decisions," *Annual Review in Automatic Programming*, Volume 6, Part 5, (1970), pages 257-288.

Petri nets are applied to compiling FORTRAN programs for a CDC 6600 computer. The FORTRAN program is converted into a Petri net showing precedence constraints between operations. This net is merged with a Petri net representing the CDC 6600 CPU. Timing information is associated with the transitions, and an exhaustive search is used to determine the sequence of operations and assignment of registers which minimizes execution time.

This paper is a revision of Section V of [Holt et al. 1968].

[Shapiro and Thiagarajan 1978] R. Shapiro and P. Thiagarajan, "On the Maintenance of Distributed Copies of a Data Base," Report ISF-78-04, Institut fur Informationssystemforschung, Gesellschaft fur Mathematik und Datenverarbeitung, Bonn, West Germany, (July 1978), 14 pages.

[Shepardson and Sturgis 1963] J. Shepardson, and H. Sturgis, "Computability of Recursive Functions," *Journal of the ACM*, Volume 10, Number 2, (April 1963), pages 217-255.

This paper develops the register machine model and shows its equivalence to Turing machines. The register machine model of computation is easy to model with an extended Petri net model, showing the equivalence of the extended Petri net model to Turing machines.

[Sifakis 1977a] J. Sifakis, "Comportement Permanent des Reseaux de Petri Temporises," AFCET Journees sur les Reseaux de Petri (AFCET Workshop on Petri Nets), Paris, France, (March 1977), pages 227-247, (In French).

[Sifakis 1977b] J. Sifakis, "Use of Petri Nets for Performance Evaluation," *Measuring, Modeling and Evaluation of Computer Systems: Proceedings of the*

Third International Workshop on Modeling and Performance Evaluation of Computer Systems, Amsterdam: North-Holland, (1977), pages 75-93.

[Sifakis 1977c] J. Sifakis, "Homomorphisms of Petri Nets: Application to the Realization of Fault-Tolerant Systems," Technical Report 90, Institute of Applied Mathematics and Computer Science, Grenoble, France, (November 1977).

[Sifakis 1978a] J. Sifakis, "Realization of Fault-Tolerant Systems by Coding Petri Nets," *Proceedings of the Eighth Annual International Conference on Fault-Tolerant Computing*, New York: IEEE, (June 1978), page 205; Also (revised) *Journal of Design Automation and Fault-Tolerant Computing*, Volume 3, Number 2, (1979).

[Sifakis 1978b] J. Sifakis, "Structural Properties of Petri Nets," *Proceedings of the Seventh Symposium on Mathematical Foundations of Computer Science 1978*, Lecture Notes in Computer Science, Volume 64, Berlin: Springer-Verlag, (September 1978), pages 474-483.

[Sifakis 1979a] J. Sifakis, "Le Controle des Systemes Asynchrones: Concepts, Proprietes, Analyse Statique," These de Docteur es Sciences, University of Grenoble, Grenoble, France, (June 1979), 233 pages, (In French).

[Sifakis 1979b] J. Sifakis, "Use of Petri Nets for Performance Evaluation," Advanced Course on General Net Theory of Processes and Systems, Hamburg, (October 1979); Also Lecture Notes in Computer Science, Berlin: Springer-Verlag, (1980).

[Slutz 1968] D. Slutz, "The Flow Graph Schemata Model of Parallel Computation," Ph.D. dissertation, Department of Electrical Engineering, Massachusetts Institute of Technology, Cambridge, Massachusetts, (August 1968), 256 pages; Also Technical Report 53, Project MAC, Massachusetts Institute of Technology, Cambridge, Massachusetts, (September 1968), 254 pages.

[Starke 1978] P. Starke, "Free Petri Net Languages," *Proceedings of the Seventh Symposium on Mathematical Foundations of Computer Science 1978*, Lecture Notes in Computer Science, Volume 64, Berlin: Springer-Verlag, (September 1978), pages 506-515.

[Stucki 1973] M. Stucki, "An Approach for Synthesizing Transition Logic Circuits," *Proceedings of the 11th Annual Allerton Conference on Circuit and System Theory*, (October 1973), pages 418-425.

[Stucki 1975] M. Stucki, "Synthesis of Level Sequential Circuits: Further Development of a Procedure Based on a Petri-Net Type of Behavioral Description," *Proceedings of the 13th Annual Allerton Conference on Circuit and System Theory*, (October 1975), pages 896-904.

[Symons 1976] F. Symons, "Modelling and Analysis of Communications Protocols Using Petri-Nets," Technical Report 140, Department of Electrical Engineering Science, University of Essex, (September 1976).

[Thieler-Mevissen 1976] G. Thieler-Mevissen, "The Petri Net Calculus of Predicate Logic," Internal Report ISF-76-09, Institut fur Informationssystemforschung, Gesellschaft fur Mathematik und Datenverarbeitung, Bonn, West Germany, (December 1976), 60 pages.

[Thomas 1976] P. Thomas, "The Petri Net: A Modeling Tool for the Coordination of Asynchronous Processes," Master's thesis, University of Tennessee, Knoxville, Tennessee, (June 1976), 118 pages.

The Petri net as a model of computation is investigated. Comparisons between the Petri net model and other models (such as in [Agerwala 1974b] or [Peterson and Bredt 1974]) are made. Also a program for simulating a Petri net is presented along with lots of output from this program tracing the execution of a Petri net.

[Thornton 1970] J. Thornton, *Design of a Computer: The Control Data 6600*, Scott, Foresman and Company, Glenview, Illinois, (1970), 181 pages.

[Tsichritzis 1971] D. Tsichritzis, "Modular System Description," Technical Report 33, Department of Computer Science, University of Toronto, Toronto, Ontario, (October 1971), 20 pages.

An introduction to the use of Petri nets to model some simple operating system problems (mutual exclusion, producer/consumer) with some thoughts about how Petri nets could be used to aid the design, evaluation, and implementation of an operating system.

[Valk 1977] R. Valk, "Self-Modifying Nets," AFCET Journees sur les Reseaux de Petri (AFCET Workshop on Petri Nets), Paris, France, (March 1977), pages 161-198; also Technical Report 34, Institut fur Informatik, University of Hamburg, (July 1977), 36 pages.

[Valk 1978a] R. Valk, "Self-Modifying Nets — A Natural Extension of Petri Nets," *Automata, Languages and Programming*, Lecture Notes in Computer Science, Volume 62, Berlin: Springer-Verlag, (July 1978), pages 464-476.

[Valk 1978b] R. Valk, "On the Computational Power of Extended Petri Nets," *Proceedings of the Seventh Symposium on Mathematical Foundations of Computer Science 1978*, Lecture Notes in Computer Science, Volume 64, Berlin: Springer-Verlag, (September 1978), pages 526-535.

[Valk and Vidal-Naquet 1977] R. Valk, and G. Vidal-Naquet, "On the Rationality of Petri Net Languages," *Proceedings of the Third GI Conference on Theoretical Computer Science*, Lecture Notes in Computer Science, Volume 48, Berlin: Springer-Verlag, (March 1977), pages 319-328.

[Valette 1977] R. Valette, "Description et Verification des Systemes Paralleles," AFCET Journees sur les Reseaux de Petri (AFCET Workshop on Petri Nets), Paris, France, (March 1977), pages 1-12, (In French).

[Valette 1979] R. Valette, "Analysis of Petri Nets by Stepwise Refinements," *Journal of Computer and System Sciences*, Volume 18, Number 1, (February 1979), pages 35-46.

[Van Leeuwen 1974] J. Van Leeuwen, "A Partial Solution to the Reachability Problem for Vector Addition Systems," *Proceedings of the Sixth Annual ACM Symposium on Theory of Computing*, New York: ACM, (April 1974), pages 303-309.

The reachability problem for dimensions of one, two, or three is shown to be decidable.

[Vaudene and Vignat 1977] D. Vaudene, and J. Vignat, "Semantique d'Enonces de Synchronisation en Termes de Reseaux de Petri," AFCET Journees sur les Reseaux de Petri (AFCET Workshop on Petri Nets), Paris,

France, (March 1977), pages 137-156, (In French).

[Vidal-Naquet 1977] G. Vidal-Naquet, "Methods pour les Problemes d'Indecidabilite et de Complexite sur les Reseaux de Petri," AFCET Journees sur les Reseaux de Petri (AFCET Workshop on Petri Nets), Paris, France, (March 1977), pages 199-206, (In French).

[Volansky 1970] S. Volansky, "Graph Model Analysis and Implementation of Computational Sequences," Ph.D. dissertation, Computer Science Department, University of California, Los Angeles, California, (June 1970), 187 pages.

[Winkowski 1976] J. Winkowski, "Formal Theories of Petri Nets and Net Simulation," CC PAS Report 242, Computation Centre, Polish Academy of Sciences, Warsaw, Poland, (1976), 23 pages.

Considers an abstract model of nets with axioms where instances of these axioms create models of the theory; concerned with the fundamental concepts of net theory, as in [Petri 1973].

[Yamamoto 1975] R. Yamamoto, "An Approach to Detecting Synchronization Errors in Concurrently Executable Programs," *Proceedings of the Ninth Asilomar Conference on Circuits, Systems, and Computers*, (November 1975), pages 154-158.

[Yoeli 1973] M. Yoeli, "Petri Nets and Asynchronous Control Networks," Technical Report CS-73-07, Department of Applied Analysis and Computer Science, University of Waterloo, Waterloo, Ontario, (April 1973), 22 pages.

[Yoeli and Barsilai 1977] M. Yoeli, and Z. Barsilai, "On Behavioral Descriptions of Communications Switching Systems," Technical Report 99, Technion — Israel Institute of Technology, Haifa, Israel, (June 1977), 24 pages; Also *Digital Processes*, Volume 3, (1977), pages 307-320.

[Yu and Murata 1978] S. Yu, and T. Murata, "PT-marked Graphs: A Reduced Model of Petri Nets," *Proceedings of the 16th Allerton Conference on Communication, Control and Computing*, (October 1978).

[Zervos 1977] C. Zervos, "Colored Petri Nets: Their Properties and Applications," Technical Report 107, Systems Engineering Laboratory, University of Michigan, Ann Arbor, Michigan, (January 1977), 317 pages.

[Zuse 1979] K. Zuse, "Petri-Nets from the Engineer's Viewpoint," Advanced Course on General Net Theory of Processes and Systems, Hamburg, (October 1979); also Lecture Notes in Computer Science, Berlin: Springer-Verlag, (1980).

Index

A

B

C

D

E

F

G

D

E

F

G

H

I

J

K

L

M

N

O

P

Q

R

S

T

U

V

W

Z